Cybernetic Avatar

Hiroshi Ishiguro · Fuki Ueno · Eiki Tachibana
Editors

Cybernetic Avatar

 Springer

Editors
Hiroshi Ishiguro
Graduate School of Engineering Science
Osaka University
Toyonaka, Osaka, Japan

Fuki Ueno
Graduate School of Informatics
Nagoya University
Nagoya, Japan

Eiki Tachibana
Graduate School of Engineering Science
Osaka University
Toyonaka, Osaka, Japan

ISBN 978-981-97-3751-2   ISBN 978-981-97-3752-9   (eBook)
https://doi.org/10.1007/978-981-97-3752-9

Cover credit: Photo by Kurima Sakai

This Springer imprint is published by the registered company Springer Nature Singapore Pte Ltd.
The registered company address is: 152 Beach Road, #21-01/04 Gateway East, Singapore 189721, Singapore

If disposing of this product, please recycle the paper.

Preface

With the advances in the Internet, AI, and robot technologies, our daily lives and professional environments have increasingly incorporated a variety of digital media. My research and development efforts have been centered around a new, emerging type of medium: avatars. Avatars, which include both teleoperated robots and computer-generated (CG) agents, enable users to operate in physically distant locations or even virtual environments, such as the metaverse. Attracted by their convenience and the broad spectrum of possibilities, various organizations, including companies and hospitals, have explored utilizing robot avatars. However, despite the initial enthusiasm, the widespread adoption experienced only a modest surge and failed to firmly establish itself. Nevertheless, the recent shift toward remote work driven by the COVID-19 pandemic has served as a catalyst for the resurgence of avatar technology, propelling its research, development, and practical application into a new phase. The large-scale societal shift toward telework has clearly highlighted avatar technology's potential to reshape how we work, communicate, and interact within a rapidly evolving digital landscape.

I am currently leading a project under the Moonshot Research and Development Program Goal 1, "The Realization of a Society in Which Human Beings Can Be Free from the Limitations of Body, Brain, Space, and Time by 2050". It was launched in 2020 by the Japan Science and Technology Agency (JST) with the aim to empower individuals from diverse backgrounds and with various responsibilities, such as caregiving or parenting, to pursue activities that align with their lifestyles, free from the constraints of their bodies, physical space, and the limitations of their brains and time. Our project, named "The Realization of an Avatar-Symbiotic Society Where Everyone Can Perform Active Roles Without Constraint," focuses on developing Cybernetic Avatars (CAs), a type of avatar that are designed to enhance people's physical, cognitive, and perceptual abilities through the integration of advanced technologies such as robotics and AI. Thus, our project aims to contribute to the realization of Goal 1, facilitating a world where people are liberated from the confines of space and time.

This book stems from the JST's Moonshot R&D project described above, presenting a vision of a future where avatars play an integral role in shaping the fabric of our interconnected society. It introduces our ongoing efforts to advance avatar technology and is structured into nine chapters. Chapter 1 discusses the potentially revolutionary impact of CAs as a new medium of communication, liberating individuals from physical barriers and creating more flexible work environments. Chapters 2–4 present developments in CAs with advanced autonomous functionality. Chapters 5 and 6 discuss the creation of a CA platform that connects multiple operators and CAs. Chapter 7 explores the physiological and neuroscientific effects of avatars and other media on operators and users. Finally, Chaps. 8 and 9 discuss the societal implementation of CAs.

Toyonaka, Osaka, Japan   Hiroshi Ishiguro
February 2024

Acknowledgment This work was supported by JST Moonshot R&D Grant Number JPMJMS2011.

Contents

About the Editors

Hiroshi Ishiguro received his Ph.D. from Osaka University, Japan, in 1991. He is a distinguished professor at Osaka University, the visiting director of Hiroshi Ishiguro Laboratories at the Advanced Telecommunications Research Institute (ATR), and a project manager there. His research interests are interactive robotics, avatars, and AVITA, Inc. In 2011, he won the Osaka Cultural Award. In 2015, he received the Prize for Science and Technology by the Minister of Education, Culture, Sports, Science, and Technology. He was also awarded the Sheikh Mohammed Bin Rashid Al Maktoum Knowledge Award in Dubai in 2015, the Tateisi Award in 2020, and an honorary doctorate from Aarhus University, Denmark, in 2021.

Fuki Ueno is a researcher in the Graduate School of Informatics at Nagoya University, Japan. She is working on the Moonshot R&D project managed by Hiroshi Ishiguro.

 **Eiki Tachibana** is a project assistant professor in the Graduate School of Engineering Science at Osaka University, Japan. He is working on the Moonshot R&D project managed by Hiroshi Ishiguro.

Chapter 1
Introduction: Cybernetic Avatar

Hiroshi Ishiguro

Abstract The chapter discusses the potentially revolutionary impact of Cybernetic Avatars (CAs) as a new medium of communication, liberating individuals from physical limitations and enabling them to work more freely. Through CAs, people with disabilities and the elderly will gain new opportunities for social engagement. Moreover, we envision a more inclusive society where humans coexist with robots and AI, fostering diversity and collaboration. The chapter also explores the transformative initiatives of our JST Moonshot R&D Goal 1, and the social and ethical considerations inherent in a society in which CAs and humans are integrated.

1.1 Avatar as a New Medium

With advances in the Internet, AI, and robot technologies, we have come to use various media in our daily lives and work. In particular, remarkable advances have been made in the media to support human communication. Starting with the radio, followed by the telephone, television, and the Internet, our lives and society have been transformed.

The reason people are so fascinated with media is that the purpose of our lives is to communicate with others. Human beings understand themselves by interacting with others, and they develop society together.

Ishiguro has focused on the research and development of avatars as a new technology for such media. Avatars are teleoperated robots or CG agents. They can be used to operate in remote locations or virtual spaces such as the metaverse.

However, the development of this avatar technology did not start until the turn of the millennium. I presented a very simple robotic avatar combining a videoconferencing system and a mobile platform in 1999 at the International Conference on Robots and Intelligent Systems (IROS) (Ishiguro and Trivedi 1999), and the first model of Geminoid, a robot avatar that looks exactly like me, was developed in

H. Ishiguro (✉)
Osaka University, Toyonaka, Osaka, Japan
e-mail: ishiguro@sys.es.osaka-u.ac.jp

© The Author(s) 2025
H. Ishiguro et al. (eds.), *Cybernetic Avatar*,
https://doi.org/10.1007/978-981-97-3752-9_1

1

2006 (Ishiguro and Libera 2018). Around 2010, several startup companies around the world were developing and selling robot avatars (Takayama et al. 2011).

However, it was a relatively small boom, and those robot avatars never took root in the world. At that time, remote work itself was not accepted. Not only the avatars, but also the idea of working daily with a videoconferencing system was not accepted by society. Therefore, companies, hospitals, and other organizations that initially introduced robot avatars with high hopes for their convenience and various possibilities gradually stopped using them.

However, the coronavirus pandemic has made remote work commonplace, and the research, development, and practical application of these avatars have become active again. During the pandemic, many people used videoconferencing systems to work remotely because face-to-face meetings were not feasible.

1.2 The Vision of Moonshot Goal 1 Avatar Symbiotic Society Project

The Moonshot Research and Development Program, which began in FY2020, consists of nine goals. Goal 1 of the project is "The realization of a society in which human beings can be free from the limitations of the body, brain, space, and time by 2050" (JST 2020). Ishiguro is working on research and development as one of the project managers for Goal 1. The following is a more detailed explanation.

The aim of Goal 1 is to enable people of various backgrounds and values, such as those who need to raise children and/or take care of the elderly, to participate in a variety of activities in accordance with their lifestyles. To achieve this, we need to create a society in which people are free from the constraints of the body, brain, space, and time. For this purpose, we need to develop Cybernetic Avatar (or CA, which means a developed avatar fused with artificial intelligence) technology, which can extend people's physical, cognitive, and perceptual abilities by highly utilizing a series of technologies known as cyborgs and avatars, and then apply it in a socially accepted manner. The research and development of CAs will be carried out while taking into account socially accepted ideas.

On the basis of Goal 1, Ishiguro aims for "The Realization of an Avatar-Symbiotic Society where Everyone Can Perform Active Roles without Constraint" (Ishiguro 2020). The goal is to create a society in which everyone can participate in a variety of work, education, medical care, and daily social activities without having to go to the workplace by freely operating multiple CAs remotely. In 2050, our lifestyles will change dramatically in terms of how we choose our locations, how we use our time, and how we expand our capabilities, and we will enable a symbiotic society with avatars that are in balance with society.

A more specific CA utilization is shown in Fig. 1.1. In education, a home class will be done by a teacher using a CA. Typical instruction can be provided by the CA's autonomous AI functions, and unexpected questions can be handled by the teacher via teleoperation, allowing him or her to operate many CAs simultaneously. However, students from all over the world can gather at the school using CAs and engage in various discussions.

The same will be true at work. Work can be done at home by inviting experts via CAs to the home, and meetings can be held at the office with members from around the world. This way of working will minimize commuting and allow people to work freely.

In medical care, simple medical examinations, such as those for a cold, will be performed at home by a doctor using a CA. This will greatly reduce the risk of spreading infectious diseases. However, in city hospitals, various specialists will examine patients using CAs, so even a small hospital in a city can function as well as a general hospital.

Fig. 1.1 Avatar symbiosis

As this happens, more and more people will use CAs as conversation partners in their daily lives, and they will attend parties, sports, travels, etc., as CAs. In other words, everyone, including the elderly and the disabled, will be able to participate freely in a variety of activities using a large number of CAs with expanded physical, cognitive, and perceptual abilities that surpass those of ordinary people. They will be able to work and study anywhere and anytime, minimizing commuting to and from work and school and allowing them to have more free time.

In this way, the Moonshot R&D project aims to create an avatar symbiosis society in which everyone can use multiple CAs to express themselves in various ways and to be active freely. CAs here refer to teleoperated robots and CG agents, and "free (自在 (jizai) in Japanese)" refers to the state in which CAs can act as the operator wishes by extending the operator's capabilities while taking into account the operator's intentions.

The concept of "free (自在 (jizai) in Japanese)" is particularly important. For a single person to use multiple CAs, the CAs must have the ability to perform tasks autonomously. The operator controls the CAs, which operate almost autonomously, and directly manipulates them only when necessary, thereby using multiple CAs simultaneously. In this case, the CAs must not ignore the operator's intentions and act autonomously but must act autonomously in response to the operator's intentions. This is "free." To manipulate CAs freely means to manipulate multiple CAs according to one's intentions, while making full use of their autonomous AI functions, etc.

1.3 How Will CAs Change Society?

I call the world in which CAs are intervening the virtualized real world (see Fig. 1.2). The virtualized real world is a world in which CAs combine the advantages of both the real world and the virtual world. Normally, we work in the real world using our physical bodies. The advantage of working in the real world is, of course, income. However, because we work in the real world, recovering from any mistakes we make is not easy. This is because failures always haunt our one and only flesh-and-blood body.

However, in the virtual worlds that have spread after the creation of the Internet, various worlds have been created, and many people can operate anonymously. Therefore, even if something goes wrong, they can easily move to another virtual world and continue their activities or continue their activities under a different name. However, the problem is that only a very limited number of people earn income from their activities in virtual worlds.

The virtualized real world is a world in which CAs integrate these advantages of earning income in the real world with the advantages of being different selves in the virtual world. In the virtualized real world with CAs, one can freely work with different CAs, becoming different selves.

This virtualized real world is expected to be achieved in Japan. In addition to Internet technologies, technologies to create humanoid robots (Moonshot calls them

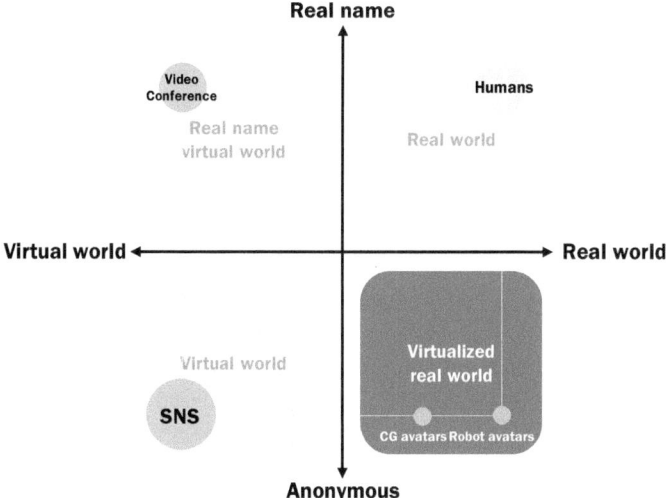

Fig. 1.2 Avatar symbiotic society

presence CAs) and CG characters (Moonshot calls them CG-CAs) are necessary to create a virtualized real world, and these technologies are Japan's strong points. In addition, Japan has a culture that easily accepts such robots and CG characters. In the West, humanoid robots and CG characters tend to be difficult to accept in daily life, but in Japan, for example, Vtubers (people who use CG characters to stream on YouTube) have already become commonplace and accepted in daily life. Therefore, the virtualized real world is expected to start in Japan. Once it takes root in Japanese society, it will be able to spread to the rest of the world. For example, just as the gaming world, such as Nintendo consoles (the Nintendo Switch and Sony PlayStation, etc.), has spread from Japan to the rest of the world, so too will the virtual world spread.

The virtualized real world is a virtual world created by CAs through their sensors and actuators, which are linked to the real world. In other words, the CAs provide the individual operator with a new existence, and a large number of CAs create a new society composed of CAs and humans. At the same time, CAs' AI functions, such as environmental recognition, can extend or replace the operator's recognition capabilities. Also, the operator can extend his or her own cognition of society (social awareness, human relationships, and the meaning of things) through the cooperation of a large number of CAs.

The virtualized real world achieved by CAs is a new world that offers a variety of possibilities and that will greatly enhance society. However, it will also cause new social problems, for example, (1) the real-world anonymization problem with CAs, (2) the ability expansion problem, (3) the multiple CA manipulation problem by a single operator, and (4) the human relationship augmentation problem. Because the virtual world of the Internet is independent of the real world, anonymity, capability expansion, and relationship expansion are not major problems. If a problem occurs, it can be solved by returning to the real world, and the problem can be handled independently between the virtual world and the real world. However, the virtualized real world, as shown in Fig. 1.2, is rooted in the real world, so these problems must be handled more carefully.

In the virtual world, various societies can exist at the same time, and rules and ethical norms can be created for each society. However, the real world is always a single society with a single set of rules and codes of ethics, and while CA technology can multiplex that single society, it is always rooted in the real world, which is a single society, and must therefore inherit some of the rules and codes of ethics of the traditional real world.

The rights and responsibilities of humans, CAs, and autonomous robots also need to be discussed. Questions such as whether or not CAs and autonomous robots can be allowed to do things that are not allowed to humans and how we can trust CAs and autonomous robots are issues that need to be considered immediately.

(1) **Real-world anonymization problem**: The first problem with CAs is that of anonymization, which has both advantages and disadvantages. Working with CAs means working with different faces, which in some cases is easier, while in others the inability to identify individuals can be a problem. If a police officer or a medical professional is not trustworthy in uniform, he or she can ask for identification when receiving services. In other words, an "avatar authentication mechanism" is needed.

(2) **Ability expansion problem**: As already mentioned, CAs can expand their abilities to work. However, this ability is not the one that the person has in the first place, and various problems may arise from this gap. The gap between the two is likely to create various problems: CAs can dramatically improve perceptual abilities. When humans interact with other humans, we do so with the belief that others have similar abilities as humans. We do not expect them to see things that we do not see. However, a CA with advanced perceptual abilities may use special sensors to see things that the other person wants to keep secret.

CAs can also be used to extend locomotion. They will be able to move faster than humans and act on the environment with more force than humans. They will be able to break things that humans cannot break. CAs may have overwhelming power that humans do not have. Such CAs cannot be used freely by everyone. Just as a driver's license is required to drive a car, a licensing system needs to be established to use CAs with superior mobility.

With CAs, we can work in a way that has a great impact on people. For example, a person can work as a famous historical figure or as a current celebrity. This augmentation of figures by CAs, like the augmentation of athletic ability, must be properly controlled and, if necessary, licensed.

(3) **Multiple existence problem**: One of the important goals of CA development in Moonshot-type R&D projects is to allow one operator to handle multiple CAs. At present, if an operator uses many CAs with different personalities, it may

cause confusion about his own personality. How serious this problem is needs to be determined by actual CA use.

If a single operator simultaneously uses a CA with the same personality that exists in several completely different locations, problems may arise for the recipient of the service. We humans live our lives assuming that people can only exist in one place at a time. However, if a single operator has access to multiple CAs, he/she can be in multiple places at the same time. The recipient of the service may feel that his/her CAs exist everywhere. If they have negative feelings about the CAs, they may feel that they are being watched at all times and in all places.

(4) **Relationship augmentation problem**: With CAs, we can operate in a different world from our normal self with a different personality. It could be said that we can start our lives over. In this case, the relationships between individuals will become more diverse. People who are already active in the virtual world, belonging to various communities, will be likely to have complex relationships to some extent, but these relationships will not necessarily be rooted in the real world. However, in the virtualized real world using CAs, truly diverse human relationships that are rooted in the real world will become possible. Such diverse relationships should expand our human potential and bring about many positive effects. However, some people may be troubled by the complexity of human relationships.

1.4 Challenges of CAs Research and Development

Solutions to these issues are not available right now, nor does this document address all of them. Most of the issues are currently under research. Through this book, we hope that readers will learn how CA systems are being researched, developed, and implemented in society and will consider how these issues might be solved in the future.

As mentioned, a variety of new social and ethical issues will arise in an avatar symbiosis society. Though a society full of great potential will be created, various problems will arise because of the new society. While carefully considering such issues, the Moonshot R&D project also aims to establish a code of ethics that embraces CAs in our society.

However, while such challenges remain in CA development, there is no doubt that it will bring about significant changes to society, and in particular CAs will be an indispensable technology for the achievement of diversity and inclusion. In the virtualized real world where CAs will be created, people will be free from physical constraints, and they will be able to work freely. People with various disabilities and the elderly who have difficulty moving their bodies will be able to work freely.

Thus, freeing people from physical limitations is critical to achieving a society of diversity and inclusion. Many kinds of discrimination that need to be eliminated from society often stem from the physical body. The history of discrimination has been that people have been mistreated for having a different skin color and for being physically disabled. Since the start of the millennium, the lack of recognition of gender diversity has also become an issue. All of these are caused by the physical body, and with CAs we can free ourselves from such physical limitations and be free to be who we want to be.

With CAs, human diversity can be achieved, and various people will be able to work together in the same world. I believe this will fully enable inclusion.

CAs, supported by AI technology, are becoming more and more capable. For example, CAs' visual and auditory abilities have already surpassed those of humans. AI technology can detect subtle emotions expressed in the facial expressions of the interlocutor, which cannot be judged by the human eye. For example, its ability to express itself using gestures is superior to that of humans. It is difficult for a person with considerable training to speak with beautiful gestures, but with AI technology a CA can add such gestures to the content of speech. In other words, anyone can easily operate CAs and expand their own capabilities through AI technology.

If the AI technology is applied to a CA's dialogue capability, it will become a CA that can perform simple dialogue autonomously. It will be a CA that speaks autonomously while following the operator's intention. However, an autonomous robot is a machine that makes its own decisions and acts on its own initiative. Here, we also refer to robots that act autonomously while following human intention as CAs.

In any case, not only CAs that are operated by humans speaking directly to them but also CAs that operate autonomously according to human intentions are expected to be active in future society. Such CAs will either be human or robotic. A CA that is operated by a human directly speaking to it may possibly be treated like a human. However, it may be difficult to distinguish whether the CA is operated by a human speaking directly or by AI following human intentions but operating autonomously. The reason is that AI technology has already become capable of human-like voices and human-like interaction.

In such a society, the boundary between humans and CAs will blur, and a society in which humans, robots, and CG characters all work together will likely be created. Such a society is the next step after diversity and inclusion. Humans will be freed from physical limitations, and a more developed form of diversity and inclusion will be possible. We will build a richer society with diverse colleagues, including not only humans but also robots and AI that support human society. I believe this is the society of the future.

References

Ishiguro H, Libera FD (2018) Geminoid studies: science and technologies for humanlike teleoperated androids. Springer, Singapore

Ishiguro H, Trivedi M (1999) Integrating a perceptual information infrastructure with robotic avatars: a framework for tele-existence. In: Proceedings 1999 IEEE/RSJ international conference on intelligent robots and systems. Human and environment friendly robots with high intelligence and emotional quotients (Cat. No.99CH36289). IEEE, pp 1032–1038

Ishiguro (2020) The realization of an Avatar-Symbiotic Society where everyone can perform active roles without constraint. https://avatar-ss.org/. Accessed 21 Jan 2024

JST (2020) Moonshot R&D: moonshot goal 1. Japan Science and Technology Agency. https://www. jst.go.jp/moonshot/en/program/goal1/. Accessed 21 Jan 2024

Takayama L, Marder-Eppstein E, Harris H, Beer JM (2011) Assisted driving of a mobile remote presence system: system design and controlled user evaluation. In: 2011 IEEE international conference on robotics and automation. IEEE, pp 1883–1889

Chapter 2
Development of Cybernetic Avatars with Humanlike Presence and Lifelikeness

Hiroshi Ishiguro, Kohei Ogawa, Yoshihiro Nakata, Mizuki Nakajima, Masahiro Shiomi, Yuya Onishi, Hidenobu Sumioka, Yuichiro Yoshikawa, Kazuki Sakai, Takashi Minato, Carlos T. Ishi, and Yutaka Nakamura

Abstract Cybernetic Avatars (CAs) are controlled by an operator through an interface that communicates movements, voice, or the intent of action. The operator can use the CA to perform activities remotely. In other words, the CA is the operator's alter ego. Therefore, the CA should have humanlike presence and lifelikeness. This chapter introduces related research, focusing on the development of a humanlike and life-like CA along with its interface technology.

H. Ishiguro (✉) · Y. Yoshikawa · K. Sakai
Osaka University, Toyonaka, Osaka, Japan
e-mail: ishiguro@sys.es.osaka-u.ac.jp

Y. Yoshikawa
e-mail: yoshikawa@irl.sys.es.osaka-u.ac.jp

K. Sakai
e-mail: sakai.kazuki@irl.sys.es.osaka-u.ac.jp

K. Ogawa
Nagoya University, Nagoya, Aichi, Japan
e-mail: k-ogawa@nuee.nagoya-u.ac.jp

Y. Nakata
The University of Electro-Communications, Chofu, Tokyo, Japan
e-mail: ynakata@uec.ac.jp

M. Nakajima
Tokyo Denki University, Adachi, Tokyo, Japan
e-mail: mizuki.nakajima@mail.dendai.ac.jp

M. Shiomi · Y. Onishi · H. Sumioka
Advanced Telecommunications Research Institute International, Seika-cho, Soraku-gun, Kyoto, Japan

© The Author(s) 2025
H. Ishiguro et al. (eds.), *Cybernetic Avatar*,
https://doi.org/10.1007/978-981-97-3752-9_2

2.1 Introduction

Cybernetic Avatars (CAs) are controlled by an operator through an interface that communicates movements, voice, or the intent of action. The operator can use the CA to perform activities remotely. In other words, the CA is the operator's alter ego. Therefore, the CA should have humanlike presence and lifelikeness. CAs were developed based on androids, which are humanlike robots. Geminoid is the most famous CA with humanlike presence. A CA with its own presence can act instead of the person controlling it. For example, a CA with the humanlike presence of a minister in a busy government ministry can perform official duties in various locations on behalf of the minister, dramatically increasing the minister's productivity.

However, a CA that closely resembles a person is not the best for every situation. Robotic and computer graphics technologies can be used to create a life-like CA that does not resemble a person; this CA can resemble cute animals or animated characters that are lifelike. Older and younger generations prefer such CAs. Humanlike CAs can sometimes be intimidating because they have a presence similar to that of humans, and CAs that only have a life-like feel, such as that of a small animal, are not intimidating and are easily accepted by many people. A life-like CA is particularly suitable as a dialogue partner for the elderly and young.

This chapter introduces related research, focusing on the development of CAs with humanlike presence and a lifelikeness, and their interface technology. In Sect. 2.2, we describe the development of a CA with humanlike presence. In Sect. 2.6, we describe the development of a life-like CA that does not have a humanlike presence, but resembles cute animals or animated characters. As an extension to these characteristics, we are developing a mobile childlike CA and a hugging CA that can be used in a mental therapy settings. These CAs are described in Sects. 2.4 and 2.5. An automatic motion generation system was required for these CAs to adequately express the intentions, emotions, and personality state; this system is described in Sect. 2.7. The construction of the system's basic learning model is described in Sect. 2.8. The basic technologies used in the interfaces of these CAs are based on cognitive science research and described in Sect. 2.3.

e-mail: m-shiomi@atr.jp

Y. Onishi
e-mail: y-onishi@atr.jp

H. Sumioka
e-mail: sumioka@atr.jp

T. Minato · C. T. Ishi · Y. Nakamura
RIKEN, Seika-cho, Soraku-gun, Kyoto, Japan
e-mail: takashi.minato@riken.jp

C. T. Ishi
e-mail: carlos.ishi@riken.jp

Y. Nakamura
e-mail: yutaka.nakamura@riken.jp

2.2 Research and Development of Cybernetic Avatar with Humanlike Presence

2.2.1 Research and Development of Human-Engaged Robots and Avatars

Until 2000, research and development in robotics focused on manipulation and navigation. Manipulation research and development began with the Unimate industrial robot developed by George Charles Devol Jr. in 1962. Navigation research began with Shakey's research and development at SRI International between 1966 and 1972. After approximately 2000, a new field of robotics research and development called interaction emerged. This research and development was triggered by the IEEE/ACM International Conference on Human–Robot Interaction (HRI), an international conference initiated by researchers worldwide, including Ishiguro, who is involved in the research and development of robots that interact with humans.

Ishiguro has been engaged in the research and development of robots, particularly androids, that have humanlike appearances. Figure 2.1 shows the flow of Ishiguro's research and development.

Ishiguro has been involved in the research and development of autonomous and teleoperated robots, such as avatars, throughout his career. In 1997, Ishiguro developed Town Robot, an autonomous robot designed to interact with people. In 2000, Ishiguro developed Robovie at the Advanced Telecommunications Research Institute International (ATR). In 2015, in collaboration with Kawahara and others at Kyoto University, Ishiguro developed ERICA, an autonomous talking android with humanlike appearance, for the JST ERATO Ishiguro Human Robot project.

In 1999, Ishiguro presented a teleoperated robot/avatar combining a video conferencing system and a mobile cart at the IEEE/RSJ International Conference on Intelligent Robots and Systems (IROS), now the largest international conference on robotics. Around 2010, many companies attempted to make this type of avatar commercially viable. One such example is Texai, which was developed by Willow Garage and funded by Google.

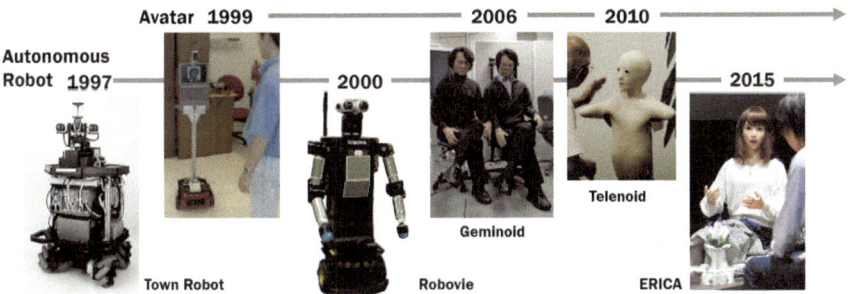

Fig. 2.1 Research and development of Ishiguro's avatars and autonomous robots

However, avatar development ended in a primary boom in 2010. By 2020, most companies stopped researching, developing, or selling robot-based avatars. One of the reasons that discouraged robot avatar research was rareness of remote work. However, during the COVID-19 pandemic, which began in 2019, remote work became more prevalent. Avatars have the potential to advance rapidly with the global of prevalence of remote work.

During the 2000s, Ishiguro continued his research and development of avatars and created Geminoid HI-1, which culminated in the development of an avatar that closely resembles his appearance in 2006. In addition, Ishiguro created a neutral-looking Telenoid that resembled a human; the Telenoid did not have a discernable gender, age, or personality.

People do not always accept avatars that closely resemble human beings. Children and the elderly tend to prefer avatars that do not closely resemble humans. To develop services that utilize avatars for people in various situations, researchers must develop different types of avatars.

Ishiguro has concurrently worked on the research and development of autonomous robots while developing avatars. This emphasis stems from the understanding that avatars and autonomous robots are complementary. Ishiguro initially researched and developed autonomous robots that interacted with people. However, to realize autonomous robots that can interact with people, we need data on how these robots engage with them. Ishiguro began his research on avatars to gather this data. Avatars could be used remotely to interact with people; then, data on how people interact with this avatar, which mimics an autonomous robot, could be collected.

Although Ishiguro began researching avatars for this purpose, he realized that avatars could be effective in assisting people with their activities. Since then, he has been involved in the research and development of both autonomous robots and avatars.

The R&D of autonomous robots is also important for the R&D of avatars. To make avatars more operator friendly, they must have various autonomous functions.

2.2.2 Operating Interface for HP-CA

A Cybernetic Avatar (CA) with humanlike presence, such as a Geminoid, an avatar with Ishiguro's appearance, is called an HP-CA.

The operational interface of an HP-CA, whose primary function is talking to people, is shown in Fig. 2.2. The operating procedure was as follows: The operator interacts with a human in front of the HP-CA while monitoring the video and audio from cameras and microphones installed in the CA's eyes and ears and from cameras and microphones in the room. The HP-CA's lip movement (Ishi et al. 2012), head movement (Sakai et al. 2016), and emotion estimation for facial expressions (Fu et al. 2023) were performed automatically. Because of the automatic estimation of the HP-CA's movement and facial expressions, the HP-CA's operator can operate the avatar without worrying about their posture.

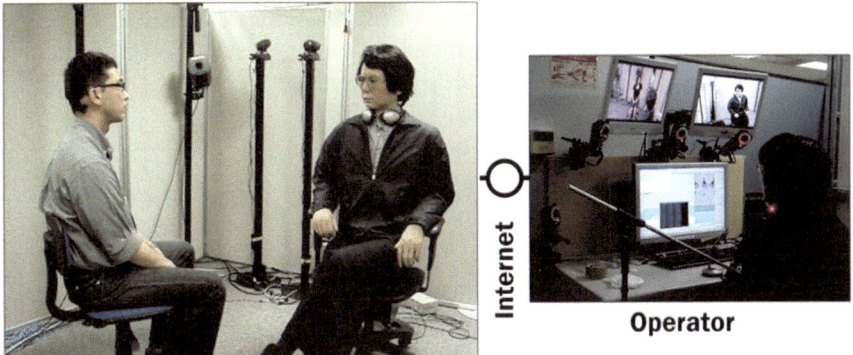

Fig. 2.2 Operating interface of the HP-CA

In general, a CA can be operated in two ways. One is to faithfully transmit the operator's movements to the CA and transmit information from sensors on the CA to the operator. This method is used when a CA is used to perform physical tasks. The other is the method used in dialogue-based tasks; such tasks convey only the operator's words and simple movements. This second method makes the operator feel like the CA's body is theirs because their voice and gestures are mimicked by the CA (Alimardani et al. 2013, 2016). Another important aspect of the second teleoperation method is that it mediates the operator's intentions.

In the teleoperation used in this section, the operator primarily communicates only their voice to the CA; the operator's voice communicates their intention. The communication of their intentions has multiple stages:

1. The operator conveys nonverbal emotional information, such as facial expressions, in addition to their voice, and the CA generates motions based on the voice and emotional information.
2. The operator conveys only their voice, and the CA estimates the emotion from the voice and generates motions based on the voice.
3. The operator conveys only the main points of their speech, and the CA generates speech, voice, and motions from these main points.
4. The operator gives the CA instructions, and the CA generates speech.
5. The CA observes the operator's behavior to infer the operator's intentions and behaves based on the inferred intentions.

A survey of the multiple phases of teleoperation reveals a problem in intent mediation. That is, does the intention to move the CA lie with the operator or CA? The first stage clearly places the intention with the CA; however, at stage 5, the CA presumes the operator's intention; thus, the CA's intention is largely reflected.

The intention of the operator or CA is also influenced by the intentions of the dialogue partner. When using a CA to provide services, the operator or CA must fully consider the intentions of the dialogue partner receiving the service. In other words, the intentions are not only those of the operator or CA but also those of the dialogue partner, and the problem of how these intentions are mediated must be dealt with. This problem is an important and interesting issue in CA research and development and must be addressed further in the future.

2.2.3 Research and Development of HP-CA in Our Moonshot Project

Current research and development of CAs is beginning to produce results in the above areas of intention communication. Research on estimating emotions and generating gestures from an operator's voice has begun to produce results. In addition, recent advances in large-scale language models have made it possible for operators to communicate only the main points of their speech to the CA.

For Ishiguro's HP-CA, Geminoid HI-6 (the 6th generation of Geminoid HI-1, which was developed in 2006), we designed eye gaze, head movements, and gestures in collaboration with experts on gestures, creating an HP-CA that has expressions similar to those of a human (Fig. 2.3). While expressive dialogue using gestures usually requires training, the behavior of this CA is generated automatically; therefore, anyone can use it to engage in an expressive dialogue.

2.2.3.1 Gesture Generation from Voice

To realize an HP-CA that is more expressive than the operator, we developed a function that generates motion according to the contents of the operator's speech. In the system developed in this study, each word was labeled as shown in Table 2.1, and appropriate motions were assigned to each label. The corresponding words listed in Table 2.1 are examples of implemented words, and the same motion may be assigned to different labels. For example, for the label "Greeting," the gesture is like bowing (Fig. 2.4a), and for the label "Realize," used when talking about the future, the gesture is that of opening and unleashing the mind (Fig. 2.4b). In addition, when CAs are

Fig. 2.3 HP-CA's gestures in dialogue (Geminoid HI-6)

Table 2.1 Examples of behavior and language correspondence

Label	Example of corresponding words
Greeting	Hello, good evening, good morning
Monitor	Figure, show
Realize	Realize, execution, carry, mellow
Body	Body, android
Impossible	Not, finish, cannot
Expect	Expect, want, wish
Me	Me, myself, Ishiguro
Develop	Development, accepting
Goodbye	Goodbye, bye bye
Fillar	Well
Excitement	Amazing, cool, surprised
Future	Deal with, future
Nod	Yes, that's right
Denial	No

talking about what they hold dear, as in the label "Expect," they use a gesture of closing their arms, symbolizing a protecting stance.

We developed HP-CA, which can present very rich gestures based on the correspondence between words and gestures generated from Ishiguro's pre-recorded speech.

Humans find it difficult to use beautiful hand gestures during presentations; they require considerable training to realize this skill. However, HP-CAs can deliver speeches while using ideal hand gestures without significant training.

Thus, the CA can be a useful tool for people with disabilities. Even if a person has a disability in his/her leg, he/she can work anywhere and anytime using a CA. The CA can also be a useful tool for able-bodied people. They can give presentations with expressive abilities beyond their own.

2.2.3.2 Lecture by HP-CA

We confirmed that Ishiguro's HP-CA can give lectures to an audience. The HP-CA has the potential to present lectures of the same or better quality than the person operating them because of psychological effects. Shimaya et al. (2019) studied communication using avatars and reported that conversations could be facilitated when the robot was operated remotely or in the same room. In addition, when the HP-CA is used as an avatar, it can provide a stronger sense of presence than video conferencing (Sakamoto et al. 2007).

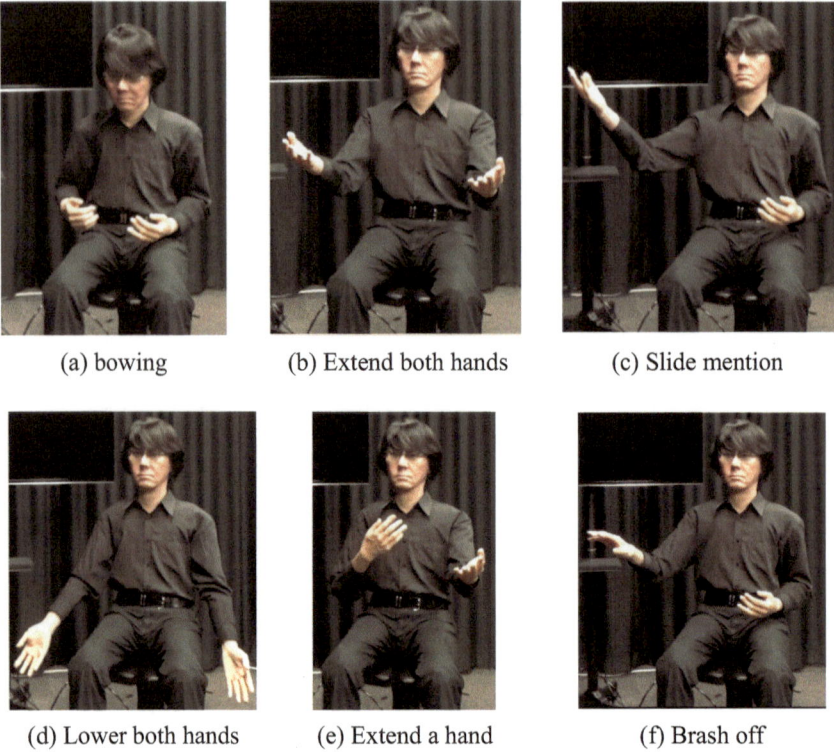

(a) bowing   (b) Extend both hands   (c) Slide mention

(d) Lower both hands   (e) Extend a hand   (f) Brash off

Fig. 2.4 Examples of gestures

Kamide et al. (2014) evaluated nonverbal information expressed by humanoid robots. According to their study, the actions of a humanoid robot giving a presentation, such as making eye contact with the audience or pointing at a screen, are important for improving the audience's understanding. In other words, the effective use of motion in robot presentations can enhance their impact.

Many other effects related to human interaction with avatars and robots have been studied (numerous research examples can be found in Ishiguro and Libera (2018)). Against the background of these studies, we evaluated an audience impressions when HP-CA was used to present lectures. Here, we administered a questionnaire after a lecture presented by an HP-CA, evaluated the impressions of the audience based on the answers obtained, and examined the HP-CA's educational and other effects. In the first hour of the lecture, the HP-CA gave a lecture with a voice and gestures based on a recorded voice and manually assigned gestures; in the next hour, Ishiguro remotely operated the HP-CA to interact with the audience. In the remote control, Ishiguro selected and expressed gestures according to the content of his speech.

The title of the lecture was "Avatar and Future Society;" the HP-CA introduced the avatars and autonomous robots that have been developed by Ishiguro so far. The lecture was given to students and their parents (approximately 900 people in total) at

Fig. 2.5 Lecture by HP-CA

the Tottori Prefectural Tottori Nishi High School; a voluntary questionnaire survey was administered to the students after the lecture. The lecture was held at the Tottori Prefectural Citizens' Cultural Hall, Rika Hall (Fig. 2.5).

Students were asked to answer questions regarding anthropomorphism, warmth, competence, discomfort, and educational effectiveness on a 5-point scale Valid responses were obtained from 245 audience members (male $= 102$, female $= 136$, non-response $= 7$) between the ages of 15 and 18 years (mean $= 16.16$, variance $= 0.898$).

In the lectures in this study, items related to anthropomorphism were rated highly for humanlike appearance. In addition, items related to warmth were rated high for sociability and kindness, whereas those related to emotions were rated low. These evaluations of emotion, kindness, and sociability are expected to be influenced by the operator's speaking style.

All items related to competence were rated high. This high rating may have been due to audience bias caused by being in a lecture and the operator being a university professor.

Discomfort was not experienced, although there were some variations in the evaluation of discomfort. This is believed to be influenced by the audience's interest in and familiarity with the robot.

The educational effectiveness score was above three for all questions. However, further investigation is needed because this factor was influenced by the audience's prior interest in the robot.

2.2.3.3 Real-Time Motion Generation by Word Prediction

We developed a function that generates gestures from voice in real-time. Because of the function's real-time nature, it requires predicting the occurrence of words in speech. Nishiguchi et al. (2017) proposed the rule that in situations where a robot converses with more than two people, the robot should prioritize actions over speech. To ensure that the avatar's behavior corresponds to its words, we developed a word prediction function that is based on a neural-network. The system predicts words and generates the avatar's behavior in the following steps: (1) Recognizing the speech of the avatar's operator. (2) Predicting the next utterance by using the speech recognition results as the input. (3) Searching for words corresponding to gestures. (4) Determining the appropriate gesture in an instance where an action corresponds to multiple words. (5) Controlling the avatar.

In this study, Bidirectional Encoder Representations from Transformers (BERT) was used to predict the avatar operator's next word. The accuracy of this prediction depends on the number of characters inputted.

The prediction accuracy increased monotonically from 10 to 50 characters and decreased slightly at 60 characters compared with the accuracy at 50 characters. Therefore, for the character counts tested in this experiment, the highest and lowest prediction accuracies were achieved with 50 (35.6%) and 10 (19.8%) characters, respectively.

To improve prediction accuracy, we are working on speech prediction using GPT, a large-scale language model. Instead of predicting words, this system recognizes the operator's utterance and predicts the subsequent sentence. This method improves real-time performance and enables the generation of motion according to the operator's intentions. We aim to develop a more general system by referring to related studies. These efforts will help us to continue improving the system.

2.3 Research on the Cognitive Aspects of a High Presence Teleoperation Interface

Teleoperated robots have become increasingly prevalent in various fields, allowing humans to remotely control robots in unsafe or impractical situations. These robots are used in search and rescue operations, scientific research, space exploration, and critical inspections (Rea and Seo 2022). However, despite their extensive use in specialized scenarios, remotely operated robots have not yet been widely adopted by the public for work and personal use (Bartneck et al. 2020). For the moment, even expert operators encounter challenges in teleoperation tasks; they struggle with basic collision avoidance and experience increased stress levels (Rea and Seo 2022). This challenge raises the question of how the average person would fare in teleoperation tasks, particularly in scenarios involving multiple robots controlled by a single operator.

The field of human–robot interaction (HRI) deals with how robots interact with humans in social settings (Bartneck et al. 2020). One of the primary obstacles to achieving effective teleoperation performance is the operator's ability, which can be constrained by technological limitations and interface design. To address this issue, researchers must focus on user-centered teleoperation interfaces that aim to reduce the cognitive load on operators and enhance their overall experience. In this section, we explore the significance of a good user experience and interface design in teleoperation systems, particularly in the context of controlling multiple robots.

Good user experience (UX) in teleoperation can be defined as a reduction in the overall load imposed on the operator during the task (Lichiardopol 2007). The overall load includes several types of load, the most significant of which is the cognitive load, which refers to the mental resources required to perform a particular task effectively. In teleoperation scenarios, reducing the cognitive load is crucial because it directly affects task performance and operator well-being. By providing an interface that minimizes cognitive load, operators can more efficiently control and monitor multiple robots, resulting in improved teleoperation performance.

2.3.1 Research Objective

The goal of creating an interface with a good user experience, that is, an interface that reduces cognitive load, can be realized in multiple ways. Notable examples include providing helpful information or making the controls of teleoperation intuitive, for example, using 3D (Regenbrecht et al. 2017) and multimodal interfaces (Triantafyllidis et al. 2020). In this section, we focus on reducing the information displayed to the operator to achieve a user experience with a reduced cognitive load.

There are several ways to achieve this objective, such as by entirely removing the background information. For our investigation, we blurred the information displayed to the operator. We used the blurring function because it is easy to deploy but effective for hiding information, which is useful for achieving the purpose of reducing the amount of information displayed. Furthermore, it provides a sense of control to the operator, as background information is still present and the degree or location of the blurring can also be adjusted. In addition, since blurring is now available for teleconferencing and other methods, people have become familiar and comfortable with it.

This section will focus on using the blurring function for when a single operator controls a single robot; however, this approach can easily be extended to a situation where a single operator operates multiple robots.

2.3.2 Proposed System Concepts

To achieve the goal of creating an interface with a good user experience, we propose a new robotic teleoperation system called COMPANIONS. COMPANIONS is an acronym for **COM**municating **PA**rtner **N** **I**ntuitive **O**perating **N**ovel **S**ystem. The system is based on the idea of mitigating information overload while preserving the intuitive aspects of the system. To achieve the desired aim, COMPANIONS incorporates the blurring of redundant background information.

When using blurring, it is assumed that a significant difference in operating ability and information overload is achieved. Even slight visual acuity and blurring can result in a marked change in a person's neuropsychological abilities for performing a task

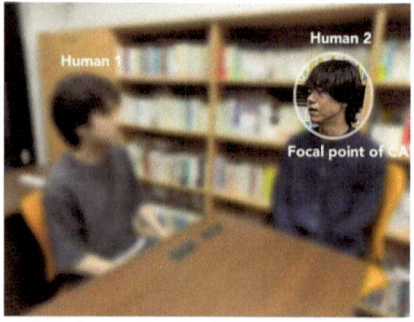

(a) Focal point on the face of active speaker

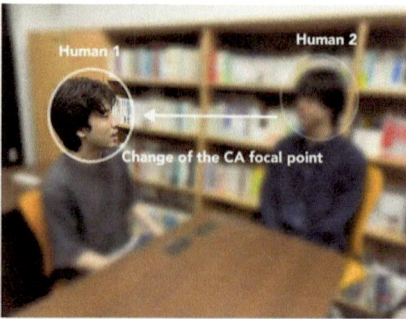

(b) Focal point changing to the new active speaker

(c) Focal point on the hand gesture of active speaker

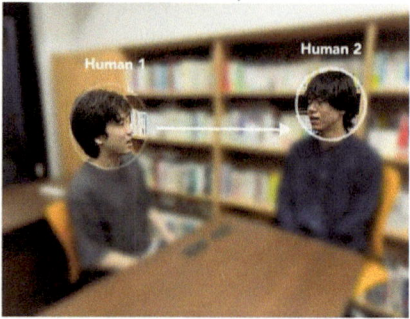

(d) Focal point on the active speaker in a quick back-and-back scenario

Fig. 2.6 COMPANIONS design for one robot—two humans interaction scenario

(Bertone et al. 2007). Taking this idea forward, it was postulated that by introducing blurring to redundant background objects, the overall quantity of objects that a user must focus on decreases. Thus, only the minimum possible information required for efficient operation would be displayed. COMPANIONS aims to use blurring to improve the operator's focus and operations through reduced cognitive strain when teleoperating a robot (Fig. 2.6).

2.3.3 Experiment

2.3.3.1 Objectives

The experiment had two main aims: (1) to test the effectiveness of information reduction on the mitigation of cognitive load for the operator, and (2) to test the subjective comfort (or preference) of the operator.

To achieve these aims, three experimental conditions were considered: No Blur, Blur, and Blackout, corresponding to full, partial, and no background displays,

Fig. 2.7 Three experimental conditions. Obtained after performing Chroma key operation and shown to the participants

(a) Blur Condition

(b) No Blur Condition

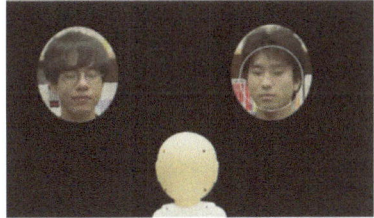

(c) Blackout Condition

respectively (Fig. 2.7). The rationale behind having three conditions is to show that the level of "background removal" matters and that blur will strike a perfect balance between showcasing the appropriate amount of information to the operator.

Thus, the experimental hypothesis is:

1. Hypothesis 1: Blur (Fig. 2.7a) and blackout (Fig. 2.7c) conditions will lead to a reduced cognitive load compared to No Blur (Fig. 2.7b).
2. Hypothesis 2: The blurred (Fig. 2.7a) condition will be more subjectively comfortable than the blackout (Fig. 2.7c) and no-blur (Fig. 2.7b) conditions.

2.3.3.2 Method

Experimental Situation

As mentioned above, there were three degrees of hidden background information that could be displayed to the operator during the experiment: full, partial, and complete. It is hypothesized that the blur condition, which strikes a perfect balance between

Fig. 2.8 Schematic of the experimental setup, with operator room (left) and interlocutor room (right)

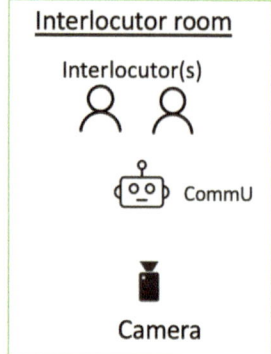

hiding information but still allows a holistic view of teleoperation, is preferred. Thus, three experimental conditions were considered: Blur, No Blur, and Blackout (hereafter referred to as black).

Experimental Procedure

A schematic of the experimental setup is shown in Fig. 2.8. There are two separate room setups: one housing the participants and the other containing the supposed interlocutors and CommU robot. The combined video was shown to the participants (operators) on a Tobii display. The operators controlled the eye gaze of the CommU robot using keyboard arrow keys while simultaneously interacting with the interlocutors.

Experimental Evaluation

Eighteen participants were recruited for the experiment. Among the 18 participants, 3 were female and 15 were male. Most participants were university students of ages 18–23.

Subjective impressions of the participants' mental workload were obtained through questionnaires using three types of psychometric studies: the Technology Acceptance Model (TAM) (Lewis 2019), NASA-Task Load index (TLX) (Stanton et al. 2017), and User Experience Questionnaire (Hinderks et al. 2018). The NASA-TLX was used to measure mental demand and provide a subjective evaluation of the cognitive load related to Hypothesis 1. The TAM and UEQ provide user impressions about the experiment, and as such, are related to Hypothesis 2.

Furthermore, the eye gaze based tracked based on the Tobii screen was used as an objective measure of the cognitive load. In conjunction with the questionnaires mentioned above, the eye-gaze data provides more varied means of evaluating the participants' cognitive loads.

There are two types of eye movement: saccades and fixations (Conklin et al. 2018). Saccades are involuntary movements between visuals and fixations refer to the fixing of the eye gaze on a particular visual (image or text). The eye movement distance was used as an objective measure of cognitive load. Research has shown that the eye movement distance can be used as a measure of cognitive load.

For Hypothesis 2, to measure the participants' subjective impression of the experimental conditions, a questionnaire employing a ranking system based on the participants' preferences was utilized. Additionally, the UEQ and TAM questionnaires were also used.

2.3.3.3 Results

The box and whisker plot shown in Fig. 2.9 illustrates the cognitive load aspect of the NASA-TLX questionnaire. We used the Bonferroni correction on the results of Friedman and Wilcoxon signed-rank tests to show that there was a significant difference between the blurred and no-blur conditions.

To analyze the results of the objective movement of an operator's gaze, we created a heat map of the gaze, as shown in Fig. 2.10. This figure shows a representative sample of a participant's gaze. The eye gaze is mostly centered on the two interlocutors; however, for the no-blur condition, the eye gaze was more scattered. The area marked by a red circle in Fig. 2.10 shows the distracted eye gaze segment.

To draw statistical insights about the eye gaze, we must perform further analyses. A larger saccadic movement between two fixation points indicates a large cognitive load. Hence, the overall eye-gaze flight across the display was recorded. Since eye movement during an activity indicate the level of cognitive load and gaze flight is indicative of mental effort, a statistical test of the normalized data can be performed.

The eye-gaze flight data are shown in Fig. 2.11. These data show that the total eye-gaze flight is different for different conditions; namely, conditions where the operator receives less information (blurred and black) have less gaze flight.

Subjective user preferences were gathered at the end of the three experimental conditions via a questionnaire that asked them to rank the three conditions in order of preference. The user preferences and mean rank obtained for each condition were

Fig. 2.9 Box and whisker plot of the cognitive load aspect of the NASA-TLX questionnaire

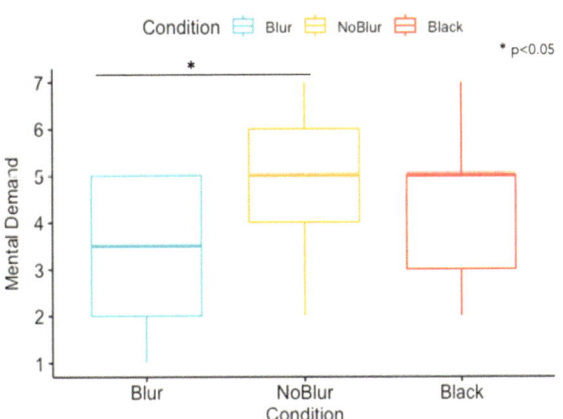

Fig. 2.10 Tobii eye gaze
heat map for the three
experimental conditions

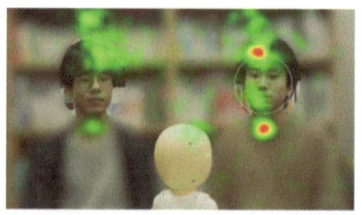

(a) Blur Condition

(b) No Condition

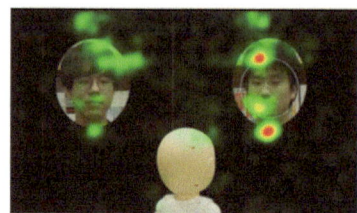

(c) Blackout Condition

Fig. 2.11 Box and whisker
plot of the normalized eye
gaze flight data for blur, no
blur, and blackout conditions

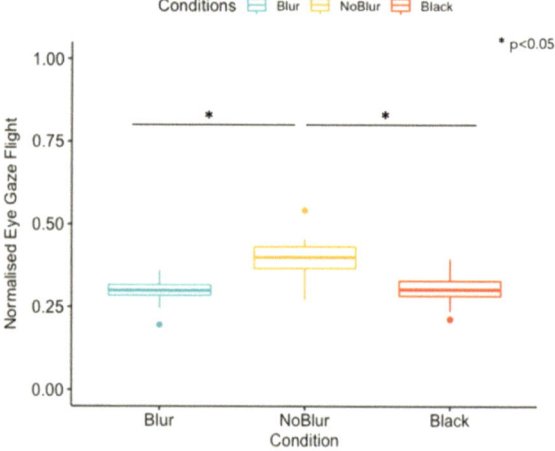

Fig. 2.12 Mean rankings of the experimental conditions

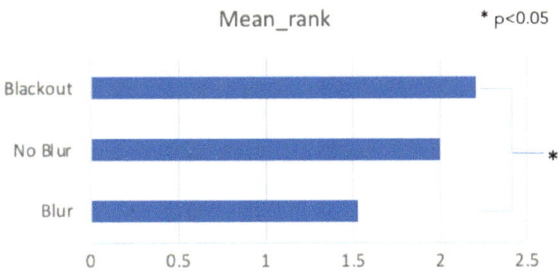

plotted. The mean rank is the weighted mean, where the mean of all the rankings for an option are weighted by the frequency with which an option is ranked at a certain position. Figure 2.12 shows that conditions where part of the operator's display is blurred are the most preferred, with a mean rank lower than the other conditions, indicating that most participants ranked this condition as ranks 1 or 2.

2.3.4 Summary

To test Hypothesis 1, three metrics were defined, and their corresponding results are shown. The blur condition reduced both the subjective (mental demand, Fig. 2.9) and objective evaluations (gaze flight, Fig. 2.11) of cognitive load for the no blur condition. Surprisingly, the blackout condition suppressed only the objective metric of the cognitive and not the subjective metric. The eye gaze flights of the blur and blackout conditions are significantly different from those of the no-blur condition. Hence, the above result partially supports Hypothesis 1; that is, the blur condition can reduce cognitive load but does not improve task performance.

For Hypothesis 2, the subjective user experience of the participants was measured using the ranking questionnaire; the results of the two questionnaires, TAM and UEQ, showed no significant difference among the three conditions; however, Fig. 2.12 shows that the blur condition ranked higher than the other conditions. In addition, statistical tests showed a significant between the blur and blackout conditions (Fig. 2.12). Thus, the participant's rankings support Hypothesis 2. The participants might have preferred the blur condition over the blackout condition because they found the blackout condition to be unfamiliar and distracting.

The partial support of Hypothesis 1 suggests that selectively blurring information displayed to an operator can help in realizing our goal of reducing the operator's cognitive load. Further, the partial support for Hypothesis 2 indicates that the goal of achieving a good user experience can also be realized with the blurring system.

Thus, our experimental results indicate that the blurring system can reduce cognitive load in the case of a single robot with a single operator; given the increased complexity and heightened mental demand, this interface design has the potential for use in a scenario where one operator controls multiple robots. The experiment described above has a few limitations: it only measured the qualitative impact of changing the amount of information displayed, it could only use fixed prerecorded video settings, and the participants were of similar demographics (age group, education level, nationality, etc.).

2.4 Research and Development of a Mobile Humanoid CA

In this section, we focus on the research and development of a mobile humanoid Cybernetic Avatar (CA) called Yui. The aim of this research and development is not only for Yui to resemble a human in appearance but also to strongly convey the presence of the operator to the interlocutor, thereby achieving natural emotional transmission. In particular, we detail the development of a head unit that allows for a human-like appearance and emotional expression by reproducing the operator's facial expressions through remote operation; we also develop an operation interface that provides a natural operating experience for the user and mirrors the operator's expressions in Yui, enhancing the conveyance of the operator's presence; finally, we investigate the significance of mobility mechanisms in humanoid CAs in facilitating natural communication between an operator and the interlocutor.

The humanoid CA, as a remotely operated robot, plays a crucial role not only in enabling remote tasks but also in deepening emotional communication between the operator and interlocutor. The CAs realistic appearance, facial expressions, gaze, and speech aim to strongly convey the presence of the operator, thus achieving the natural transmission of emotions. The proposed operation interface could allow the operator to feel the presence of the remote interlocutor and possibly vice versa, suggesting that both parties could feel as though they are conversing face-to-face.

Mobility in humanoid CAs significantly expands the range of tasks that the operator can perform remotely and enables more natural communication by allowing the adjustment of the CA's position relative to the interlocutor. Here, we propose a CA that employs a wheeled mobility system to move around. Despite being wheeled, the full-body movement of the humanoid allows for more effective interaction between the operator and interlocutor.

2.4.1 Realistic Head Unit for the Humanoid CA: Development and Features

Traditional remote-operation robot technology focuses on operability and enhances the operator's sense of presence (Darvish et al. 2023). However, in remote communication, the interlocutor's perception of the operator's presence is equally important. From this perspective, our project initially developed the head unit shown in Fig. 2.13a for the humanoid CA Yui (Nakajima et al. 2024). This unit strengthens emotional communication and interaction between the operator and interlocutor.

Yui's head unit adopts a childlike and gender-neutral design to facilitate communication with a broad audience. By adopting a design that minimizes distinctive features, an approach similar to that used in Telenoid (Ogawa et al. 2011) is adopted. The goal of this head unit is to evoke the presence of the operator through the behaviors and facial expressions of the CA. The head, including the three degrees of freedom in the neck, has 21 degrees of freedom and controls the eyeballs, skin, and overall posture. The adoption of electric motors, considering their quietness and responsiveness, facilitates natural movement. Some motors are used for directly driving joints, whereas others utilize both forward and reverse rotations to actuate two different points by pulling wires; this configuration enables the nuanced movements of the skin. This unique assignment of motor rotations increases the total number of actuation points, resulting in 28 actuation points.

The head unit is equipped with prosthetic eyes that contain wide-angle lens cameras (Fig. 2.13b) that balance a broad field of view with a humanlike appearance. Additionally, built-in microphones in both ears (Fig. 2.13c) provide the operator with stereo acoustics of the remote environment and directional information about the interlocutor, enabling communication as if the operator were in the same space as the interlocutor.

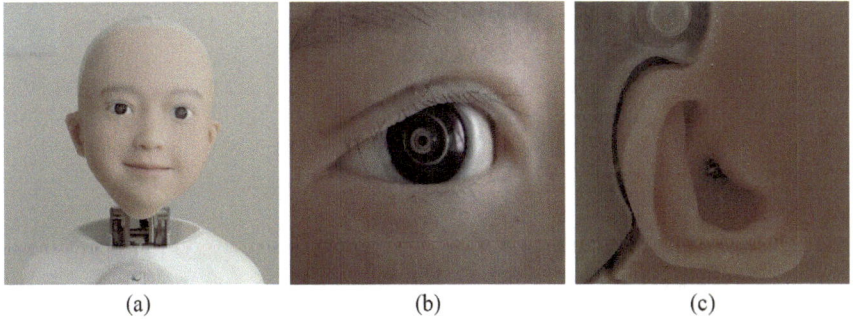

(a)   (b)   (c)

Fig. 2.13 Head unit for the humanoid CA Yui: **a** head unit; **b** prosthetic eye containing wide-angle lens camera; **c** microphone embedded in the prosthetic ear

2.4.2   Operation Interface for the Humanoid CA Operation

Each of Yui's eyeballs can move independently from side-to-side, while they move in unison when moving up and down, giving a total of three degrees of freedom. Synchronizing the movement of Yui's eyeballs with those of the operator enhanced the intuitive perception of the distance and relative positions of remote objects. In our experiment (Shinkawa and Nakata 2023), using the android Ibuki (Nakata et al. 2022), which has the same eyeball degrees of freedom, we synchronized the operator's eye movements with the android's to observe surrounding objects. The results confirmed that synchronizing eye movements reduces the operator's task execution time and workload. The reduction of these metrics indicates that facilitating gaze transitions with low operational loads during teleoperation can potentially streamline communication.

The microphones embedded in the humanoid CA's ears capture stereo sounds and present them as stereo acoustics to the operator. The shape of the ears allows the determination of the direction of sound sources, including the front and back, enabling the operator to experience a sense of physical presence.

The operator's facial expressions, recognized by cameras mounted on a Head-Mounted Display (HMD), are reflected in the movements of the humanoid CA's skin. This skin movement allows the operator's expressions, such as smiles, to be transmitted to the interlocutor in real time. In addition, the operator's speech, captured by the HMD microphone, is played through the CA's chest speaker.

2.4.3   Mobility Mechanisms in Humanoid CA

The introduction of mobility mechanisms in humanoid CAs is crucial for enhancing the operator's sense of presence. Past studies have shown that feedback resulting from avatar actions increases an operator's sense of presence (Ma and Kaber 2006). Mobility mechanisms may allow the operator to interact more dynamically with a remote environment, further enhancing the sense of presence. Furthermore, this mobility enables a more comprehensive observation of remote locations and more proactive interactions with people and objects, expanding what the avatar can do.

The integration of mobility mechanisms also significantly affects nonverbal communication. In particular, the adjustment of distance and positioning relative to the interlocutor is an essential element in communication (Hall et al. 1968; Kendon 1990); it facilitates more natural and effective dialogue.

Our humanoid CA adopts a wheeled mobility system. However, since our humanoid CA must closely resemble natural human movement, it must have human-like behavior during movement. Therefore, we installed a linear actuation mechanism on a differential two-wheeled cart to simulate the knee flexion and extension associated with walking; this mechanism causes the upper body to oscillate up and down (Nakata et al. 2022). This oscillation results in humanlike movement. In addition, it

Fig. 2.14 Mobile android Ibuki with a person

can allow people walking alongside the android to synchronize their walking cycles with the oscillation of the android's upper body (Yagi et al. 2022). Research on human relationships indicates that people tend to synchronize their walking cycles more easily when they have better first impressions (Cheng et al. 2020). If we can adjust the behavior of the moving humanoid CA, it could potentially facilitate communication between the operator and the interlocutor through full-body movements, promoting more natural interactions. Figure 2.14 shows the mobile android Ibuki with a person walking alongside it.

2.4.4 Potential Applications of Humanoid CA

The images in Fig. 2.15 represent various potential applications of humanoid CAs. These include roleplay scenarios conducted using the head unit in actual operation.[1] For instance, deep dialogue with students during supervision and discussion may be achieved, and emotional connections with patients may be deepened during remote medical consultations. Furthermore, gatherings with close friends, and emotionally engaging reporting, e.g., detailed reports of delicious food, comparable to face-to-face interactions, might be experienced. In language learning, practical skill improvement can be realized through practice, akin to face-to-face sessions. In addition, stronger emotional bonds with family members may be fostered in family conversations, even when the people having the conversation are not at the same place. These applications showcased the potential of communication through humanoid CAs.

---

[1] Yui: Android Avatar (English Subtitles) https://youtu.be/D0R2R-64RKU Accessed 17 Nov 2023.

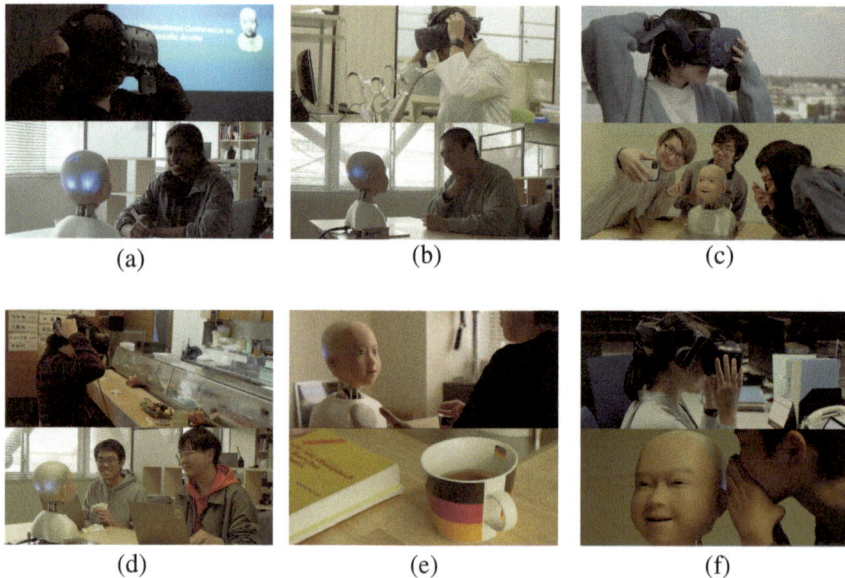

(a)   (b)   (c)

(d)   (e)   (f)

Fig. 2.15 Potential applications of humanoid CA: **a** supervision and discussion; **b** medical consultations; **c** gatherings with close friends; **d** emotionally engaging reporting; **e** language learning; **f** family conversations

While these applications are currently hypothetical, technical advancements in humanoid CAs have increased their feasibility. Future research should focus on evaluating and improving the practicality of such scenarios. In particular, the combination with mobility mechanisms can enrich user experience. In the long term, we anticipate new possibilities for humanoid CAs in fields such as education, healthcare, and social interaction.

2.4.5 Concluding Remarks

In this section, we discussed the development and features of the mobile humanoid CA Yui. Specifically, we introduced a head unit that reflects the operator's facial expressions; further, we introduced an operation interface that offers a natural experience for the operator along with an effective mobility mechanism. These additions showcase the potential of humanoid CAs. These technologies not only enhance the quality of interaction during remote communication but also pave the way for practical applications in everyday life and business. Future research should focus on refining these technologies and their applications to promote the practicality and adoption of humanoid CAs.

2.5 Development of a Huggable CA

2.5.1 Background of a Huggable CA

The development of hardware systems capable of interacting with individuals, operators, and both physical and virtual agents is pivotal in CA research. To achieve this capability, researchers have introduced various platforms, including humanlike androids (Glas et al. 2016; Shiomi et al. 2020b; Nakata et al. 2022), robot devices (Minami et al. 2021; Asaka et al. 2023), computer graphic-based agents (Yoshikawa et al. 2023; Moriya et al. 2023), and other platforms (Horikawa et al. 2023). These systems enable people to interact autonomously with CAs and operators through teleoperation. Other studies have demonstrated the importance of conversational interaction in various contexts: educational support for children (Kawata et al. 2022), online job interview training (Yoshikawa et al. 2023), and attentive listening in psychiatric daycare (Ochi et al. 2023).

For interactions between humans and CAs in particular human–robot interaction contexts, researchers have focused on touch interaction because of its effectiveness in natural and acceptable interactions. Human–robot touch interaction offers numerous benefits for individuals, akin to human-to-human touch. These advantages include mental therapy effects (Shibata 2004), pain reduction (Geva et al. 2020), encouragement of self-disclosure (Shiomi et al. 2021), improvement in motivation (Shiomi and Hagita 2016; Higashino et al. 2021), and stress buffering (Shiomi and Hagita 2021).

While touch interactions are effective in human-avatar interaction contexts, the current design of CAs primarily focus on verbal interaction, neglecting the potential for touch interactions between individuals, operators, and CAs. The CAs developed above successfully advanced conversation-based interactions, although their capabilities for touch interactions were limited. Rather than extending the capabilities of CAs that were not originally designed for touch interaction, we chose to develop CAs designed for touch interaction because such an approach is more useful for conducting research on human-avatar touch interactions.

Considering these factors, we developed huggable CAs that physically interact with people. We developed two types of huggable CAs: Moffuly-MS and Avatar Hiro-chan; the types of CAs allow for different types of touch interactions between humans and CAs. To facilitate huggable interactions with humans, we outfitted fabric-based capacitance-type touch sensors with Moffuly-MS and Avatar Hiro-chan. In the remainder of this section, we discuss the specifics of huggable CAs.

2.5.2 *Moffuly-MS: A Huggable CA with a Large Stuffed-Toy Appearance*

2.5.2.1   Hardware and Software

Moffuly-MS is a modified version of Moffuly-II (Onishi et al. 2023), which is a large teddy-bear-type robot that can hug people and touch their heads and backs while hugging. The Moffuly-MS resembles a gigantic rabbit (Fig. 2.16, left), and its hardware settings resemble those of Moffuly-II: approximately 200 cm tall with 110 cm arms and three DOFs for each arm (two DOFs for each elbow and one DOF for each wrist). We covered its metal frame and cotton cushions with fabric that acts as skin. Fabric-based capacitance-type touch sensors were installed on the skin to detect contact with people. We also equipped the CA with a removable mouth to reduce the risk of infection because people's faces come into contact the mouth during hugs.

The Moffuly-MS control device also resembles that of Moffuly-II. We used Raspberry Pi 4 to control each motor (Dynamixel MX-106R, 8.4 N m), read the sensor values on the fabric skin, play sounds during interaction, and enable teleoperation through a network. By reading the sensor values, such as motor joint angles and torque, the system controls the maximum number of joints and their speed during hug interactions to ensure safe contact with people.

We implemented two hugging behaviors: autonomous and teleoperated. In the autonomous setting, the Moffuly-MS closes its arms when an interacting person hugs it until the touch sensors on its arm detect contact. After detecting contact, Moffuly-MS stops its arm movements and maintains the hug. Moffuly-MS opens its arms autonomously after the person releases it.

In the teleoperated setting, the operator can control the hug behavior of the Moffuly-MS using a network. Moreover, using online conference systems such

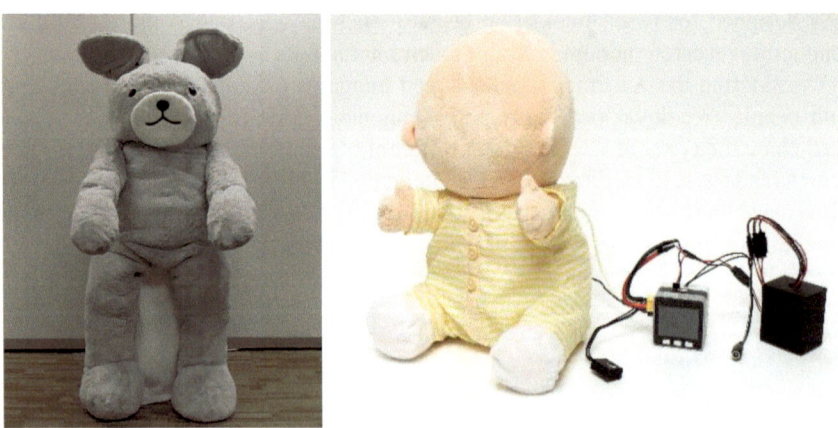

Fig. 2.16 Moffuly-MS (left) and avatar Hiro-chan (right)

Fig. 2.17 Squeezing gesture during hugging

as Zoom, the operator can directly talk to people through Moffuly-MS's hug interactions.

2.5.2.2 Intra-hug Gesture Designs

Intra-hug gestures, such as rubbing and squeezing, play essential roles in both human–robot (Block et al. 2023) and human–human hug interactions (Goodwin 2020; Mondada et al. 2020; Onishi et al. 2021; Dueren et al. 2021). For example, individuals in relationships touch each other's faces and heads to convey intimate emotions. Therefore, we implanted two intrahug gestures (rubbing and squeezing) in two areas (the head and back). Based on previous studies, the robot firmly held the interacting individual for three seconds (squeezing gesture) or vertically moved its hand for three seconds (rubbing gesture). In this implementation, we used Moffuly-MS's left hand or each gesture; its right hand continued to hold the lower back of the person it was hugging for both behaviors. The speed of the arm movement was set at approximately 10 cm/s to efficiently stimulate the C-tactile fibers using a rubbing gesture (Essick et al. 1999). Figures 2.17 and 2.18 show examples of squeezing and rubbing gestures during the hug interaction.

2.5.3 Avatar Hiro-Chan: A Huggable CA with a Baby-Like Stuffed-Toy Appearance

2.5.3.1 Hardware and Software

Avatar Hiro-chan is a modified version of HIRO (Sumioka et al. 2021), a baby-like stuffed device whose design is based on a minimal human design approach for interactions with seniors with dementia. Avatar Hiro-chan is identical to a faceless baby-like

Fig. 2.18 Rubbing gesture during hugging

robot named HIRO, which is approximately 30-cm tall. We chose this design to avoid the incongruence between faces and voices in the context of emotional representation. For example, if the robot's face expresses sad emotions but its voice suggests happiness, seniors with dementia may become confused during the interactions and discontinue them. As a previous study reported that senior citizens could imagine facial expressions by interacting with a conversational robot (Sumioka et al. 2014), we also excluded facial features from Avatar Hiro-chan.

The hardware settings of HIRO and Avatar Hiro-chan have different control systems. Avatar Hiro-chan uses M5Stack for teleoperation through a network, has one DOF for each arm for touch interactions with people, and can connect fabric-based capacitance-type touch sensors, such as Moffuly-MS, to detect touch interactions. Avatar Hiro-chan uses audio modalities to interact with people. Like HIRO, it uses recordings captured at the age of one year. We installed 91 voices: 20 positive, 25 weakly positive, 17 weakly negative, and 29 negative. Moreover, in the weakly positive voices, we installed three different babbling sounds, denoting meaningless articulations such as "mama."

For interactions between Avatar Hiro-chan and users, we again implemented autonomous and teleoperated modes. In the autonomous mode, similar to HIRO, Avatar Hiro-chan uses acceleration sensor outputs. When the values on the acceleration sensors have frequent significant changes (i.e., users who physically interact with Avatar Hiro-chan like caring for a baby), Avatar Hiro-chan's internal state becomes positive and then it uses positive voices in interactions. Because a past study found that such negative voices as crying discourage users from interacting with HIRO (Yamato et al. 2022), Avatar Hiro-chan uses fewer negative voices in its interaction even though the acceleration sensor values have not changed (i.e., users are not physically interacting with Avatar Hiro-chan).

2.5.4 Potential Use Cases

2.5.4.1 Moffuly-MS

In this subsection, we describe possible use cases for Moffuly-MS by considering the existing interaction studies between humans and CAs. One possible case is to investigate the appearance effects of huggable CAs by comparing them with an original Moffuly-II and other CAs with humanlike appearances, such as Geminoid (Sakamoto and Ishiguro 2009), ERICA (Glas et al. 2016), and SOTO (Shiomi et al. 2020b). Because appearance effects have an essential role in human–robot touch interaction (Shiomi and Hagita 2021), comparing touch interaction effects with different CAs provides interesting knowledge, including gender effects. Similar to this context, interaction with CAs may be influenced by such operator characteristics as gender and voice. Investigating the combination effects among the genders of users, the operators, and the CAs would be interesting.

Analyzing touch interaction behaviors using touch sensors is a typical type of study related to huggable CAs. In physical interaction between people and CAs, people may change their touch behaviors based on their attitudes and perceived feelings. Since Moffuly-MS has fabric-based capacitance-type touch sensors on its skin, it can easily gather people's touching behavior data. Understanding touching behaviors and their implied attitudes by analyzing their characteristics will facilitate both autonomous and teleoperated interactions to appropriately respond to user actions.

Another use case is investigating the effects of hug interactions on individuals with autism spectrum disorder (ASD). Several studies reported the effectiveness of using CAs in the context of conversational training for them (Kawata et al. 2022; Yoshikawa et al. 2023; Ochi et al. 2023), although these studies focused less on physical interaction between robots and such individuals. On the other hand, some studies reported that tactile feeling increases perceived comfort during communication with ASD individuals (Tatsukawa et al. 2016; Miguel et al. 2017). Related to this effect, researchers investigated the effects of a hug machine, which provides pressure to such individuals to calm the sensitive (Grandin 1992; Edelson et al. 1999). Following these studies, hug interaction with Moffuly-MS may provide positive effects toward conversational task-based training to ASD individuals.

Investigating the effects on operators is another interesting use case because existing studies on interaction between humans and CAs have mainly focused on the effects on users who interact with CAs. However, we believe that interaction by CAs may influence operators, e.g., receiving positive impressions toward the users due to the interaction contents by CAs. Providing touch interaction through CAs to users may also influence the perceptions of operators.

2.5.4.2   Avatar Hiro-Chan

In this subsection, we describe potential use cases for Avatar Hiro-chan based on current interaction studies between humans and CAs. One use case is to conduct a similar experiment with HIRO (i.e., one- or two-week trials at nursing homes) using Avatar Hiro-chans controlled by specialists and gather teleoperation logs to understand how professional people interact with dementia seniors. Such trials might provide important guidelines for designing robot behaviors for dementia seniors and a useful dataset to train autonomous controlling systems for Avatar Hiro-chan. For example, gathering voice timing and their contents would be useful for understanding how professionals forge interaction between them.

Analyzing caring behaviors toward Avatar Hiro-chan using both touch and acceleration sensors is another typical study related to huggable CAs. In particular in interaction with dementia seniors and Avatar Hiro-chan, understanding how the former treat it is essential for long-term acceptable interaction. Due to behavioral and psychological symptoms of dementia (BPSD) (Cerejeira et al. 2012), such aggressive behaviors as physical tantrums may occur. An autonomous recognition system for such situations may calm dementia seniors by changing Avatar Hiro-chan's behaviors and summoning staff to support such dementia seniors.

Another possible use case is to investigate the effects of multiple Avatar Hiro-chans for interaction with one user because of their size and simple teleoperation. Past studies in human–robot interaction showed the effectiveness of multiple robot conversations in the context of understandability and attracting people's intention (Sakamoto et al. 2009; Shiomi and Hagita 2016; Iio et al. 2021; Shiomi et al. 2023). Using multiple Avatar Hiro-chans might also provide fruitful knowledge in the context of multiple robot interaction with dementia seniors.

Moreover, focusing on both senior citizens and children is another interesting perspective in the Avatar Hiro-chan context. Our preliminary trials showed that children greatly accepted HIRO in after-school care environments; although we designed Avatar Hiro-chan for dementia seniors, it can be used to investigate how children interact with huggable CAs. For example, grandparents can operate Avatar Hiro-chan and interact with their grandchildren by teleoperation. Such childcare support trials through a teleoperated robot are an active research field in Japan. In fact, robotics researchers developed a teleoperated robot in the context of childcare support and investigated the social acceptance of such a robot system (Abe et al. 2018; Nakanishi et al. 2022). Similar to this context, analyzing operators' control logs while interacting with children offers interesting opportunities for comparing the differences in the behavior sequences of dementia seniors and children.

2.5.5   Summary

Existing studies related to interaction between CAs and people are mainly focused on conversational interaction, although physical interaction has an essential role in human–human interaction. Enabling physical interaction between people and CAs is critical for more acceptable and effective services by CAs. Moreover, adding such a modality increases our understanding of the effects of physical interaction on the perceptions of not only users but also operators. Therefore, this chapter introduced two kinds of huggable CAs designed to physically interact with people. The first is Moffuly-MS, which has a large, huggable rabbit-like appearance. The second is Avatar Hiro-chan, which has a small, huggable baby-like appearance. Both CAs have the capability to interact with people physically through fabric-based touch sensors. We also developed a teleoperated system that enables operators to interact with users by CAs and discussed possible use cases using two different kinds of huggable CAs.

2.6 Development of Life-Like CA and Mechanisms for Collaborative Conversation

CAs, which are capable of looking after the lives of children (Nakanishi et al. 2022), the elderly (Noguchi et al. 2023), and individuals requiring special attention, are posited to provide a sense of security to both guardians and dependents by offering personalized services through dialogue. However, if a CA possesses an overly strong presence that is difficult to ignore, users may become overly reliant on it, incurring increased costs for generating dialogue content capable of meeting such excessive expectations. Furthermore, the presence of an overly dependable CA may inadvertently impede user initiative in their activities. In response, the development of small-sized, life-like CAs, that do not have human-level presence but resemble cute animals or animated characters that are lifelike, is expected to serve roles similar to those of pets. Such entities can integrate seamlessly into living spaces without being imposing while still engaging deeply with users when necessary.

There are two possible mechanisms that enable artificial agents to communicate with humans: automatic and teleoperated dialogue systems. Considering the cost of human resources, caring for children and the elderly necessitates the implementation of autonomous dialogue systems. However, given the high individualization (and associated costs) required and the difficulties in building highly adaptive general artificial intelligence, such systems should ideally be hybrid in nature, engaging in human intervention only in scenarios demanding emergency responses or attentive listening. However, this implies that human operators must manage multiple CAs. Remaining aware of dialogues provided simultaneously by various CAs at different locations is challenging; hence, a mechanism that assists operators in swiftly grasping dialogue history and initiating appropriate conversations upon transition is necessary.

Recently, another approach considered for providing a flexible ability to sustain conversations with humans involves enabling multiple robots to talk collaboratively (Arimoto et al. 2018; Nishio et al. 2021) and increase their influence (Shiomi et al. 2020a; Okada et al. 2023). Accordingly, this study aims to develop a life-like CA empowered by collaborative dialogue with paired CAs. The goal is to support operators in simultaneously managing dialogue services through multiple pairs of CAs in various locations. This section introduces a system of multiple cooperative CAs that can not only autonomously provide dialogue services in pairs but also allows an operator to smoothly participate in a conversation with a user via a CA, with the second CA serving an assistive role in the understanding of dialogue history and initiating conversations. Two field experiments using the proposed system demonstrated its feasibility.

Furthermore, technology that controls multiple CAs and supports operators in communicating with others was expanded to create a new online communication environment called CommU-talk. Each user can use a 3D model of a small humanoid robot named CommU as their CA to talk to other avatars and the proxies of other participants. Notably, the control signal from each user to their avatar is also utilized to partially control other avatars, which facilitates user communication by displaying supportive nonverbal responses toward their respective avatars. Finally, future work is discussed based on current accomplishments.

2.6.1 Collaborative Dialogue System of Multiple CAs for Simultaneous Dialogue Services at Multi-locations

If CAs capable of autonomous dialogue services in multiple locations are deployed, with human operators only intervening to control the CAs when autonomous operations fail, a small number of operators could use these CAs to provide high-quality dialogue services at multiple sites. For this dialogue service, where one operator manages N CAs (a 1:N CAs system), each CA must sustain dialogue, and the operator should seamlessly take when necessary.

However, sustaining dialogue is generally not a simple task. Autonomous robots are expected to sustain dialogue and gather information, even after some failures in conversation. Moreover, transitioning to remote dialogue is not straightforward. It is challenging for operators to instantly understand the history of a conversation, which makes it difficult to develop appropriate dialogue flow after sudden operator participation. In other words, operators face the challenge of not only understanding the context of a conversation but also immediately and simultaneously initiating hospitable dialogue. In response, our project considers the utilization of a method that achieves dialogue continuation through the collaboration of multiple robots (Arimoto et al. 2018; Nishio et al. 2021), which resolve two challenges: continuing dialogue and handing over conversations between CAs and the operator (Fig. 2.19).

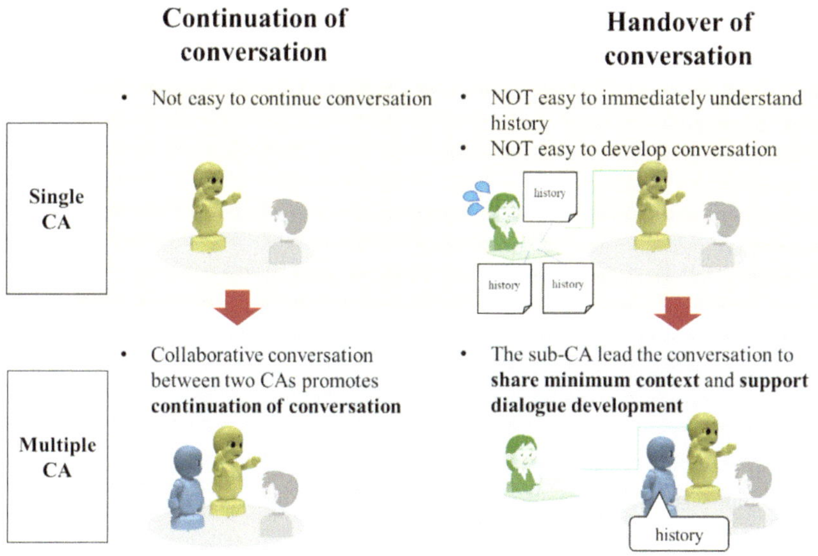

Fig. 2.19 Challenges in 1:N CAs control systems (continuing dialogue and switching conversations between automatic and teleoperated modes), and the expected advantage of using multiple CAs for the systems

2.6.1.1 Dialogue Continuation Through Multiple CAs Collaboration

We have devised a method called "Dialogue without Voice Recognition" that continues dialogue even when human speech cannot be recognized; this method coordinates multiple robots that actively prompt speech from humans and engage in conversations with each other or take turns speaking (Yoshikawa et al. 2017). To test our method, we constructed situations in which robots continued to respond evasively without basing their statements on human speech. Past studies have shown that using multiple robots for this type of dialogue is less likely to make people feel ignored compared to when a single robot conducts "Dialogue without Voice Recognition." Further, using multiple robots improves the perceived sense of conversation (Arimoto et al. 2018).

Thus, we developed a system that offers prolonged dialogue opportunities to the elderly by deepening the dialogue based on recognized results when possible, and otherwise, maintains the dialogue without failure through "Dialogue without Voice Recognition." Here, the expectation is that having one robot respond when a person does not speak gives the person a sense of reassurance that they do not need to speak, allowing for dialogue continuity. Indeed, in a demonstrative experiment conducted at an elderly care facility, it was reported that over 50% of the elderly did not cease dialogue for more than 30 min, indicating that a significant level of dialogue continuation was achieved (Nishio et al. 2021). The same strategy was applied to build a recommendation dialogue system by repeating and answering questions and demonstrate its practical feasibility through a field experiment to recommend food at a food court in a shopping mall (Sakai et al. 2022). The same recommendation dialogue system was adopted for the multiple CA dialogue system introduced in this section.

2.6.1.2   Handover of the Conversation Between CAs and the Operator

Using multiple CAs in one location has more advantages than the continuation of conversation in the automatic mode. When switching, the operator must not only understand the context, but also immediately start a hospitable conversation. To facilitate this, if the second robot automatically generates a summary of the past conversation, the operator not only becomes aware of it but can also share it with the guest, establishing it as the focal point for further development in the conversation. In addition, even after starting a conversation with the operator, the second robot can support the operator in advancing the conversation through their main avatar.

2.6.1.3   Implementation of Multiple CAs Dialogue System with Supportive Handover

Figure 2.20 illustrates a rough sketch of the proposed system, which has two modes. The first is the autonomous mode on the left side. The second is the teleoperation mode on the right side. In the autonomous mode, the Q-A-based multi-robot dialogue system (Sakai et al. 2022) maintains dialogue by asking the human guest questions. Questions, responses, and comments were provided alternately by the two CAs. When either CA spoke, their voice was generated with text-to-speech (TTS) software or a pre-recorded sound file was played with the loudspeaker associated with the speaking CA; the movements of the speaking CA's mouth (open-close) and arms (drumming gesture) were produced in synchrony with the CA's utterance to make the interlocutor feel as if the sound was produced by the CA's mouth. Note that one of the CAs is prepared to speak with the same voice as the operator to conceal the unnaturalness of the handover between the autonomous and teleoperation modes.

In the teleoperation mode, the operator talks through one of the CAs by talking in front of the microphone associated with the CA's laptop interface. The captured voice is converted by a voice-changer software to sound like a CA and is produced from the loudspeaker associated with the corresponding CA. Note that the captured voice is also analyzed to detect formants that generate synchronized mouth (open-close) and arm (drumming gesture) based on the simplified version of Ishi's method (Ishi et al. 2018).

The most distinguishable feature of the method is that it triggers words that share dialogue history, which ensures successful switching. In the system used in the field experiments introduced in this section, the sub-CA asks the main CA, which is controlled by an operator, questions to summarizes or refers to the answers obtained from the guest user interacting with the CAs. The operator then starts developing

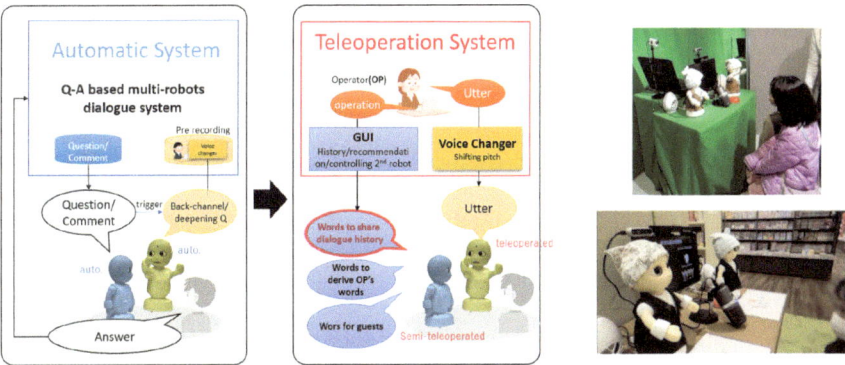

Fig. 2.20 Multiple-CA dialogue system with supportive handover (left) and its field experiments in an amusement park (right top) and children's bookstore (right bottom)

conversation to recommend something in accordance with the context. For example, the sub-CA may ask "Well, I understand that the guest likes white tigers very much. Do you know something to recommend to a person who loves white tigers?" When the sub-CA asks the main CA, the operator must perceive that they must answer the questions through the main CA and should be able to naturally develop a conversation for the recommendation by answering the question; this process might reduce the difficulty and cognitive load of handing over the conversation.

Meanwhile, the operator can also trigger utterances from the second robot through a simple button-type GUI, which is implemented to derive the necessary and convenient utterances for task accomplishment. In the recommendation conversation, the sub-CA generates FAQs about the recommended item, which prompts the operator to provide information included in the answer of the FAQs such as "Where can the guest find the recommended item in the shop?" or "How much is it?" This function allows the operator to convey recommendations without being too pushy.

2.6.1.4 Evaluation of Improved Conversation Quality in Souvenir Recommendation in Amusement Park

A field experiment was conducted in an aquarium and zoo (Nifrel, EXPOCITY) frequented by a significant number of children. Two conversational booths, each housing a pair of small desktop humanoid robots, called CommU CAs, were installed near the park exit gates. These installations aimed to provide a dialogue service that introduces visitors to souvenir shop products located beyond the exit.

Children were invited to sit in front of the CAs and initiate autonomous dialogue by pressing a button. In this mode, the CAs inquired about the children's experiences in the park and asked them questions about their preferences; their responses were used for souvenir recommendations and to deduce the children's preferred park experiences. After several preferences were determined and potential recommendations were narrowed down, an operator started a conversation that introduced the selected souvenir items. After arriving at the booth, the operator chooses from a list of potential items to introduce. The CA without an operator prompted the operator to refer to the user's responses, facilitating the introduction of products based on the autonomous dialogue. This system was designed to enable the operator to listen to a summary of the prior conversation; the summary enables the operator to introducing souvenirs that the child might prefer and gain insight into the rationale behind product recommendations. This approach allows for a seamless transition from autonomous to remote operated dialogue.

We compared the days when only the autonomous mode was used both to gather information from the children and recommend an item to when the operator performed this task. On days when the operator managed recommendations, there were occasions when only one pair of robots had a guest. However, in many cases, both pairs had two guests simultaneously. In such cases, the robots must maintain conversations before switching to the operator. When switching in, the operator should be able to start a conversation immediately. In approximately 68% of cases where the operator took over and completed services, both booths were occupied, following the proposed method for transitioning from autonomous to teleoperated modes. On days when the operator managed the recommendations, the conversation was rated better than those when only the autonomous mode was used.

These results suggest that the proposed handover method successfully transitioned from the autonomous mode to the teleoperated mode without introducing a significant discontinuity that could disrupt the recommendation conversation. Meanwhile, supplementing autonomous dialogue with a human touch through the teleoperated conversation improved the overall impression.

2.6.1.5   Evaluation of Reduction of Teleoperator's Load to Handover in Children's Bookstore

Similarly, a pilot study was conducted in the picture book section of a bookstore (TSUTAYA, EXPOCITY store) visited by several children. Two conversational booths, each featuring a pair of small desktop humanoid robots called CommU CAs, were set up at two locations within the picture-book section; these booths provided a dialogue service that introduced picture books to the guest children.

The procedure for engaging the children in dialogue was akin to that of Nifrel, as explained in the previous subsection. Upon pressing a button, an autonomous dialogue began; during this conversation, the CAs asked the children about their preferences based on picture books they previously read previously. This information was used to recommend new picture books the children might enjoy. Once preferences were collected and recommendations were narrowed down, an operator was prompted to take over the conversation and introduce the picture book that would be recommended to the child. Following the same protocol as Nifrel, the nonoperator-controlled CA assists the transition to the teleoperated mode by referring to the user's responses to the recommended item.

During the transition between the autonomous and teleoperated modes, we prepared conditions both with and without implementing the proposed collaborative handover to assess its impact on operator support in the dialogue system transition. Specifically, the ease with which the operators entered the conversation was assessed. The NASA Task Load Index (NASA TLX), a tool used to evaluate the burden of performing a task, was also used in this evaluation.

This study compared operator evaluations on days when customer service was provided with and without the proposed collaborative handover. In this experiment, two pairs of twin CAs were used by a single operator. The operator changed every hour and was asked to evaluate the load felt during the operation. Seventeen operators used the system with the proposed method while another seventeen operators used the same system without the proposed transition method. The proposed method was confirmed to be superior in terms of the operators' workloads, including factors such as time pressure. This superiority suggests that the proposed method may reduce the cognitive and psychological burden on operators during the management of and entry into dialogue. Moreover, although an increase in the purchasing rate of recommended products was not tested for the possible conditions of this investigation, an increase in the purchasing rate of recommended products compared to the period before the experiment was observed. This indicates that using CAs for recommendations may aid in the selection of picture books.

2.6.2 Extending Multiple CAs Dialogue System to Online Conversation

Online conferencing systems such as Zoom, Skype, Microsoft Teams, and Google Meet have become globally ubiquitous, a trend that has significantly accelerated with measures taken to combat the spread of COVID-19. Recent advancements in avatar technology have provided users with the option of employing computer-generated avatars (CG avatars) instead of using their actual faces in these interfaces. Concurrently, the concept of the metaverse—a virtual realm grounded in virtual/augmented reality technologies—has gained prominence. In this metaverse, social activities, including conversations, are expected to occur through avatars that run parallel to the physical world. Avatar-mediated communication is expected to foster a sense of security among users by enabling them to conceal their true identities. Close alignment of movements between users and their avatars can provide users with a heightened sense of body ownership and agency (Daprati et al. 1997; Blakemore et al. 1999; Nishio et al. 2012).

However, it remains unclear how users can attain a strong sense of participation in conversations and behave naturally in the virtual world. This ambiguity stems partly from a lack of precise methods for perceiving and acting in a virtual environment. This deficiency inhibits the emergence of natural, timely, and socially appropriate responses between avatars, which is essential for cultivating the aforementioned sense of immersion. To overcome this challenge, we hypothesize that the technology introduced in this subsection, which involves the control of multiple CAs by an operator to facilitate operator-to-operator communication, would be advantageous for enhancing participation in the virtual world. This technology empowers the operators to harness other avatars to shape the communication scene according to their desired outcomes, as demonstrated in this subsection.

2.6.2.1 Implementation of Semi-autonomous Social Avatar Room: CommU-Talk

The method of controlling multiple CAs to support operator communication was extended to create a new online communication environment called CommU Talk (Fig. 2.21). An online virtual conferencing system was used with computer graphics robots (CG robots) representing the participant's proxy agents. A three-dimensional CG model of a humanoid robot called CommU was used to draw CG robots that talked to each other; therefore, this conferencing system was called CommU-Talk. Each user used a 3D CG robot as a proxy agent to talk to other avatars and participants. Notably, the signal of each user used to control his/her avatar was also used to partially control other avatars to support user communication by exhibiting supportive nonverbal responses toward his/her own avatar.

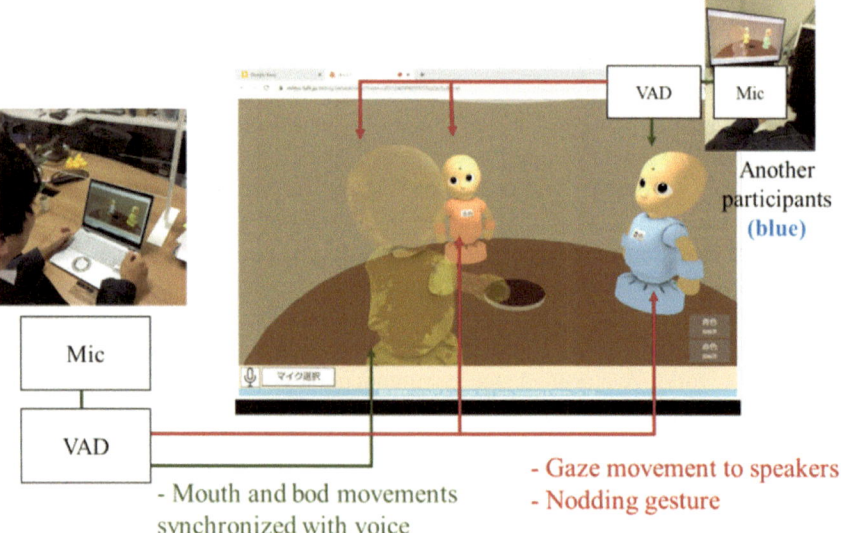

Fig. 2.21 Information flow in CommU-Talk

In CommU-Talk, the participants' voices were captured using microphones and listened to each other. Furthermore, the captured voices were utilized to automatically produce non-verbal behaviors of CG robots so that they looked like they were actively speaking and attentively listening to each other; the speaking CG robot moved its lips and hand gestures in synchrony with its voice, while the listening robots directed their gaze and nodded toward the speaking robot. The automatic function to produce nonverbal behavior synchronized with voices is expected not only to simplify the control of CG robots but to also provide users with a rich sense of agency; additionally, it ensures attention from others, independent of how the participants typically achieved it.

2.6.2.2 Applications of CommU-Talk

CommU-Talk is expected to be used to alleviate communication hurdles that individuals may experience with other participants; this is achieved by concealing their identities and subtly representing the emotions of all participants while maintaining essential social signals to foster the senses of being listened to and looked at. Therefore, it is anticipated to encourage participants who might feel hesitant about certain people, such as their boss or less familiar people, to talk with an improved perception of social presence, co-presence, and overall meeting experience (Yasuoka et al. 2022). Such a reduction in hurdles has also been utilized to build a less intimidating easier environment for practicing communication for individuals with communication challenges, such as those with ASD. Yoshikawa et al. showed that CommU-Talk can be used by people with ASD. They asked three groups of five individuals with ASD to conduct a mock interview in which two interviewers interviewed an interviewee (Yoshikawa et al. 2023). In total, 25 sessions were conducted, each consisting of interview practice and group discussions to reflect on the interviewee's performance using the recording of the interview, which was taken in the avatar room. The results demonstrate that the participants tolerated the repetition of conversations, and their experiences were sufficiently real, demonstrated by an improved interview performance in physical interviewers after the virtual test interviews.

The center window shows the interface for the participant depicted on the left and includes his yellow avatars as well as the blue and red avatars operated by other participants. The operator's voice activity detected by the microphone (VAD) is used to generate his avatar's synchronized mouth, body, and gaze movements as well as the nodding gestures of the other avatars.

2.6.3 Summary

In this section, we introduce a multiple CA dialogue system that allows an operator to control multiple CAs and establish successful conversations with an interlocutor interacting with the CAs. The proposed method empowers the operator to provide dialogue services simultaneously at multiple locations, leveraging the advantages of collaborative conversation strategies between two pairs of CAs. Specifically, it enables the continuation of conversation in the autonomous mode and seamless conversation handover during the transition from the autonomous to the teleoperation mode. We demonstrated the feasibility of these capabilities through field experiments and evaluated their performances in terms of recommendation quality and operator workload. To effectively manage multiple dialogue services simultaneously, future research should focus on technological improvements such as summoning the operator when necessary, summarizing past conversations for seamless handovers, and suggesting the use of a secondary CA during conversations to reduce the cognitive burden on the primary CA.

The concept of a multiple CA dialogue system, designed to assist operators in communication, is further extended to create a semi-autonomous social avatar room called "CommU-Talk;" the aim of this room is to enhance online conversations through avatars. Methods that can be used to enhance a user's psychological experience in the semi-autonomous social avatar room must be specified. This can be achieved by modeling how humans recognize and engage in multiparty conversations, considering the influence of cultural and clinical backgrounds. These considerations are essential for advancing the adoption and application of these technologies.

2.7   A Study on Generating Natural CA Motion Without Being Aware of Teleoperation

CAs with humanlike presence are humanlike in appearance, which makes their interlocutors experience them as having humanlike presence, even in remote locations. However, the control required to generate their humanlike presence, like mimicking human motion, is complex. For the operator to control the CA as intended and provide the presence of the CA through its movements, it is impractical for the operator to fully control the CA movements according to his/her own state. To realize a CA with humanlike presence, a function that recognizes the operator's intentions, emotions, and personal characteristics and reflects them in the CA's movements must be developed. Such a function would enable the operator to freely operate the CA without being aware of the teleoperation.

This section introduces methods for generating CA motion to operate CAs at will. The method employs deep learning and statistical models based on interpersonal interaction data. The key ideas for achieving humanlike behavior in this approach are synchronizing multimodal expressions (motion and voice) and determining the situational dependence of behavior. In addition, to enable free operation, motion that reflects the mental state and personality of the operator must be automatically generated. Therefore, this section introduces methods for generating motion that reflects the mental state and personality of the operator by extending the aforementioned motion generation methods. In interpersonal communication, people change their characteristics depending on the situation, and their motion depends on these characteristics. For example, it is easy to imagine that people have different change speech manners and gestures when they are with their friends compared to when they are with their parents. Therefore, a model of the CA's movements must consider its character to ensure its free-operation. This section also introduces the modeling of characters and the effects of expressing characters in CA. Finally, techniques for estimating the internal state of an operator are introduced. The teleoperation system automatically generates CA motion based on the estimated internal state of the operator, allowing the operator to easily operate the CA as intended.

2.7.1 Methods to Generate Humanlike CA Motions

Gestures are crucial for increasing the human qualities of a CA to achieve smoother interactions with humans. An effective system to model human gestures that match speech must be embedded in CAs. We first propose a gate recurrent unit (GRU)-based autoregressive generation model for gesture generation, which is trained with a convolutional neural network (CNN)-based discriminator in an adversarial manner using a Wasserstein generative adversarial network (W-GAN)-based learning algorithm. The W-GAN algorithm was used to avoid generating similar gestures for the same input. The model was trained to output the rotation angles of the joints in the upper body and implemented to animate a CG character. The motions synthesized by the proposed system were evaluated via an objective measure and subjective experiment, showing that the proposed model outperformed a baseline model trained by a state-of-the-art GAN-based algorithm using the same dataset. This result reveals that it is essential to develop a stable and robust learning algorithm for training gesture-generation models (Wu et al. 2021).

Data-driven approaches also have greater generalizability in various contexts than rule-based methods. However, most studies have no direct control over the human impressions of robots. The main obstacle is that creating a dataset covering various impression labels is not trivial. Based on previous findings in cognitive science on robot impressions, we proposed a heuristic method to control the impressions of different levels of personality (extroversion) without manual labeling; we also demonstrated its effectiveness on a CG character and partially on a humanoid robot through subjective experiments (Wu et al. 2022).

Figure 2.22 shows an overview of the proposed model. The model is primarily conditioned by prosodic features extracted from the speech signal and has an additional conditional input for gesture traits for both the generator and discriminator.

Although the dataset we chose contained gesture data for the entire body, we only used the upper-body joints to train our model. We define three classes of gesture traits: low, neutral, and high. Low indicates that the speed and amplitude of a specific gesture sample are below the dataset average, neutral indicates that they equal to the dataset average, and high indicates that they are above the dataset average. Our proposed

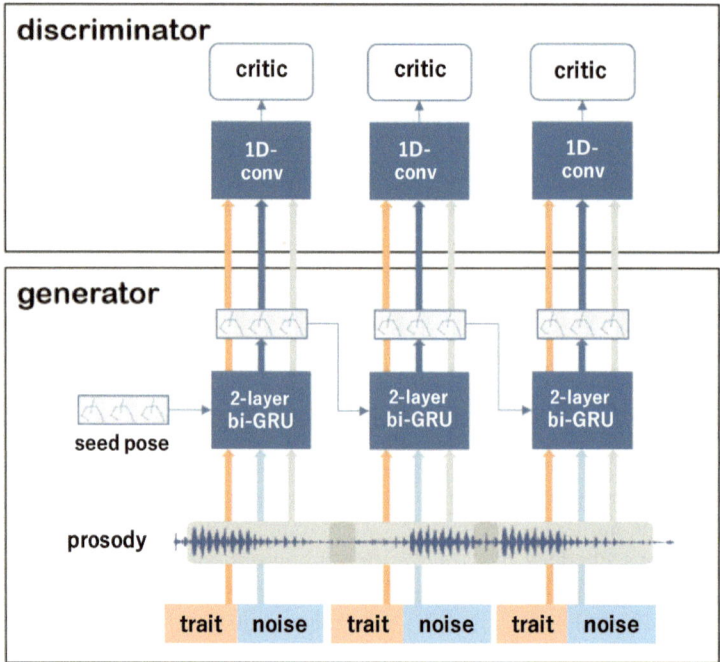

Fig. 2.22 Overview of the proposed gesture generation model

model can then generate gestures for different label inputs, that is, low, neutral, and high, conditioned on the speech input.

In total, five groups were evaluated: ground truth, which was the motion-captured data (Gg); a baseline model based only on prosody (Gb); and the proposed gesture traits, low (Gl), mid (Gm), and high (Gh). We first randomly chose three utterance samples from the test set and used the speech signals to generate gestures using the baseline and proposed models with different trait labels. Videos were recorded for both the CG character (using Unity software) and the humanoid robot CommU (Fig. 2.23). Snapshots of upper-body motions synthesized in a CG character are shown.

Figure 2.24 shows the perceived extroversion of the CG character and CommU for all five conditions. For the CG characters, the gradation of perceived extroversion for the proposed model matches the purpose of the proposed pseudo-labeling method. For CommU, the perceived extroversion was lower in the low model, but no significant differences were found between the neutral and high models. This may be due to the fewer DOFs and CommU's actuator constraints, which make the gestures of CommU less expressive.

We proposed a conditional GAN-based co-speech gesture generation model that exploits cognitive heuristics while maintaining the flexibility of data-driven methods. The experimental results on the CG characters/humanoid robots showed that our

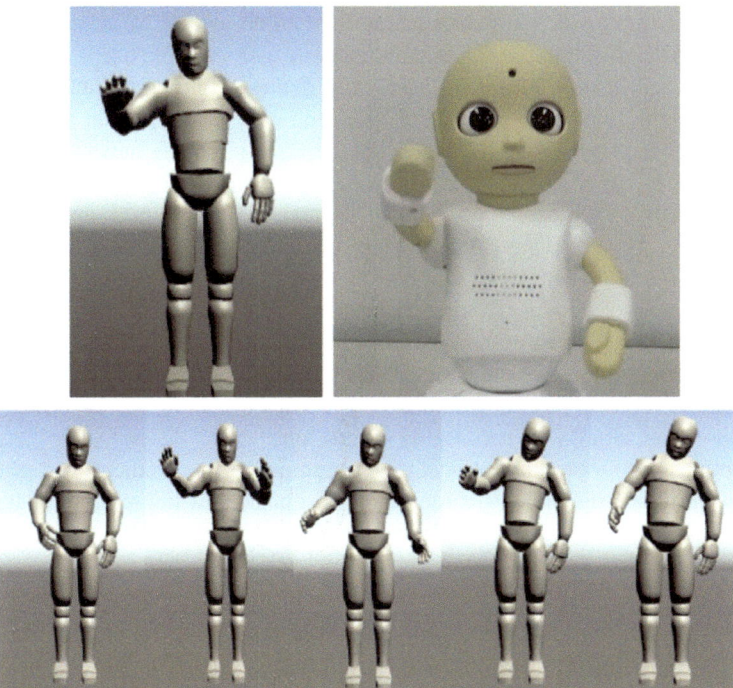

Fig. 2.23 CG character and the humanoid robot CommU (top). Snapshots of gestures generated in the CG character (bottom)

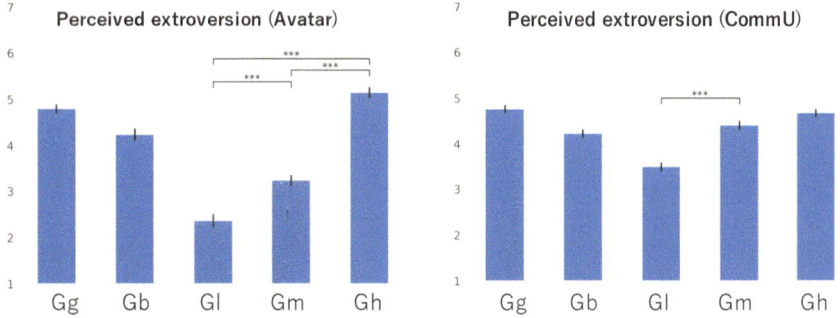

Fig. 2.24 Perceived extroversion on the CG character (left) and CommU (right)

model controlled perceived extroversion, and these findings are consistent with those of human cognitive behavior. Furthermore, our proposed model highlights the potential of fine control with a "black box" data-driven method. The burden of collecting big data can be moderated by adopting findings from cognitive science.

In an in-person dialogue with multiple participants, people naturally gaze toward others or avert their gazes according to their dialogue roles and mental states. We aimed to develop a CA that can generate humanlike eye movements to achieve smoother and more engaged dialogue interactions with multiple users.

We analyzed gaze behavior in three-party dialogue data, accounting for turn-taking, dialogue roles, and gaze aversion during conversations (Ishi and Shintani 2021). For this purpose, we used multimodal three-party dialogue data and first analyzed the distributions of (1) the gaze target (toward dialogue partners or gaze aversion), (2) gaze duration, and (3) eyeball direction during gaze aversion. Each distribution was modeled based on the analysis results, and a gaze motion generation system was implemented. Gaze movements were generated using CommU, which provides degrees of freedom for the head and eyeballs. Subjective evaluation experiments have shown that natural behaviors are achieved by the proposed gaze-control system, which accounts for dialogue roles and eyeball movement control (Shintani et al. 2021). Further details on gaze data analysis and evaluation are described in the next subsection, along with personality expression.

We improved our proposed model to generate the gaze movements of a dialogue robot in multiparty dialogue situations and investigated how impressions change for models created by the data of speakers with different personalities (Shintani et al. 2022).

The following roles of the dialogue participants are considered in three-party interactions:

- Speaker (Sp): The person who takes the turn and holds the dialogue floor.
- Addressee (AD) or main listener (ML): The target person who the speaker primarily addresses
- Side participant (SD) or sublistener (SL): A dialogue participant who is neither a speaker nor an addressee.

We analyzed the percentages of gaze targets, that is, how often the gaze was directed to a target person or averted to a direction other than a person's face. For a three-party dialogue, gaze targets are categorized as either dialogue partners (Sp, AD, and SD) or gaze aversion (GA). Figures 2.25 and 2.26 show the distributions of the gaze targets for more extroverted (A) and more introverted (B) persons, respectively. For example, different distributions can be observed depending on the dialogue roles and whether a person belongs to class A or B. Figures 2.25a and 2.26a show that when the dialogue role is "Speaker," Person A looks at the Addressee, the Side Participant, or averts their gaze at the same rates. In contrast, Person B had a higher percentage of gaze aversion and a very low percentage of looking at the Side Participant. Overall, it can be observed that Person A looks more at the dialogue partner, while Person B has a higher percentage of gaze aversion.

Similar distributions were obtained for the gaze duration and eyeball direction for each person and dialogue role. These distributions were parameterized and used to generate gaze movements.

We conducted subjective experiments to evaluate how impressions of personality changed when using the gaze models of different speakers to generate a robot's gaze behavior. In this study, we adopted the android Nikola to evaluate gaze behavior. Nikola has 32 DOFs for the face and 3 DOFs for the neck, allowing the generation

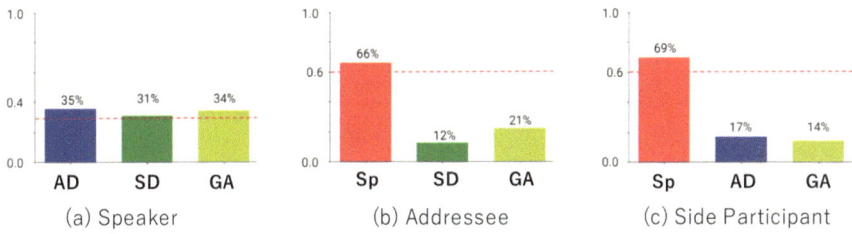

Fig. 2.25   Distributions of gaze targets during each dialogue role for person A (extroverted)

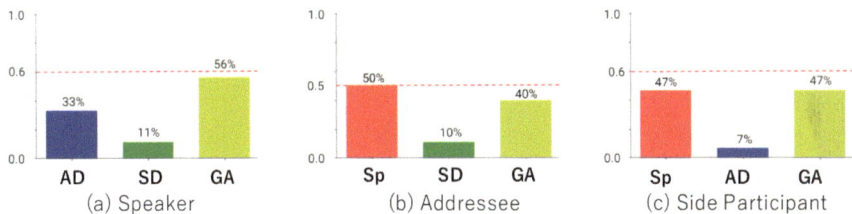

Fig. 2.26   Distributions of gaze targets during each dialogue role for person B (introverted)

of several facial expressions that convey different emotions (Sato et al. 2022). To control gaze behavior in a dialogue scenario, we controlled the following actuators: eyeball pitch, left eyeball yaw, right eyeball yaw, upper eyelids, head pitch, and head yaw. The lip actuators were controlled based on the speech signal of the target speaker, and the lip corner and cheek actuators were controlled to generate smiling faces during laughter.

Two motion types were generated by the proposed gaze-generation system using the model parameters of Persons A (PA-M) and B (PB-M). For comparison, another motion type was generated by reproducing the gaze behavior of the speakers (PA-R and PB-R), mapping the measured head angles of the speakers, and moving their eyeballs according to the gaze labels. Video clips were created for each type of motion. Figure 2.27 shows a snapshot of the video used in the experiment. Figure 2.28 shows the subjective perceptual scores (mean and standard errors) of the personalities for the different conditions. Significant differences through repeated measures of one-way ANOVA were achieved between the motions generated by models PA-M and PB-M for the voice of Person A, as shown in Fig. 2.28. We found that changing the gaze model parameters of subjects with distinct personalities was, to some extent, effective in changing their impressions of personality traits.

In conclusion, based on the analysis of gaze behaviors in three-party dialogue data, we observed differences in the distributions of gaze model parameters (gaze targets, gaze durations, and gaze patterns during gaze aversion) for speakers with different personalities. The gaze behaviors generated by the android using the models of the speaker types A and B were perceived to have different extraversion levels. In future work, we plan to investigate the effects of other factors, such as prosodic features in

Fig. 2.27 Gaze control in android Nikola

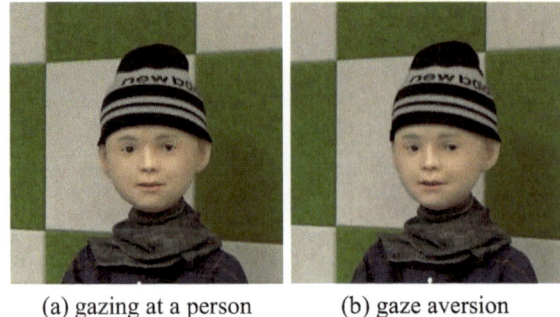

(a) gazing at a person   (b) gaze aversion

Fig. 2.28 Subjective personality results

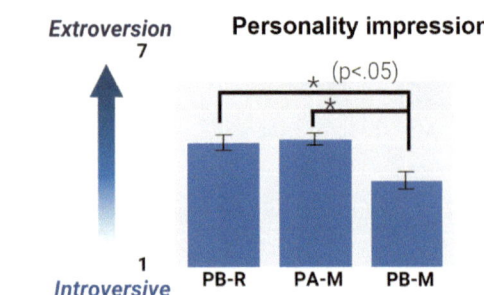

the voice, and the possibility of generalizing gaze model parameters to control the expression of personality.

2.7.2 Modeling Motions Expressing the Operator's Character

As a case study, we investigated how negative expressions by CA, such as dissatisfaction and anger, would affect people's impressions in persuasive tasks. Specifically, we explored how to furnish an android robot with socially motivated expressions geared toward eliciting adherence to the COVID-19 guidelines. We analyzed how different behaviors associated with social expressions in such situations occur in human to human interactions and designed a scenario in which a robot utilizes context-inspired behaviors (polite, gentle, displeased, and angry) to enforce social compliance. We then implemented these behaviors in an android robot and subjectively evaluated how effectively it expressed them; additionally, we investigated how they were perceived in terms of their appropriateness and effectiveness in enforcing social compliance with COVID-19 guidelines. We also considered how the subjects' values regarding compliance awareness affected the robot's behavioral impressions. Our evaluation results indicated that participants generally preferred polite behaviors by the robot, although participants with higher levels of compliance awareness manifested higher preferences toward the appropriateness and effectiveness of social compliance enforcement through negative robot expressions (Ajibo et al. 2021).

To make telepresence CAs automatically express the operator's emotions, we also need technology that can recognize the operator's emotional state. We also developed deep-learning-based techniques for this purpose.

In a multimodal emotion recognition task based on speech and facial images, we propose an adversarial training-based classifier with isolated Gaussian regularization to regularize the distribution of latent representations and further smooth the boundaries among different categories. Four emotion categories (neutral, joy, anger, and sadness) were classified with 70% accuracy (Fu et al. 2023).

We also addressed the emotion recognition problem by considering gesture information as an additional modality. For this purpose, we proposed an attention-based convolutional neural network that uses skeletal data as input to predict the speaker's emotional state. A graph-attention-based fusion method was also proposed to combine our model with models using other modalities to provide complementary information in the emotion classification task and effectively combine multimodal cues. The combined model utilized audio signals, text information, and skeletal data. The model significantly outperforms the bimodal model and other fusion strategies, proving its effectiveness (Shi et al. 2021).

2.8 Development of an Interaction Behavior Learning Method for CAs

When using an avatar system for remote communication, the quality of the information obtained is reduced owing to the technical limitations of the observation system and communication channels. For example, the camera's narrow angle of view limits the area that the operator can perceive, and delays in communication make timing difficult, thus making the operation less straightforward. On the other hand, if the operator is "on the spot," they can easily recognize the situation through their five senses and behave appropriately.

Multiple degrees of freedom must be manipulated to manipulate an avatar with a high level of expressiveness. This complicates the avatar manipulation, whereas using a simple avatar reduces expressiveness. For daily avatar use, it is crucial to develop a teleoperation system that is easy for the teleoperator to operate while providing rich expressions for users who interact with it.

The objective of this study is to develop a semi-autonomous remote control method for avatars, allowing highly expressive interactions through various modalities by conveniently operating specific modalities. Specifically, we are developing semi-autonomous avatars that behave according to the sensor information obtained in the field in addition to the information sent by the remote operator. These avatars are expected to provide richer interactions than those that operate remotely. For this purpose, we are working on modeling human–human interactions using a deep generative model and developing mechanisms for rich human–robot interactions using this model (Okadome and Nakamura 2023a).

2.8.1 Modeling Dyadic Interaction

Interaction-behavior generation is an important issue in the development of robots that interact naturally with humans. Recently, interactive systems that use natural language dialogue, such as smart speakers, have become useful tools for providing various types of information. However, dialogue with such systems lacks the fluency and natural flow observed in human–human interaction due to a limited level of expressiveness, including gestures and well-paced dialogue turns. To achieve rich human–robot interaction, methods that address these issues are needed (Forlizzi 2007).

When operating through a semi-autonomous avatar, the operator receives limited information on the remote site, and the operable modality is limited owing to the complexity of the process. Under these conditions, the same problems arise with autonomous dialogue systems when striving for a dynamic and engaging conversation between the avatar and its dialogue partner. Therefore, the automatic generation of modalities independent of operator guidance and the adjustment of instructions to suit the context, that is, through semi-autonomous avatars, are expected to enhance the user's experience.

In human-to-human dialogue, the behavior of dialogue participants is influenced by the behavior of others; therefore, it is necessary to consider the behavior of all dialogue participants simultaneously, not just that of individuals (Kwon et al. 2015). Note that "behavior" here includes both movement and speech. In other words, they act in accordance with the interlocutor's behavior, such as backchanneling during the interlocutor's utterances and expressing gestures when explaining something. To realize such a "full-duplex communication" interaction, we modeled human–human interaction behavior.

In this study, we developed motion generation models for behaviors during dyadic interactions and investigated their characteristics. First, the videos and audios of the two participants in the dialogue were recorded simultaneously (referred to as Persons R and L, respectively, as shown in Fig. 2.29). Features such as poses and speech segments were extracted from recorded images and audio, and the time series of features at each time point was treated as quantified "behavior." Specifically, the data were processed in discrete time and modeled for pairs of time series $X_L(t)$ and $X_R(t)$ of features of constant time length T, where $X_L(t) = [x_L(t - T + 1), x_L(t - T + 2), \ldots, x_L(t)]$ and $X_R(t) = [x_R(t - T + 1), x_R(t - T + 2), \ldots, x_R(t)]$ are the behaviors of persons L and R, respectively. $x_{(\cdot)}(t)$ is a feature of person $(\cdot)$ at time t.. In other words, by treating changes in feature values (history) over a fixed time window as behavior, we modeled the spatiotemporal structure of the behavior of dialogue participants.

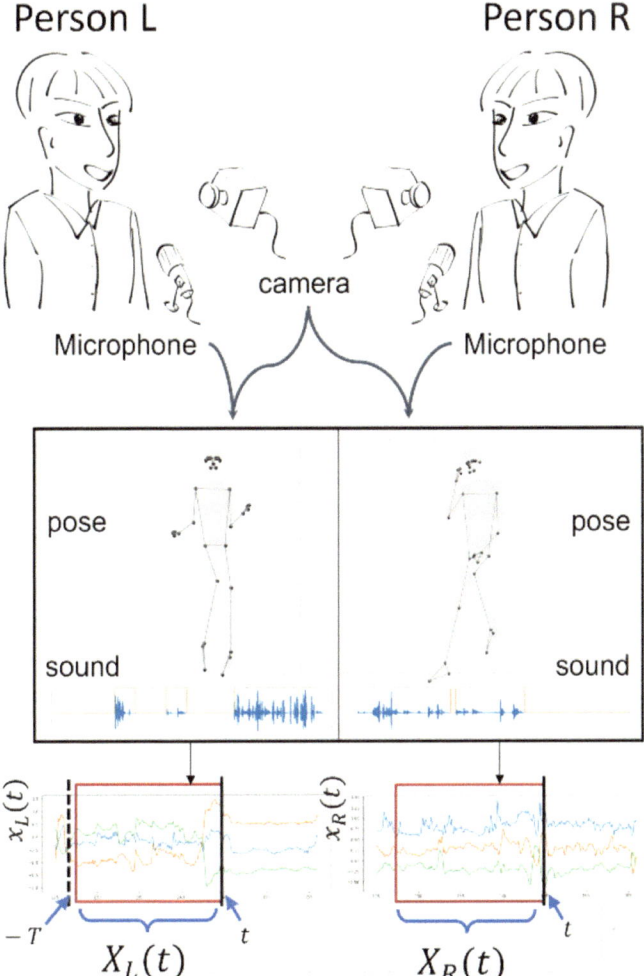

Fig. 2.29 Measurement of the behavior of dialogue pairs and extraction of features

Because human behaviors during dialogue influence each other, these features are expected to influence each other. For example, in a situation where R is the speaker, the features computed from R's speech signal would indicate that he is speaking, whereas the features corresponding to L's nodding behavior would change at some point during R's speech. These relationships were modeled using data-driven methods. Using this "full-duplex" behavioral model of humans, the goal is to develop semi-autonomous avatars that can be easily manipulated to produce rich expressions.

2.8.2 Generative Model for Dyadic Interaction

Gesture generation methods using deep generative models have been studied to generate the humanlike behavior of humanoids or CG-agents, agents. Deep generative models are methods of sampling real-sample-like data from random number generators and have recently been used in various fields other than modeling motion, such as imaging. Notable methods include autoencoders and their variants, such as variational autoencoders (VAE), generative adversarial networks (GAN), and diffusion models (Goodfellow et al. 2016; Ho et al. 2020). In this study, as a framework for semi-autonomous avatars, we model interaction behavior using VAEAC (Variational autoencoder with arbitrary conditioning) (Ivanov et al. 2019) and diffusion models that can recover missing data ('inpainting' in the field of image processing), that is, they can be generated under arbitrary conditions. The generative model was trained to generate samples that were indistinguishable from the training data. For the behavior during the interaction, the measured behaviors $X_L(t)$ and $X_R(t)$ are the training data. If we can obtain samples from the distribution $p(X_L, X_R)$ of the data, the mimicking data can be sampled; that is, we can obtain samples similar to the true data. Using a generative model, the sample $\left(\widehat{X}_L, \widehat{X}_R\right)$ can be approximated. Using generative models that can be generated under arbitrary conditions, given some elements $\left(\overline{X}_L, \overline{X}_R\right)^+$ in a sample, the remaining elements $\left(\widehat{X}_L, \widehat{X}_R\right)^-$ can be generated to be consistent with the given conditions, and mimicking data $\left(\widehat{X}_L, \widehat{X}_R\right)$ can be synthesized from $\left(\overline{X}_L, \overline{X}_R\right)^+$ and $\left(\widehat{X}_L, \widehat{X}_R\right)^-$.

They can be generated under arbitrary conditions and have various applications. If elements before a certain time are given as a condition, this corresponds to a later prediction of behavior. In addition, if only one feature value is unknown, the data are complemented such that their behavior is consistent with that of the other modalities. Given only person R's behavior X_R as a condition, it can infer the behavior of person L with whom it is interacting. Furthermore, future behaviors, including variability, can be evaluated by generating many samples (Takashiro et al. 2021). In this section, we present the deep generative models we have been working on for head movements and gestures.

A generative model using the VAEAC was constructed for head movements during dialogue. The facial images and audio of the two participants during the dialogue were recorded, and landmarks of the head (feature points indicating the positions and shapes of the facial contours, eyes, eyebrows, nose, and mouth) were extracted using OpenFace (Baltrusaitis et al. 2018). In addition, speech segments were detected from the speech signal to compose behaviors X_L and X_R. While the landmarks obtained here are represented by the coordinates of 68 feature points, such as facial contours and eye shapes, each point does not move independently but can be expected to move with a low degree of freedom (i.e., move synchronously). Therefore, feature points were first divided into the mouth and other parts (hereafter referred to as face), and dimensionality reduction was performed on the features each time using the VAE.

As a result, (X_L, X_R) became a 3-dimensional array with $2 \times 7 \times T$ by compressing the mouth and face into a 3-dimensional vector and adding the speech signal.

An evaluation of impressions was conducted to assess the performance of the generative model constructed using this method, considering both measured and generated movements. Specifically, the evaluators watched video clips created under three conditions and ranked them based on whether they adopted the behavior depicted in each clip. Note that the evaluators were graduate students and researchers involved in research on interactive robots and programmed the movements of interactive robots daily. The video clips produced were based on the following procedure:

Positive: Feature points are extracted from the recorded video, and a photorealistic head image is synthesized using the feature points as input. A bilayer model was used for synthesis. Because the original video clips were considered to have a better image quality than the synthesized clips, they were synthesized in comparison with the other conditions. Generate: Conditional on the speech signals of L and R and the images of R, the motion of L is generated. Subsequently, a photorealistic head image is synthesized. Negative: Composite head motion of person L extracted from randomly selected video clips at other times instead of the current video clip. A comparison of these results showed that, in some cases, the video clips generated by the proposed method were highly evaluated compared to the positive ones (Takashiro et al. 2021).

For upper-body movements or gestures during dialogue, we built a generative model using a diffusion model, which has recently attracted attention for generating high-quality samples, particularly in the field of image generation (Okadome and Nakamura 2023a). Because DDPM, which requires many iterations, is not suitable for real-time generation owing to computational time issues, DDIM (Song et al. 2020), which can generate samples with a small number of iterations, was used. We developed a learning method that considered the mask shape used during inference. The following three-generation tasks were used in the evaluation:

- Prediction: The second half of the time-series data were generated conditional on the first half.
- Completion: The first and last halves of the time-series data were used as conditions for the generation of the intermediate part of the time-series data. In this case, it is necessary to generate data that matches the conditions before and after the intermediate part.
- Reaction: Generate one person's behavior, conditional on another person's behavior. The model used here generates 10 s behavior and has been confirmed to generate a greater variety of actions in each task than VAEAC. The system could generate motions with a computation time of approximately 28 ms, which is sufficiently fast for the system to generate robot motions (Okadome and Nakamura 2023a).

2.8.3 Feature Space of the Interaction

Behaviors during interactions are reciprocal. For example, behaviors such as backchanneling or smiling in response to the utterances of an interaction partner can be observed. Therefore, we investigated a feature extraction method that focuses on the temporal structure.

Methods for analyzing behavioral entrainment include calculating the correlation between time-series data obtained from microphones and cameras. However, it is difficult for humans to determine whether the behaviors of a pair of induvial a video clip are synonymous. Therefore, we have been studying a framework for self-supervised learning that focuses on the temporal structure of human–human interaction. The framework improves the accuracy of predicting future behavior, specifically the occurrence of turn taking. Additionally, a feature space for data with a spatiotemporal structure can be obtained. The impression evaluation suggested a relationship between the position of the sample in the feature space and the impression of the video clip.

In this study, we used an operation called "lag operation" to construct the feature space (Okadome and Nakamura 2023b). The lag operation uses a composite sample of one person's behavior $X_L(t)$ at time t and the other person's behavior $X_R(t + \tau)$ at $t + \tau$. A deep neural network is pre-trained to infer the lag τ. This method was derived from the idea that the effects of the mutual influences of behaviors during interaction are thought to decrease as the time difference increases. This framework (Fig. 2.30) automatically constructs a teacher signal and does not require labeled data, which are typically used in supervised learning tasks. Not using labeling data makes it possible to perform pre-training using only sensor data, which is relatively easy to obtain.

In the experiment, five types of data were prepared as time deviations $\tau \in [-1.0, -0.5, 0.0, 0.5, 1.0]$, and the network was trained to infer the delay class. Here, the case of $\tau = 0.0$ is the measured data. After learning, samples with different τ were distributed in different areas in the feature space. However, there were many areas of overlap in their distributions. In other words, there were samples in which the effects of the lag operation were indistinguishable.

When the lag-free samples $(X_L(t), X_R(t))$ are mapped into the feature space, some data are projected into regions where they are judged to be almost $\tau = 0.0$, while others are projected into regions where the magnitude of the time shift is difficult to distinguish, i.e., where samples of different classes are mixed within the feature space. Picking some of the video clips in each domain, we found some characteristic behaviors; for example, the former included simultaneous smiling behavior, while the latter continued to show the speaker's thinking and speaking without looking at the listener.

Therefore, we propose an index $\mathcal{R}$ that indicates how clearly a sample is classified by calculating the degree of class overlap at each point in the feature space. We investigated how this difference in $\mathcal{R}$ changed the impression of the interaction between the video clips. Specifically, the participants watched video clips with different R

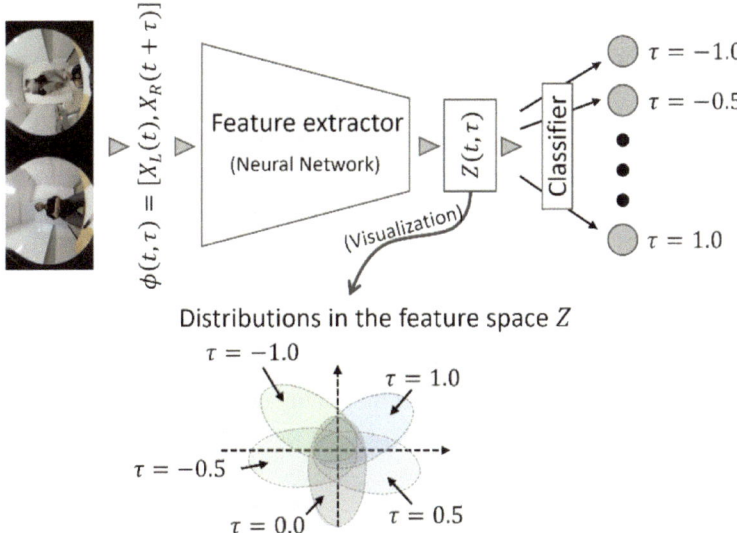

Fig. 2.30 Feature extraction using lag operation

values of $\mathcal{R}$ extracted from the recorded video and provided their impressions of the video clips. We found that the answers to questions such as "Is the pace of the conversation good?" or "Is the conversation lively?" and the indicator $\mathcal{R}$ are related. Specifically, video clips with good tempo or excitement are projected onto the area with a large $\mathcal{R}$, while video clips featuring a speaker speaking thoughtfully and appearing inattentive to the interlocutor are projected onto the area with a small $\mathcal{R}$.

In the future, we plan to conduct research on the refinement of the generative model with reference to the knowledge obtained from the construction of the feature space and the use of $\mathcal{R}$ as an indicator of the quality of interaction between the robot and humans. In the experiment, five types of data were prepared as time deviations $\tau \in [-1.0, -0.5, 0.0, 0.5, 1.0]$, and the network was trained to infer the delay class. Here, the case of $\tau = 0.0$ is the measured data. After learning, samples with different τ were distributed in different areas in the feature space. However, there were many areas of overlap in their distributions. In other words, there were samples in which the effects of the lag operation were indistinguishable.

When the lag-free samples $(X_L(t), X_R(t))$ are mapped into the feature space, some data are projected into regions where they are judged to be almost $\tau = 0.0$, while others are projected into regions where the magnitude of the time shift is difficult to distinguish, i.e., where samples of different classes are mixed within the feature space. Picking some of the video clips in each domain, we found some characteristic behaviors; for example, the former included simultaneous smiling behavior, while the latter continued to show the speaker's thinking and speaking without looking at the listener.

Therefore, we propose an index $\mathcal{R}$ that indicates how clearly a sample is classified by calculating the degree of class overlap at each point in the feature space. We investigated how this difference in $\mathcal{R}$ changed the impression of the interaction between the video clips. Specifically, the participants watched video clips with different R values of $\mathcal{R}$ extracted from the recorded video and provided their impressions of the video clips. We found that the answers to questions such as "Is the pace of the conversation good?" or "Is the conversation lively?" and the indicator $\mathcal{R}$ are related. Specifically, video clips with good tempo or excitement are projected onto the area with a large $\mathcal{R}$, while video clips featuring a speaker speaking thoughtfully and appearing inattentive to the interlocutor are projected onto the area with a small $\mathcal{R}$.

In the future, we plan to conduct research on the refinement of the generative model with reference to the knowledge obtained from the construction of the feature space and the use of $\mathcal{R}$ as an indicator of the quality of interaction between the robot and humans.

2.8.4 Semi-autonomous Teleoperation System for Cybernetic Avatars

Using the generative model described in the previous subsection, we built prototype systems for the teleoperation of CAs (Nishimura et al. 2023). In recent years, remote interaction technologies using computer graphic (CG) avatars and other devices have become widely used. In such systems, the behavior of the avatar is usually manipulated by the operator at every step. However, it is not easy for the avatar operator to recognize the situation in a remote environment and behave accordingly, owing to the limitations of the sensor system. It would also be cumbersome to control all avatar behaviors, especially if the avatar is highly expressive. Of course, there are ways to achieve this, such as building an immersive VR space; however, this is not a framework that can be used easily. In this study, we constructed a semiautonomous teleoperation system that automatically completes a part not indicated by the operator using a generative model.

Figure 2.31 shows the proposed framework. In this model, the avatar is controlled using a generative model and supplemented with information that is not controlled by the avatar operator. Thus, the avatar is controlled not only by the avatar operator's instructions but also by considering sensor information from the remote environment. By selecting the information (modality) to be manipulated, various forms of manipulation can be considered, including extreme cases where the avatar behaves autonomously if the operator does not perform any operations. Currently, for CG avatars, we constructed a system that controls head movements (facial orientation and landmarks) using the VAEAC. In this system, sensor information on the behavior (image sequence and voice) of User B, interacting face-to-face with the avatar in a remote environment, and the voice of User A, the avatar operator, is received. It then

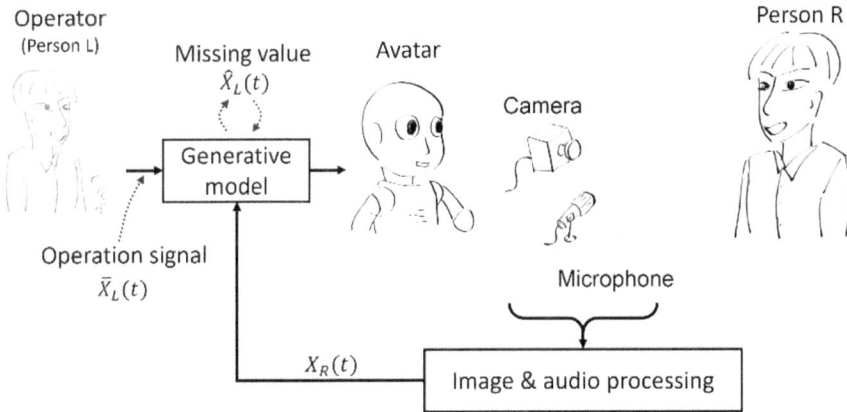

Fig. 2.31 Semi-autonomous teleoperation system for a CA

outputs the image sequence of the CG avatar (Fig. 2.31). We are also working on building a semi-autonomous teleoperation system for androids.

2.8.5 Summary

In this section, we introduced generative models of a person's behavior during dialogue and a framework for a semi-autonomous teleoperation system that uses the generative model. We are currently implementing and evaluating the proposed system. This framework stands out for its ability to generate avatar behaviors aligned with both the situation (including the interaction partner's behavior) and the operator's instructions. It achieves this by "filtering" the avatar operator's instructions through a generative model, which tailors the avatar behaviors to the specific situation. In the future, we aim to build an operational system with a low workload in which the remote operator can operate the avatar by indicating the necessary information at the necessary time, and a system with a low cognitive load in which one user can operate the avatar while performing other tasks.

The method presented in this section uses a generative model of human interaction to construct a semi-autonomous teleoperation system. However, because human reactions are influenced by appearance and other factors, models that mimic human behavior are not always effective. As a demonstration experiment, we are currently building a framework for learning the behavior of a receptionist robot at a shopping mall. We studied the learning behaviors using inverse reinforcement learning (Chen et al. 2023), which learns human operations, and (forward) reinforcement learning, which is an autonomous learning method. The results have been successful, with autonomous androids transforming the behavior of passersby more effectively than human-operated androids. A future challenge is to build a semi-autonomous teleoperation system that can both imitate human behavior and learn to operate effectively through autonomous learning.

References

Abe K, Shiomi M, Pei Y et al (2018) ChiCaRo: tele-presence robot for interacting with babies and toddlers. Adv Robot 32:176–190. https://doi.org/10.1080/01691864.2018.1434014

Ajibo CA, Ishi CT, Ishiguro H (2021) Advocating attitudinal change through android robot's intention-based expressive behaviors: toward who covid-19 guidelines adherence. IEEE Robot Autom Lett 6:6521–6528. https://doi.org/10.1109/LRA.2021.3094783

Alimardani M, Nishio S, Ishiguro H (2013) Humanlike robot hands controlled by brain activity arouse illusion of ownership in operators. Sci Rep 3:2396. https://doi.org/10.1038/srep02396

Alimardani M, Nishio S, Ishiguro H (2016) Removal of proprioception by BCI raises a stronger body ownership illusion in control of a humanlike robot. Sci Rep 6:33514. https://doi.org/10.1038/srep33514

Arimoto T, Yoshikawa Y, Ishiguro H (2018) Multiple-robot conversational patterns for concealing incoherent responses. Int J Soc Robot 10:583–593. https://doi.org/10.1007/s12369-018-0468-5

Asaka T, Seppelfelt GDC, Nagai T, Yukizaki S (2023) HumanoidBot: framework for integrating full-body humanoid robot with open-domain chat system. Adv Robot 37:1171–1186. https://doi.org/10.1080/01691864.2023.2256386

Baltrusaitis T, Zadeh A, Lim YC, Morency LP (2018) OpenFace 2.0: facial behavior analysis toolkit. In: 2018 13th IEEE international conference on automatic face & gesture recognition (FG 2018). IEEE, pp 59–66

Bartneck C, Belpaeme T, Eyssel F et al (2020) Human-robot interaction. Cambridge University Press, Cambridge

Bertone A, Bettinelli L, Faubert J (2007) The impact of blurred vision on cognitive assessment. J Clin Exp Neuropsychol 29:467–476. https://doi.org/10.1080/13803390600770793

Blakemore S-J, Frith CD, Wolpert DM (1999) Spatio-temporal prediction modulates the perception of self-produced stimuli. J Cogn Neurosci 11:551–559. https://doi.org/10.1162/089892999563607

Block AE, Seifi H, Hilliges O et al (2023) In the arms of a robot: designing autonomous hugging robots with intra-hug gestures. ACM Trans Hum Robot Interact 12:1–49. https://doi.org/10.1145/3526110

Cerejeira J, Lagarto L, Mukaetova-Ladinska EB (2012) Behavioral and psychological symptoms of dementia. Front Neurol 3. https://doi.org/10.3389/fneur.2012.00073

Chen Z, Nakamura Y, Ishiguro H (2023) Outperformance of mall-receptionist android as inverse reinforcement learning is transitioned to reinforcement learning. IEEE Robot Autom Lett 8:3350–3357. https://doi.org/10.1109/LRA.2023.3267385

Cheng M, Kato M, Saunders JA, Tseng C (2020) Paired walkers with better first impression synchronize better. PLoS ONE 15:e0227880. https://doi.org/10.1371/journal.pone.0227880

Conklin K, Pellicer-Sánchez A, Carrol G (2018) Eye-tracking. Cambridge University Press, Cambridge

Daprati E, Franck N, Georgieff N et al (1997) Looking for the agent: an investigation into consciousness of action and self-consciousness in schizophrenic patients. Cognition 65:71–86. https://doi.org/10.1016/S0010-0277(97)00039-5

Darvish K, Penco L, Ramos J et al (2023) Teleoperation of humanoid robots: a survey. IEEE Trans Rob 39:1706–1727. https://doi.org/10.1109/TRO.2023.3236952

Dueren AL, Vafeiadou A, Edgar C, Banissy MJ (2021) The influence of duration, arm crossing style, gender, and emotional closeness on hugging behaviour. Acta Psychol (Amst) 221:103441. https://doi.org/10.1016/j.actpsy.2021.103441

Edelson SM, Edelson MG, Kerr DCR, Grandin T (1999) Behavioral and physiological effects of deep pressure on children with autism: a pilot study evaluating the efficacy of Grandin's hug machine. Am J Occup Ther 53:145–152. https://doi.org/10.5014/ajot.53.2.145

Essick GK, James A, McGlone FP (1999) Psychophysical assessment of the affective components of non-painful touch. NeuroReport 10:2083–2087. https://doi.org/10.1097/00001756-199907 130-00017

Forlizzi J (2007) How robotic products become social products. In: Proceedings of the ACM/IEEE international conference on human-robot interaction. ACM, New York, pp 129–136

Fu C, Liu C, Ishi CT, Ishiguro H (2023) An adversarial training based speech emotion classifier with isolated Gaussian regularization. IEEE Trans Affect Comput 14:2361–2374. https://doi. org/10.1109/TAFFC.2022.3169091

Geva N, Uzefovsky F, Levy-Tzedek S (2020) Touching the social robot PARO reduces pain perception and salivary oxytocin levels. Sci Rep 10:9814. https://doi.org/10.1038/s41598-020-669 82-y

Glas DF, Minato T, Ishi CT et al (2016) ERICA: the ERATO intelligent conversational android. In: 2016 25th IEEE international symposium on robot and human interactive communication (RO-MAN). IEEE, pp 22–29

Goodfellow I, Bengio Y, Courville A (2016) Deep learning. MIT Press

Goodwin MH (2020) The interactive construction of a hug sequence. In: Cekaite A, Mondada L (eds) Touch in social interaction. Routledge, New York, pp 27–53

Grandin T (1992) Calming effects of deep touch pressure in patients with autistic disorder, college students, and animals. J Child Adolesc Psychopharmacol 2:63–72. https://doi.org/10.1089/cap. 1992.2.63

Hall ET, Birdwhistell RL, Bock B et al (1968) Proxemics [and comments and replies]. Curr Anthropol 9:83–108. https://doi.org/10.1086/200975

Higashino K, Kimoto M, Iio T et al (2021) Tactile stimulus is essential to increase motivation for touch interaction in virtual environment. Adv Robot 35:1043–1053. https://doi.org/10.1080/016 91864.2021.1967780

Hinderks A, Schrepp M, Domínguez Mayo FJ et al (2018) UEQ KPI value range based on the UEQ benchmark

Ho J, Jain A, Abbeel P (2020) Denoising diffusion probabilistic models. In: Advances in neural information processing systems, pp 6840–6851

Horikawa Y, Miyashita T, Utsumi A et al (2023) Cybernetic avatar platform for supporting social activities of all people. In: 2023 IEEE/SICE international symposium on system integration (SII). IEEE, pp 1–4

Iio T, Yoshikawa Y, Ishiguro H (2021) Double-meaning agreements by two robots to conceal incoherent agreements to user's opinions. Adv Robot 35:1145–1155. https://doi.org/10.1080/ 01691864.2021.1974939

Ishi CT, Shintani T (2021) Analysis of eye gaze reasons and gaze aversions during three-party conversations. In: Interspeech 2021. ISCA, pp 1972–1976

Ishi CT, Liu C, Ishiguro H, Hagita N (2012) Evaluation of formant-based lip motion generation in tele-operated humanoid robots. In: 2012 IEEE/RSJ international conference on intelligent robots and systems. IEEE, pp 2377–2382

Ishi CT, Machiyashiki D, Mikata R, Ishiguro H (2018) A speech-driven hand gesture generation method and evaluation in android robots. IEEE Robot Autom Lett 3:3757–3764. https://doi.org/ 10.1109/LRA.2018.2856281

Ishiguro H, Libera FD (2018) Geminoid studies. Springer, Singapore

Ivanov O, Figurnov M, Vetrov D (2019) Variational autoencoder with arbitrary conditioning. In: 7th international conference on learning representations, ICLR 2019

Kamide H, Kawabe K, Shigemi S, Arai T (2014) Nonverbal behaviors toward an audience and a screen for a presentation by a humanoid robot. Artif Intell Res 3. https://doi.org/10.5430/air.v3n 2p57

Kawata M, Maeda M, Yoshikawa Y et al (2022) Preliminary investigation of the acceptance of a tele-operated interactive robot participating in a classroom by 5th grade students. In: Social robotics: 14th international conference, ICSR 2022, Florence, Italy, 13–16 Dec 2022, Proceedings, Part II, pp 194–203

Kendon A (1990) Spatial organization in social encounters: the F-formation system. In: Conducting interaction: patterns of behavior in focused encounters. Cambridge University Press, Cambridge, pp 209–237

Kwon J, Ogawa K, Ono E, Miyake Y (2015) Detection of nonverbal synchronization through phase difference in human communication. PLoS ONE 10:e0133881. https://doi.org/10.1371/journal. pone.0133881

Lewis JR (2019) Comparison of four tam item formats: effect of response option labels and order. J Usability Stud 14:224–236

Lichiardopol S (2007) A survey on teleoperation. Technische Universiteit Eindhoven 2007.155

Ma R, Kaber DB (2006) Presence, workload and performance effects of synthetic environment design factors. Int J Hum Comput Stud 64:541–552. https://doi.org/10.1016/j.ijhcs.2005.12.003

Miguel HO, Sampaio A, Martínez-Regueiro R et al (2017) Touch processing and social behavior in ASD. J Autism Dev Disord 47:2425–2433. https://doi.org/10.1007/s10803-017-3163-8

Minami A, Takahashi H, Nakata Y et al (2021) The neighbor in my left hand: development and evaluation of an integrative agent system with two different devices. IEEE Access 9:98317–98326. https://doi.org/10.1109/ACCESS.2021.3095592

Mondada L, Monteiro D, Tekin BS (2020) The tactility and visibility of kissing. In: Cekaite A, Mondada L (eds) Touch in social interaction. Routledge, New York, pp 54–80

Moriya S, Shiono D, Fujihara R et al (2023) Aoba_v3 bot: a multimodal chatbot system combining rules and various response generation models. Adv Robot 37:1392–1405. https://doi.org/10. 1080/01691864.2023.2240883

Nakajima M, Shinkawa K, Nakata Y (2024) Development of the lifelike head unit for a humanoid cybernetic avatar 'Yui' and its operation interface. IEEE Access 12:23930–23942. https://doi. org/10.1109/ACCESS.2024.3365723

Nakanishi J, Baba J, Ishiguro H (2022) Robot-mediated interaction between children and older adults: a pilot study for greeting tasks in nursery schools. In: 2022 17th ACM/IEEE international conference on human-robot interaction (HRI). IEEE, pp 63–70

Nakata Y, Yagi S, Yu S et al (2022) Development of 'ibuki' an electrically actuated childlike android with mobility and its potential in the future society. Robotica 40:933–950. https://doi.org/10. 1017/S0263574721000898

Nishiguchi S, Ogawa K, Yoshikawa Y et al (2017) Theatrical approach: designing human-like behaviour in humanoid robots. Rob Auton Syst 89:158–166. https://doi.org/10.1016/j.robot. 2016.11.017

Nishimura Y, Takashiro S, Okadome Y et al (2023) Development of a semi-autonomous teleopera-tion system for a CG avatar using a deep generative model. In: The 37-th annual conference of the Japanese society for artificial intelligence (JSAI 2023)

Nishio S, Watanabe T, Ogawa K, Ishiguro H (2012) Body ownership transfer to teleoperated android robot. In: Ge SS, Khatib O, Cabibihan J-J et al (eds) Social robotics: 4th international conference, ICSR 2012, Chengdu, China, 29–31 Oct 2012, Proceedings. Springer, Berlin, pp 398–407

Nishio T, Yoshikawa Y, Iio T et al (2021) Actively listening twin robots for long-duration conversation with the elderly. ROBOMECH J 8:18. https://doi.org/10.1186/s40648-021-002 05-5

Noguchi Y, Kamide H, Tanaka F (2023) How should a social mediator robot convey messages about the self-disclosures of elderly people to recipients? Int J Soc Robot 15:1079–1099. https://doi. org/10.1007/s12369-023-01016-x

Ochi K, Inoue K, Lala D et al (2023) Effect of attentive listening robot on pleasure and arousal change in psychiatric daycare. Adv Robot 37:1382–1391. https://doi.org/10.1080/01691864.2023.2257264

Ogawa K, Nishio S, Koda K et al (2011) Telenoid: tele-presence android for communication. In: ACM SIGGRAPH 2011 emerging technologies. ACM, New York, pp 1–1. https://doi.org/10.1145/2048259.2048274

Okada Y, Kimoto M, Iio T et al (2023) Two is better than one: apologies from two robots are preferred. PLoS ONE 18:e0281604. https://doi.org/10.1371/journal.pone.0281604

Okadome Y, Nakamura Y (2023a) Diffusion model with MASKed input for generating gestures during dyadic conversation. In: IBISML2023–19, pp 121–128 (in Japanese)

Okadome Y, Nakamura Y (2023b) Extracting feature space for synchronizing behavior in an interaction scene using unannotated data. In: Iliadis L, Papaleonidas A, Angelov P, Jayne C (eds) Artificial neural networks and machine learning—ICANN 2023: 32nd international conference on artificial neural networks, Heraklion, Crete, Greece, 26–29 Sept 2023, Proceedings, Part VIII. Springer, Cham, pp 209–219

Onishi Y, Sumioka H, Shiomi M (2021) Increasing torso contact: comparing human-human relationships and situations. In: Li H, Ge SS, Wu Y et al (eds) Social robotics: 13th international conference, ICSR 2021, Singapore, Singapore, 10–13 Nov 2021, Proceedings. Springer, Cham, pp 616–625

Onishi Y, Sumioka H, Shiomi M (2023) Designing a robot which touches the user's head with intra-hug gestures. In: Companion of the 2023 ACM/IEEE international conference on human-robot interaction. ACM, New York, pp 314–317

Rea DJ, Seo SH (2022) Still not solved: a call for renewed focus on user-centered teleoperation interfaces. Front Robot AI 9. https://doi.org/10.3389/frobt.2022.704225

Regenbrecht J, Tavakkoli A, Loffredo D (2017) A robust and intuitive 3D interface for teleoperation of autonomous robotic agents through immersive virtual reality environments. In: 2017 IEEE symposium on 3D user interfaces (3DUI). IEEE, pp 199–200

Sakai K, Minato T, Ishi CT, Ishiguro H (2016) Speech driven trunk motion generating system based on physical constraint. In: 2016 25th IEEE international symposium on robot and human interactive communication (RO-MAN). IEEE, pp 232–239

Sakai K, Nakamura Y, Yoshikawa Y, Ishiguro H (2022) Effect of robot embodiment on satisfaction with recommendations in shopping malls. IEEE Robot Autom Lett 7:366–372. https://doi.org/10.1109/LRA.2021.3128233

Sakamoto D, Ishiguro H (2009) GEMINOID: remote-controlled android system for studying human presence. KANSEI Eng Int 8:3–9. https://doi.org/10.5057/ER081218-1

Sakamoto D, Kanda T, Ono T et al (2007) Android as a telecommunication medium with a human-like presence. In: Proceedings of the ACM/IEEE international conference on human-robot interaction. ACM, New York, pp 193–200

Sakamoto D, Hayashi K, Kanda T et al (2009) Humanoid robots as a broadcasting communication medium in open public spaces. Int J Soc Robot 1:157–169. https://doi.org/10.1007/s12369-009-0015-5

Sato W, Namba S, Yang D et al (2022) An android for emotional interaction: spatiotemporal validation of its facial expressions. Front Psychol 12. https://doi.org/10.3389/fpsyg.2021.800657

Shi J, Liu C, Ishi CT, Ishiguro H (2021) 3D skeletal movement-enhanced emotion recognition networks. APSIPA Trans Signal Inf Process 10. https://doi.org/10.1017/ATSIP.2021.11

Shibata T (2004) An overview of human interactive robots for psychological enrichment. Proc IEEE 92:1749–1758. https://doi.org/10.1109/JPROC.2004.835383

Shimaya J, Yoshikawa Y, Kumazaki H et al (2019) Communication support via a tele-operated robot for easier talking: case/laboratory study of individuals with/without autism spectrum disorder. Int J Soc Robot 11. https://doi.org/10.1007/s12369-018-0497-0

Shinkawa K, Nakata Y (2023) Gaze movement operability and sense of spatial presence assessment while operating a robot avatar. In: 2023 IEEE/SICE international symposium on system integration (SII). IEEE, pp 1–7. https://doi.org/10.1109/SII55687.2023.10039342

Shintani T, Ishi CT, Ishiguro H (2021) Analysis of role-based gaze behaviors and gaze aversions, and implementation of robot's gaze control for multi-party dialogue. In: 2021 9th International conference on human agent interaction (HAI). pp 332–336. https://doi.org/10.1145/3472307. 348465

Shintani T, Ishi CT, Ishiguro H (2022) Expression of personality by gaze movements of an android robot in multi-party dialogues. In: 2022 31st IEEE international conference on robot and human interactive communication (RO-MAN). IEEE, pp 1534–1541

Shiomi M, Hagita N (2016) Do synchronized multiple robots exert peer pressure? In: Proceedings of the fourth international conference on human agent interaction. ACM, New York, pp 27–33

Shiomi M, Hagita N (2021) Audio-visual stimuli change not only robot's hug impressions but also its stress-buffering effects. Int J Soc Robot 13:469–476. https://doi.org/10.1007/s12369-019-00530-1

Shiomi M, Okumura S, Kimoto M et al (2020a) Two is better than one: social rewards from two agents enhance offline improvements in motor skills more than single agent. PLoS One 15:e0240622. https://doi.org/10.1371/journal.pone.0240622

Shiomi M, Sumioka H, Sakai K et al (2020b) SŌTO: an android platform with a masculine appearance for social touch interaction. In: Companion of the 2020 ACM/IEEE international conference on human-robot interaction. ACM, New York, pp 447–449

Shiomi M, Nakata A, Kanbara M, Hagita N (2021) Robot reciprocation of hugs increases both interacting times and self-disclosures. Int J Soc Robot 13:353–361. https://doi.org/10.1007/s12 369-020-00644-x

Shiomi M, Hayashi R, Nittono H (2023) Is two cuter than one? Number and relationship effects on the feeling of kawaii toward social robots. PLoS ONE 18:e0290433. https://doi.org/10.1371/journal.pone.0290433

Song J, Meng C, Ermon S (2020) Denoising diffusion implicit models

Stanton NA, Salmon PM, Walker GH et al (2017) Mental workload assessment methods. In: Stanton NA, Salmon PM, Rafferty LA et al (eds) Human factors methods. CRC Press, London, pp 301–364

Sumioka H, Nishio S, Minato T et al (2014) Minimal human design approach for sonzai-kan media: investigation of a feeling of human presence. Cognit Comput 6:760–774. https://doi.org/10.1007/s12559-014-9270-3

Sumioka H, Yamato N, Shiomi M, Ishiguro H (2021) A minimal design of a human infant presence: a case study toward interactive doll therapy for older adults with dementia. Front Robot AI 8. https://doi.org/10.3389/frobt.2021.633378

Takashiro S, Nakamura Y, Nishimura Y, Ishiguro H (2021) Development of a generative model for face motion during dialogue. In: IEICE technical report, pp 12–16

Tatsukawa K, Nakano T, Ishiguro H, Yoshikawa Y (2016) Eyeblink synchrony in multimodal human-android interaction. Sci Rep 6:39718. https://doi.org/10.1038/srep39718

Triantafyllidis E, McGreavy C, Gu J, Li Z (2020) Multimodal interfaces for effective teleoperation

Wu B, Liu C, Ishi CT, Ishiguro H (2021) Probabilistic human-like gesture synthesis from speech using GRU-based WGAN. In: Companion publication of the 2021 international conference on multimodal interaction. ACM, New York, pp 194–201

Wu B, Shi J, Liu C et al (2022) Controlling the impression of robots via GAN-based gesture generation. In: 2022 IEEE/RSJ international conference on intelligent robots and systems (IROS). IEEE, pp 9288–9295

Yagi S, Nakata Y, Nakamura Y, Ishiguro H (2022) Spontaneous gait phase synchronization of human to a wheeled mobile robot with replicating gait-induced upper body oscillating motion. Sci Rep 12:16275. https://doi.org/10.1038/s41598-022-20481-4

Yamato N, Sumioka H, Ishiguro H et al (2022) Interactive baby robot for the elderly with dementia—realization of long-term implementation in nursing home. Trans Digital Pract 3:14–27 (in Japanese)

Yasuoka M, Zivko M, Ishiguro H et al (2022) Effects of digital avatar on perceived social presence and co-presence in business meetings between the managers and their co-workers. In: Wong

L-H, Hayashi Y, Collazos CA et al (eds) Collaboration technologies and social computing: 28th international conference, CollabTech 2022, Santiago, Chile, 8–11 Nov 2022, Proceedings. Springer, Cham, pp 83–97

Yoshikawa Y, Lio T, Arimoto T et al (2017) Proactive conversation between multiple robots to improve the sense of human-robot conversation. In: AAAI 2017 fall symposium on human-agent groups: studies, algorithms and challenges, pp 288–294

Yoshikawa Y, Muramatsu T, Sakai K et al (2023) A new group-based online job interview training program using computer graphics robots for individuals with autism spectrum disorders. Front Psychiatry 14. https://doi.org/10.3389/fpsyt.2023.1198433

Chapter 3
Spoken Dialogue Technology for Semi-Autonomous Cybernetic Avatars

Tatsuya Kawahara, Hiroshi Saruwatari, Ryuichiro Higashinaka, Kazunori Komatani, and Akinobu Lee

Abstract Speech technology has made significant advances with the introduction of deep learning and large datasets, enabling automatic speech recognition and synthesis at a practical level. Dialogue systems and conversational AI have also achieved dramatic advances based on the development of large language models. However, the application of these technologies to humanoid robots remains challenging because such robots must operate in real time and in the real world. This chapter reviews the current status and challenges of spoken dialogue technology for communicative robots and virtual agents. Additionally, we present a novel framework for the semi-autonomous cybernetic avatars investigated in this study.

T. Kawahara (✉)
Kyoto University, Kyoto, Kyoto, Japan
e-mail: kawahara@i.kyoto-u.ac.jp

H. Saruwatari
The University of Tokyo, Bunkyo, Tokyo, Japan
e-mail: hiroshi_saruwatari@ipc.i.u-tokyo.ac.jp

R. Higashinaka
Nagoya University, Nagoya, Aichi, Japan
e-mail: higashinaka@i.nagoya-u.ac.jp

K. Komatani
Osaka University, Ibaraki, Osaka, Japan
e-mail: komatani@sanken.osaka-u.ac.jp

A. Lee
Nagoya Institute of Technology, Nagoya, Aichi, Japan
e-mail: ri@nitech.ac.jp

© The Author(s) 2025
H. Ishiguro et al. (eds.), *Cybernetic Avatar*,
https://doi.org/10.1007/978-981-97-3752-9_3

3.1   Introduction

Speech is the most natural means of human–human communication. Therefore, it has been investigated for effective human–computer interaction and human–robot interaction. Although text-based communication is dominant in computers and smartphones, automatic speech recognition (speech-to-text) and speech synthesis (text-to-speech) are crucial for using speech interfaces. Speech interfaces have become widely available for smartphone applications, including simple voice search and more elaborate assistant software. Since 2010, with the introduction of deep learning and the accumulation of large datasets, speech recognition performance has improved significantly. This technology has also been applied to smart speakers wherein speech is the only available interface. Speech interfaces have been used in various applications such as car navigation and appliance control systems. However, these systems are modeled as reactive human–machine interfaces that respond to short, well-formed commands or queries. The content and style of speech are much different from those of human–human communication, where we speak many long utterances in turn, during which an interlocutor provides feedback. In general, current speech interfaces are designed and implemented as half-duplex communication, rather than full-duplex communication.

This raises a major question: What kind of communication style is appropriate for humanoid robots? The answer depends on the type of task and the roles that robots are expected to play. The affordances and outlook of a robot are also important considerations (Bartneck et al. 2020). If a robot is a porter or cleaner, a command-and-control interface is sufficient. A receptionist needs to engage in some dialogue, but most user inputs are simple queries. In these tasks, the goal of the dialogue is definite and observable. However, a guide (at a museum or tourist spot) or an attendant (at an event) must handle more complex interactions. Furthermore, with recent advancements in AI, more complex dialogue tasks such as interviews and counseling are also within the scope of autonomous systems. There is also increasing demand for communicative robots that can speak with people, particularly seniors living alone. Such interactions are similar to human-like conversations.

Text-based chatbots have been developed for many years. With the development of large language models such as GPT (Brown et al. 2020), chatbots have become very natural and human-like. Furthermore, the automatic speech recognition performance of conversational telephone speech has been claimed to be of human parity with a word error rate of 5% (Saon et al. 2017; Stolcke and Droppo 2017). The quality of speech synthesis has also been claimed to be almost indistinguishable from that of human speech (Shen et al. 2018). These developments raise further questions. If we combine these speech recognition, speech synthesis, and chatbot systems, can we realize a human-level spoken dialogue system? If we implement such a system in a human-looking android, will it pass a multimodal Turing test?

We can easily find an answer (or identify problems) by implementing such a system. One apparent issue is the timing or latency of responses. It usually takes several seconds before getting a response from a GPT server. When the interval becomes two seconds or longer, users often make new utterances, which can conflict with delayed system responses. Another apparent problem is simply reading out a very long response, rather than speaking naturally. Although the synthesized speech of each sentence may be natural, the speech of an entire discourse could be monotonous and unengaging. Spoken dialogue is significantly different from text-based dialogue because it is a real-time interaction and speech can convey rich information beyond text content.

We have investigated human-like spoken dialogue in this context (Kawahara 2019), which will be described in Sect. 3.4. However, several challenges must be addressed before realizing or replacing human communication. First, the current system cannot respond to all queries or establish rapport with users through empathy. More importantly, system responses may not align with the intent of the designer or controller when the dialogue task is not definite. Therefore, we introduce a novel human–avatar interaction framework using a semi-autonomous Cybernetic Avatar (CA). The CA is a hybrid of an autonomous dialogue system and a human-operated avatar. This concept is described in Sect. 3.2 and the corresponding dialogue control strategy is described in Sect. 3.5.

Communicative robots and CAs are intended to be used in the real world. When serving the general public, they are often placed in noisy environments such as shopping malls and exhibition halls. This poses another challenge for realizing spoken dialogue with communicative robots. Unlike smartphones and smart speakers, it is difficult to constrain the timing of user utterances and detect the end of utterances. User speech is occasionally mixed with other sounds such as another person's speech and background music. Therefore, we need acoustic processing to extract and enhance target speech robustly, as described in Sect. 3.3.

The system infrastructure based on CG-CA is described in Sect. 3.6, and implementations and experiments are presented in Sect. 3.7.

3.2 System Concept and Overview

This section provides an overview of a framework for semi-autonomous CAs (Kawahara et al. 2021), including a discussion of advantages and technical challenges.

3.2.1 Architecture and Process Flow

The spoken dialogue system for semi-autonomous CAs is illustrated in Fig. 3.1. In this framework, one remote operator serves many users in parallel using CAs, each of which can be a robot or virtual agent with an autonomous dialogue system. Because users and CAs may be in noisy environments such as shopping malls and exhibition halls, user speech must be separated and enhanced. When a user is talking

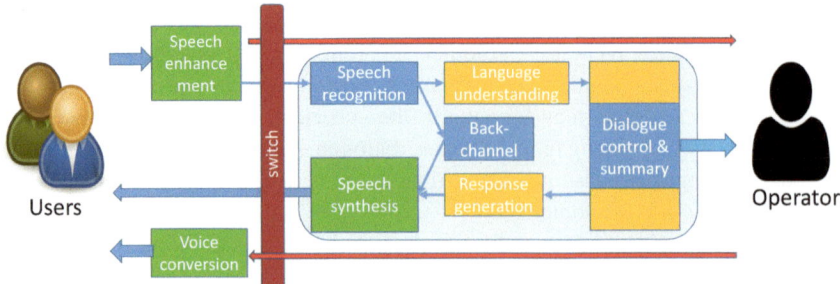

Fig. 3.1 System architecture of a semi-autonomous CA

to the operator, their speech is directly passed to the operator (top red line). Otherwise, the speech is processed by an autonomous system (middle blue box). First, it is transcribed by an automatic speech recognition module and then processed by a language understanding module. When a response is generated, it is outputted by a speech synthesis module. This flow is essentially the same as that of conventional spoken dialogue systems. Backchannels are generated in real-time to make the dialogue more engaging. Partial repeats and elaborate questions are also generated in the attentive listening mode, as described in Sect. 3.4.

When there is a problem during a dialogue session (e.g., some user queries cannot be handled by the system or the engagement of the user decreases (Inoue et al. 2018; Oertel et al. 2020)), the human operator intervenes in the session. In this case, the user's speech is directly fed to the operator, and the operator's speech is converted into the avatar's voice to make the entire dialogue coherent (bottom red line). Because one operator handles many users simultaneously, there may be cases in which intervention is required for more than one user simultaneously. To solve this problem, it is necessary to define the priority of a session and to maintain the remaining dialogue through chatting.

3.2.2 Advantages and Goals

The proposed system provides effective and efficient dialogue-based services. In contrast to a conventional human avatar, which can only interact with one user at a time, the proposed system makes it possible to serve many users in parallel. If the majority of dialogue can be handled in an autonomous manner such as explanation or listening, then one operator is expected to be able to serve three or more users simultaneously. Additionally, this framework provides a new perspective on spoken dialogue system research. Autonomous systems do not have to achieve perfect performance or human-level experiences, but they can turn to real humans when necessary, thereby expanding the applicability of such systems.

The goal of the proposed system is to achieve human-parity performance, which can be achieved through one-on-one dialogues with an avatar for all users in parallel. The proposed system should significantly outperform an autonomous system acting alone. Another important factor is smooth and natural switching between the autonomous system and the human operator, which should be imperceptible.

3.2.3 Technical Challenges

There are many technical challenges associated with realizing the proposed system. The most crucial aspect is the seamless integration of autonomous and teleoperated avatars. This requires improving autonomous systems to behave in a human-like manner. The proposed model and its implementation are discussed in Sect. 3.4.

Voice conversion is a key component of seamless switching. There are two ways to this conversion: converting the operator's speech into the system's voice or converting the system's speech into the operator's voice. Typically, an avatar's voice and face are collectively designed for the target task and should be consistent over time and location. Therefore, the operator's speech is converted into the avatar's voice in our system.

Additionally, dialogue context must smoothly switch to that of a human operator. The system should detect when it cannot continue the dialogue and must turn to a human operator. This is a difficult problem as many AI or pattern recognition systems do not know when they are making errors. There have been many investigations on the automatic evaluation of dialogue responses (Lowe et al. 2017) and detection of dialogue breakdown (Higashinaka et al. 2021), but the performance is far below the pragmatic level. Although there have been attempts to develop semi-autonomous teleoperated robots, the detection of dialogue breakdown has been performed manually or based on predefined phrases such as "that is not right" (Glas et al. 2008; Shiomi et al. 2008; Kanda et al. 2010). When the system switches to a human operator, it must provide the operator with the dialogue context in an efficient manner so that they can promptly catch up with the conversation and recover from the error. Furthermore, the system should recognize the personalities and preferences of users to generate user-adaptive responses. These issues are challenging and may depend on the nature of the dialogue task. These issues are addressed in Sect. 3.5.

The usable software platform developed in this project is described in Sect. 3.6.

3.3   Acoustic Processing for Real-World Human–Avatar Interaction

3.3.1   Acoustic Environment for CA's Ear

The environments in which we communicate can vary significantly, ranging from quiet settings such as libraries to crowded areas such as train stations. We are not typically consciously aware of such changes as humans; however, we have the ability to hear an interlocutor's voice under varying conditions. For example, even in extremely noisy construction environments, one can listen to and understand what an interlocutor says by focusing on their voice. This selective hearing is known as the **cocktail party effect** (Arons 1992).

Similar to this effect in humans, the technology used to extract individual audio sources (e.g., an interlocutor's voice) from mixed signals is called **audio source separation**. CAs should eventually perform functions similar to those of humans and act as symbiotic entities in society, which will require selective hearing abilities (Fig. 3.2). Therefore, it is necessary to develop audio source separation technology for CAs.

Audio source separation for CAs must meet the following requirements.

- **Adaptability to various acoustic environments**: A CA should be able to handle environmental noise variation, for example, when someone talks loudly in a quiet library, the noise has a clear directionality. Conversely, various sounds intermingle in noisy shopping malls. The former type of noise is called **directional noise**, which originates from a certain direction. The latter is referred to as **diffuse noise**, which is a sound that spreads in various directions. Audio source separation for CAs must consider cases in which an interlocutor's voice is mixed with such varied types of noise. It is assumed that no prior information regarding noise is available.
- **Capability for streaming and low-latency separation**: To achieve natural speech communication with humans, audio source separation must be performed in a streaming and low-latency manner. This implies that separation should be executed sequentially each time mixed signals are observed. The corresponding

Fig. 3.2 CAs should be able to hear an interlocutor's voice in a variety of acoustic environments

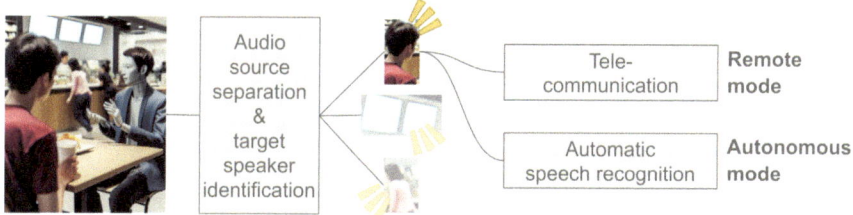

Fig. 3.3 CA auditory system. The CA must (1) identify the target speaker's voice after separating mixed signals and (2) transfer the voice to a human operator in the remote mode or to the automatic speech recognition module in the autonomous mode

technology must minimize the waiting time before an interlocutor's voice is recognized.

- **Ability to identify an interlocutor's voice after separation**: A semi-autonomous CA should have both autonomous and remote modes (Fig. 3.3). In the remote mode, an interlocutor's voice is transmitted to a remote operator, whereas in the autonomous mode, a module recognizes an interlocutor's voice. Therefore, audio source separation in CAs must identify which separated signal is the voice of the relevant interlocutor after separating mixed signals. Because the interlocutor is not fixed, it is assumed that no prior information regarding the interlocutor's voice is available.

- **Minimizing the loss of the interlocutor's voice**: In telecommunication, the quality of communication is often degraded by poor connection quality, which could cause an interlocutor's voice to be heard intermittently in the presence of background noise. If audio source separation is erroneous, residual noise may remain in the separated voice or parts of the voice may be missing. To ensure high communication quality, audio source separation must reduce these types of errors.

3.3.2 Audio Source Separation: Basic Theory

3.3.2.1 Blind Source Separation

Blind source separation (BSS) involves audio source separation without information regarding the recording environment, mixing system, or source locations. In a determined or overdetermined scenario (i.e., the number of microphones is greater than the number of sources), independent component analysis (ICA) (Comon 1994), independent vector analysis (IVA) (Hiroe 2006; Kim et al. 2007), and independent low-rank matrix analysis (ILRMA) (Kitamura et al. 2016) are the most commonly used methods for solving the BSS problem, and many studies on BSS have been reported in the area of acoustic signal processing (Sawada et al. 2019). Such overdetermined BSS methods generally assume independence between sources and estimate a "demixing

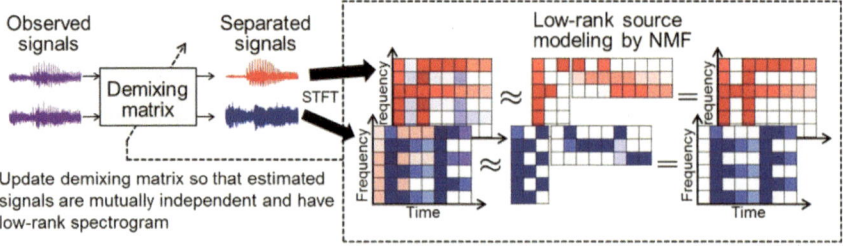

Fig. 3.4 Overview of ILRMA

matrix" that should be the inverse of the mixing system according to the results of maximum likelihood estimation. In particular, ILRMA assumes low-rankness for the power spectrogram of each source using non-negative matrix factorization (NMF) (Lee and Seung 1999) in addition to statistical independence between sources (Fig. 3.4), allowing it to achieve efficient and accurate separation. Additionally, most methods assume a rank-one spatial model, and the frequency-wise acoustic path of each source can be represented by a single time-invariant spatial basis, which is often called a steering vector. Under this assumption, the overdetermined BSS problem is reduced to a simple estimation of the demixing matrix for each frequency. This approach leads to increases in the stability of parameter optimization and computational efficiency. Recent advances in ILRMA include spatial regularization (Mitsui et al. 2018) and extension of the source statistical model (Mogami et al. 2020).

On the other hand, for underdetermined scenarios (i.e., the number of microphones is less than the number of sources), multichannel NMF (MNMF) (Ozerov and Fevotte 2010; Sawada et al. 2013) has received considerable attention. Instead of estimating a demixing matrix (rank-one spatial model), in MNMF, a full-rank spatial covariance matrix (SCM) (Duong et al. 2010) is estimated for each source. A full-rank SCM can represent not only the acoustic path but also the spatial spread of each source or diffuse noise. The target source can then be enhanced using the SCM via a multichannel Wiener filter. MNMF is more advantageous compared to ILRMA for realistic BSS tasks, especially in highly reverberant and noisy environments. However, its optimization has a high computational cost and lacks robustness against variations of the initialization (Kitamura et al. 2016). FastMNMF has been proposed to accelerate parameter estimation (Ito and Nakatani 2019; Sekiguchi et al. 2019), although its performance still depends on initial parameter values. To increase stability, an ILRMA-based initialization method can be utilized for MNMF (Kitamura et al. 2016; Shimada et al. 2018).

Additionally, rank-constrained SCM estimation (RCSCME) has been proposed for extracting directional target speech efficiently and robustly (Kubo et al. 2020; Kondo et al. 2022). BSS for extracting a specific target source from an observed mixed signal is often referred to as blind speech extraction (BSE). Although directional target speech can be expressed using a rank-one (rank-constrained) SCM, diffuse noise requires a full-rank SCM because of its spatial spread. To achieve robust and computationally efficient extraction in this BSS, one can utilize the demixing matrices obtained by BSS methods such as ILRMA. It utilizes the fact that the demixing filters for diffuse noise can cancel the directional target speech in BSS methods based on rank-one spatial models (Takahashi et al. 2009), resulting in the accurate estimation of a rank-$(M - 1)$ diffuse noise SCM, where M denotes the number of microphones. RCSCME can restore one lost spatial basis of diffuse noise via maximum a posteriori estimation without requiring a huge number of computations. Recently, Nishida et al. (2023) proposed NoisyILRMA, which can be regarded as an integration of ILRMA and RCSCME.

3.3.2.2 Supervised Source Separation

Deep neural networks (DNNs) have shown promising performance for both single-channel (Grais et al. 2014; Nakamura et al. 2021) and multichannel audio source separation (Nugraha et al. 2016; Tu et al. 2017; Qian et al. 2018). Makishima et al. (2019) unified the ILRMA-based blind estimation of a demixing matrix and DNN-based supervised updating of the source spectrogram model (Fig. 3.5). In this method, the demixing matrix (spatial model) is efficiently optimized as with ILRMA. Because this method utilizes a time–frequency spectrogram matrix estimated by a DNN to optimize the spatial model, it is called independent deeply learned matrix analysis (IDLMA). IDLMA assumes a semi-supervised scenario in which a solo-recorded dataset can be prepared for a single source in a mixed signal, where there is no solo-recorded dataset for the other sources. In this scenario, because a DNN source model for other sources cannot be prepared in advance, a new data augmentation scheme is introduced in which augmented data are used to retrain the DNN source model iteratively for all sources while optimizing the spatial model. Recently, as an extension of IDLMA, a combination of unsupervised NMF and supervised DNN source models was proposed to mitigate the training-inference mismatch problem inherent to DNNs (Hasumi et al. 2023). In (Misawa et al. 2021), IDLMA was utilized as a preprocessing step for RCSCME under diffuse noise conditions.

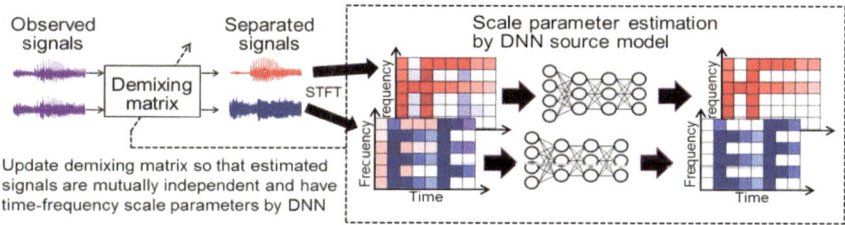

Fig. 3.5 Overview of IDLMA

3.3.3   Audio Source Separation for Human–Avatar Interaction

3.3.3.1   Architecture for Real-Time Processing

Audio source separation methods are typically designed as offline methods. For example, the ILRMA, RCSCME, and IDLMA approaches described in Sect. 3.3.2 begin separation after an entire mixed signal has been inputted. Therefore, to apply these approaches to CA systems, they must be extended to handle streaming inputs with low latency. In this section, we describe the real-time extension of an RCSCME-based BSE method.

A simple approach to implementing such an extension is to perform the entire process frame-by-frame. This approach is effective for speech extraction methods with low computational costs such as delay-and-sum beamformers. However, compared to such methods, the RCSCME-based BSE method is more sophisticated and computationally expensive, which hinders real-time execution when using this approach. One solution to this problem is to adopt a parallel computing approach called the block-wise batch approach (Mukai et al. 2004; Mori et al. 2006), where computationally expensive steps of the process are executed across multiple frames and the remaining parts are executed in a frame-by-frame manner.

The entire process of the RCSCME-based BSE method can be divided into two main steps: ILRMA and RCSCME. ILRMA consists of a large number of matrix operations and requires a sufficient number of iterations to achieve the desired separation performance. In contrast, RCSCME mainly consists of scalar operations and can operate in real-time on a modern computer.

To run these two components in parallel, dependency is an issue that must be addressed because RCSCME uses the demixing matrices estimated by ILRMA. The demixing matrix represents the acoustic environment and does not tend to change significantly unless the acoustic environment changes drastically. Therefore, using a previously estimated demixing matrix should not significantly degrade separation performance at the current time. As a result, when a CA system is operated in a relatively consistent acoustic environment (e.g., where an interlocutor and CA have face-to-face interactions), the real-time extension of the RCSCME-based BSE method can achieve sufficient separation performance.

Figure 3.6 presents a schematic diagram of the computational flow of this approach. ILRMA runs over multiple frames and estimates a demixing matrix from the most recently observed signals over a specific time window and subsequently passes the estimation results to RCSCME. RCSCME runs at each time interval of a hop size and must finish its execution within that same interval. It first separates the most recently observed signals over a specific time period using the latest demixing matrix provided by ILRMA. It then identifies one of the separated signals as the voice of the target interlocutor. Finally, it enhances the signal of the interlocutor before outputting the final signal.

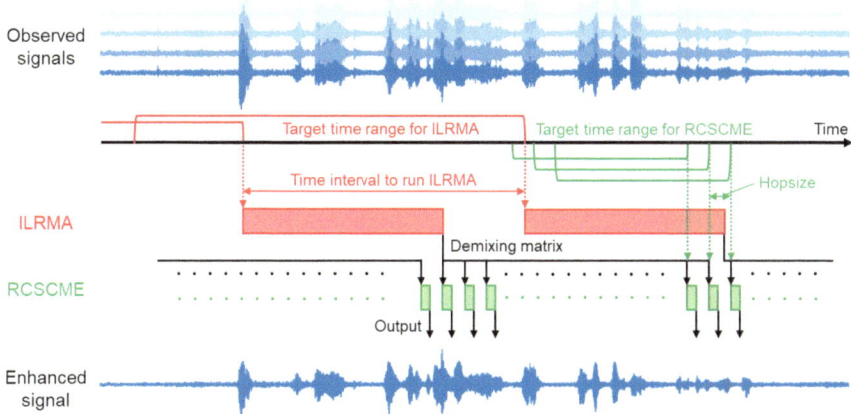

Fig. 3.6 Schematic illustration of the computational flow of the RCSCME-based BSE method. The red and green rectangles represent the processing times of the ILRMA and RCSCME parts, respectively

The algorithmic latency of the real-time extension described above is the sum of the frame and hop sizes. In practice, the total latency also includes the recording latency, which depends on the recording device, and the processing time of the BSE method, which depends on the computational resources available. A GPU can accelerate the matrix operations of ILRMA and the parallelizable scalar operations of RCSCME. Based on this notable acceleration, we experimentally confirmed that a Python implementation could operate in real time.

3.3.3.2 Use of Prior Information Regarding Interlocutor Direction

When using CAs, prior information regarding an interlocutor's voice is unavailable. However, the approximate direction of an interlocutor is often available in advance. For example, in a dialogue system with a CA, the interlocutor and CA often face each other and do not move significantly. When a camera is installed in a CA, an interlocutor's direction can often be estimated. Even if it contains minor errors, this information is useful for the BSE method and can improve separation performance. In this section, we present a method for utilizing the direction information of an interlocutor in the real-time BSE method described in the previous section.

In the RCSCME component of the real-time RCSCME-based BSE method, we must choose one of the separated signals obtained using the demixing matrix. For this purpose, an algorithm based on higher-order statistics similar to the offline BSE method can be used (Fujihara et al. 2008). However, this approach sometimes fails for various reasons. For example, the speech duration of an interlocutor included in the observed signals may be too short to compute higher-order statistics. These

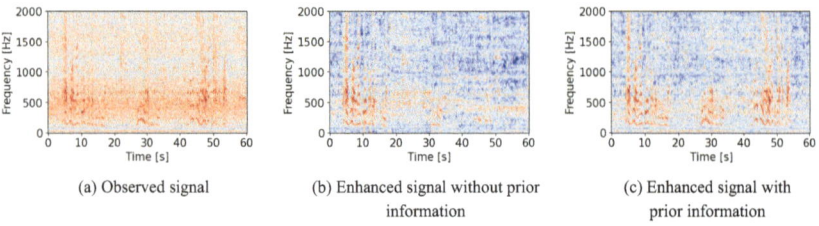

(a) Observed signal   (b) Enhanced signal without prior   (c) Enhanced signal with
                              information   prior information

Fig. 3.7 Examples of the magnitude spectrograms of observed and enhanced signals

failures lead to a partial loss of the interlocutor's voice in the enhanced signal, which significantly lowers the communication quality with the CA system.

To reduce such failures, prior information regarding an interlocutor's direction can be used to improve demixing filter estimation in ILRMA. For example, we can design the initial values of the demixing matrix using this information and assign a specific channel in the direction of the target interlocutor. The ideal steering vector in the direction of the interlocutor can be computed analytically for a plane wave. In most cases, the sounds arriving from the direction of an interlocutor can be approximated with reasonable accuracy by the plane wave.

Figure 3.7 presents examples of magnitude spectrograms of observed and enhanced signals obtained with and without prior information. The enhanced signal obtained without prior information partially lost the interlocutor's voice (see the time from approximately 25 to 55 s in Fig. 3.7b). In contrast, the enhanced signal obtained with prior information (see Fig. 3.7c) captured the entire duration of the interlocutor's voice.

3.3.4 Future Directions

We have described audio source separation as a technology for realizing human-like CA ears. Below, we describe some technologies that will be necessary to realize more advanced CA ears.

- **Semantic hearing:** In this section, we discussed audio source separation for an interlocutor's voice in a one-on-one, face-to-face conversation. However, the targets of selective hearing are diverse. For example, if an emergency vehicle such as an ambulance passes by a CA, the CA must listen to the sound of the ambulance to take appropriate action. Furthermore, in conversations involving multiple people, the target interlocutor may change frequently. The sounds to be listened to selectively vary depending on the context of communication. Therefore, semantic audio source separation that can adapt to different contexts is necessary.
- **Spatial context**: In addition to identifying the target signal ("what"), it is also important to identify its spatial location ("where"). Depending on the location of an interlocutor, a CA may need to change its posture, which can also help

operators understand the acoustic environment of the operating area. However, because the ears of a CA are significantly different from human ears (e.g., humans have only two ears on either side of their head, whereas the ears of a CA are not limited in this manner), some ingenuity is required to convey the spatial location information captured by a CA to humans.

3.4 Speech Processing for Real-Time Human–Avatar Interaction

This section addresses speech processing for creating human-like dialogue systems, which is the basis of natural human–robot interaction and semi-autonomous CAs. Based on the perspective that spoken dialogue is a real-time interaction that includes non-textual reactions, we describe several important components below.

3.4.1 Automatic Speech Recognition

The performance of automatic speech recognition has been improved using deep learning models and large datasets. Some models have achieved an accuracy of 95% for conversational speech and the robust recognition of distant speech in queries to smart speakers. Recently, end-to-end modeling (Li 2022) and self-supervised learning (Baevski et al. 2020; Hsu et al. 2021) using Transformers have provided further improvements. However, it is still challenging to deal with conversational speech without close-talking microphones because users may be unaware of the presence of a microphone. This scenario also applies to humanoid robots. The incorporation of acoustic processing as described in Sect. 3.3 is crucial and is currently being investigated.

Another challenge is that the major users of communicative robots are seniors and children, whose speech is not covered well by conventional speech recognition models.

3.4.2 Speech Synthesis

The performance of speech synthesis has also been improved by deep learning, particularly by end-to-end modeling (Shen et al. 2018). Recently, non-autoregressive models based on Transformers, which are called JETS, (Lim et al. 2022) have been widely used. The synthesis of various types of speech, including emotional speech, has also been investigated. However, it is difficult to generate natural backchannels, fillers, and laughter, as described in the following sections. Therefore, these types of voices are often pre-recorded. Although the synthesized speech of an utterance may be natural, prosody across sentences over discourse has not been modeled effectively, so conversion often becomes monotonous.

3.4.3 Turn-Taking

In conventional human–machine interfaces such as smartphones and smart speakers, the detection of utterance endpoints is obvious because such systems assume a single query or command. The start of an utterance is explicitly designated by users by tapping (push to talk) or by speaking a predefined keyphrase (wake word). This type of interface is acceptable for many applications, even when there is some latency before the system responds.

However, smooth turn-taking is critical for realizing human-like conversation. It has been observed that the average turn-switch time in human–human dialogue is less than 500 ms (Lala et al. 2018). In fact, an interval longer than 2 s feels excessively long for many users, often causing them to speak again, which can conflict with system responses. It is challenging to ensure that a system will respond within one second.

Conventional voice activity detection for automatic speech recognition is typically performed by detecting a pause approximately 400 ms after an utterance. This implies that there is already an inherent latency of 400 ms. This pause is then followed by several downstream processes such as speech recognition, language/dialogue processing, and response generation. Many of these modules based on Transformers (Vaswani et al. 2017) require the entire input for processing and cannot operate in a real-time pipeline. When these modules are operated on a cloud server, latency due to Internet communication is also inevitable. To realize prompt responses, it is necessary to deploy modules directly on robots or connected PCs.

A turn-taking model is also necessary for determining whether the current user is talking or the system can take a turn. Turn-taking has been modeled using recurrent neural networks with prosodic and lexical features trained on natural human-to-human dialogue datasets (Lala et al. 2019a).

3.4.4 Backchannel Generation

Backchannels play a significant role in providing human-like feedback behavior during dialogue. Backchannels can be classified into two categories of continuers and assessments. Continuers such as "right" and "*hai*" provide feedback for smooth talking by providing a signal of listening, understanding, and agreeing. They play a role in turn management by suggesting that the current speaker can continue. On the other hand, assessments such as "wow" and "*he-*" express the listener's reactions, such as surprise, interest, and empathy, typically at the end of utterances. Assessments also produce a sense of rhythm and synchrony.

The generation of backchannels requires the prediction of the timing, morphemes, and prosody of each occurrence (Kawahara et al. 2016). As continuers are generated in small volumes, the choices of morpheme and prosody are relatively unimportant, but timing is critical. Continuers should ideally be generated around the end of a phrase or utterance. This implies that it is too late when we conduct conventional voice activity detection, but we need to predict during user utterances. In a previous study, a machine learning model was trained with prosodic and lexical features using a natural human–human dialogue dataset (Lala et al. 2017).

The generation of assessments is more challenging because false generations significantly hamper dialogue. Furthermore, it is difficult to conduct machine learning because there are few such occurrences. However, assessment can be combined with emotional recognition. When a user exhibits positive or strong emotions, a corresponding assessment can be generated.

3.4.5 Filler Generation

Fillers such as "well" and "*ano-*" are used to signal thinking or hesitation during an utterance. They are not necessary for system responses, but they exhibit human-like behaviors. By placing a filler before speaking, one can alert the audience and show politeness. Fillers are also used to suggest taking turns. This approach is effective, especially when the turn keep or switch is ambiguous, and can thus be used in humanoid robots. The use of fillers for turn-taking has been investigated (Lala et al. 2019b).

3.4.6 Shared Laughter Generation

Laughter plays an important role in human–human communication. Laughter is not necessarily generated as a reaction to feeling funny; a large majority of laughter is used for enhancing socialization. Laughter can be effective for ice-breaking and relaxation. In particular, shared laughter, which follows an interlocutor's laughter, is useful for displaying empathy. On the other hand, laughter is sometimes used in a masochistic manner. In such cases, shared laughter should not be generated.

Inoue et al. (2022) trained a machine learning model to detect and classify user laughter into three categories of mirthful laughter, social laughter, and masochistic laughter. Corresponding shared laughter was generated for the first two cases.

3.4.7 Attentive Listening System

An attentive listening system (Inoue et al. 2020) was developed by integrating the aforementioned modules. Attentive listening is useful for seniors who need to be heard and maintain their communication skills. This system listens and encourages the subject to speak more often. The system needs to handle open-domain dialogue and respond to any utterances, but does not require a large knowledge base or a large language model. In other words, the system requires fundamental communication skills, that is, listening and understanding with interest and empathy.

The system configuration is illustrated in Fig. 3.8. While a user is speaking, the system continuously predicts the generation of backchannels using prosodic features to indicate that it is listening. Automatic speech recognition and simple natural language processing of focus word detection and sentiment analysis are conducted to generate partial repeats, elaborating questions, and assessments. These modules operate on a single laptop PC without a GPU. Examples of generated responses are presented in the figure.

A partial repeat is a simple repetition of a focus word. This function can be performed for any utterance and is effective for demonstrating that the system is understanding and encouraging the dialogue to continue. Elaborating questions are generated by connecting the most likely "wh-" question marker to the focus word. Such questions suggest a level of interest in the dialogue. Assessments are generated when positive or negative sentiments are detected in a user's utterances. Assessments are expected to show empathy. When none of the responses described above are generated, formulaic responses are generated. Their prosody can be tuned according to sentiment. The selection of responses is performed heuristically with a priority order of assessments, elaborating questions, partial repeats, and formulaic responses.

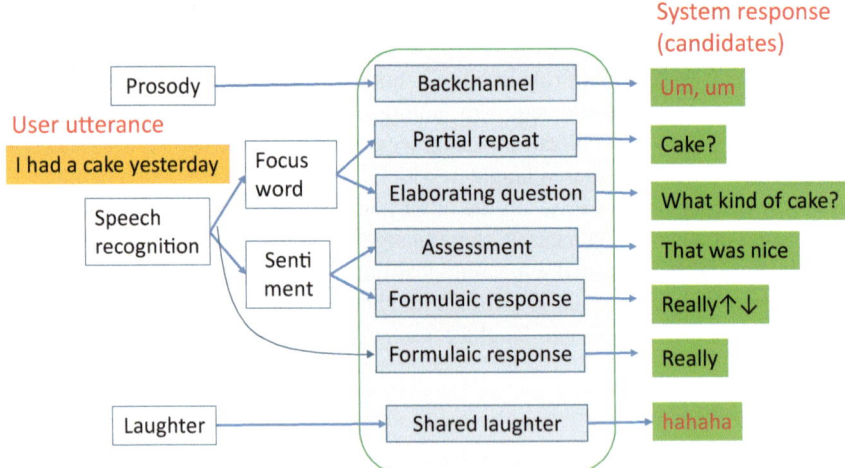

Fig. 3.8 Configuration of an attentive listening system

This order was determined according to the difficulty or infrequency of response types. Shared laughter is generated when the system detects laughter from the user.

The system can be used alone or incorporated into the semi-autonomous CAs introduced in Sect. 3.2. The details of the evaluation are provided in (Inoue et al. 2020). The system can also be combined with sophisticated chatbots using large language models, such as GPTs, by generating a simple response on the fly before obtaining an elaborate response from the server.

3.5 Dialogue Processing for CAs

3.5.1 Handover in Spoken Dialogue

Parallel conversation refers to a framework of interaction in which a small number of human operators provide dialogue services to users through CAs. CAs conduct basic interactions and if problems arise, human operators take over the dialogue to resolve these issues. Such a framework allows for high-efficiency dialogue services. For example, even if only a few service providers with specialized skills exist, it is possible to utilize their expertise to benefit many users. An important issue in this context is the handover of dialogue. Handover refers to the continuation of dialogue between the CA and the user by a human operator, who takes over from the CA.

3.5.2 Spoken Dialogue Technology for the Handover of Dialogue

Dialogue handover has traditionally been studied in the context of call routing. For example, in call centers, the process involves listening to requests from users and transferring them to the appropriate operators (Gorin et al. 1997). This technology is known as call routing and has been approached as a classification problem in machine learning. However, in call routing, operators do not take over the dialogue midway. Although there have been studies on systems that determine if a dialogue with a human has encountered problems and should be handed over to a human operator (Walker et al. 2002), such studies typically assume that human operators restart the dialogue from the beginning and do not examine how to hand over ongoing dialogues effectively. There have been a few studies in which robots have conducted most of the dialogue and only handed over certain parts to human operators. However, in these studies, the part of the dialogue handled by humans was predetermined, transforming the problem into a division of roles, rather than a true handover (Glas et al. 2012).

In a parallel conversation, a dialogue system should conduct the main conversation and when problems arise, an operator should take over the dialogue seamlessly. A corresponding framework is illustrated in Fig. 3.9. The necessary spoken dialogue

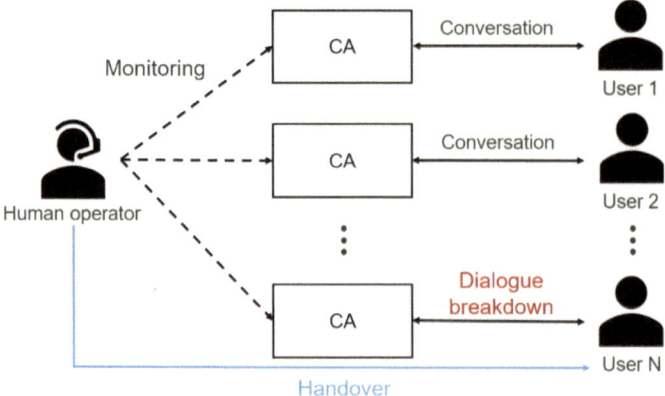

Fig. 3.9 Architecture for parallel conversation

technology primarily involves autonomous dialogue systems. For efficient dialogue services in parallel conversations, most conversations must be conducted by a CA. Additionally, dialogue breakdown detection technology is required to detect problems in a dialogue. Both speech and multimodal information can be used to detect dialogue breakdown. Furthermore, dialogue summarization technology is required. This involves creating summaries that allow human operators to understand the content of the dialogue conducted thus far between the CA and the user. Additionally, it is not always guaranteed that human operators will be available to take over immediately. In such cases, technology for buying time until a human operator is available is also necessary.

In the following subsections, we discuss autonomous dialogue systems, dialogue breakdown detection technology, dialogue summarization technology, and techniques for buying time.

3.5.2.1 Autonomous Dialogue System Technology

With the advancement of large language models (LLMs) such as ChatGPT, the performance of autonomous dialogue systems has improved significantly. Such systems can respond with high accuracy for dialogues that do not require much context such as casual conversations or guidance. High-precision dialogue has also been reported in both task- and non-task-oriented dialogues (Iizuka et al. 2023).

Here, we describe an experiment conducted at a facility called Nifrel, which is a combination of a zoo and aquarium in Osaka, Japan. In this experiment, we placed six robots (SOTA) inside the facility and used GPT-3 to provide guidance. Additionally, we prepared two operators who intervened and took over the dialogue if they judged that there were problems, thereby resolving dialogue issues (Mochizuki et al. 2023).

Figure 3.10 presents the interface used by the operators. During the month-long experiment, we found that approximately 90% of the interactions were successfully

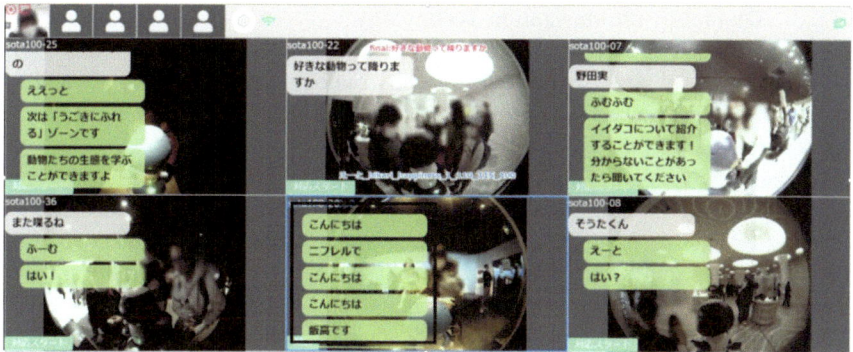

Fig. 3.10 Interface used by operators. This figure includes the images from the six robots, as well as user and system utterances

completed without any issues. This indicates that current LLMs can achieve sufficient levels of interaction. It was also noted that operator speech tended to contain fewer technical terms than system speech. With appropriate prompting, LLMs can achieve more advanced dialogue than humans depending on the field.

3.5.2.2 Dialogue Breakdown Detection Technology

Dialogue breakdown detection technology (Higashinaka et al. 2016) has primarily been studied in text-based dialogue systems. Various machine learning methods have been explored for the Dialogue Breakdown Detection Challenges. In the context of spoken dialogue such as in call centers, technologies have been developed to detect problematic dialogues using acoustic and language features (Walker et al. 2002). In future autonomous dialogue systems, the key technology will be the detection of dialogue breakdown in multimodal dialogues.

An analysis of problematic dialogues in the Nifrel experiment revealed that most issues were caused by a lack of understanding of the dialogue context, which included multimodal information. According to our analysis of problematic dialogues (Mochizuki et al. 2023), there were few errors originating from language or speech recognition, and most issues were caused by the system's inability to understand and respond to different scenarios. For example, errors such as a user's voice being too quiet or the system's inability to recognize a user's face, which prevented the start of an interaction, were observed. To address such dialogue breakdowns, technology that can accurately understand problems based on the dialogue context and multimodal information is necessary.

3.5.2.3 Dialogue Summarization Technology

To understand the content of a dialogue instantly, dialogue summarization technology is necessary. This raises an interesting question regarding what constitutes a desirable form of dialogue summarization.

We conducted experiments to investigate this issue by creating scenarios in which humans take over dialogues from each other and examined the types of information exchanged in such scenarios. The results indicated that a format connecting utterances was desirable (Yamashita and Higashinaka 2022). This format is useful because it allows for an understanding of who responded to whom and what was said.

We refer to the summarization of dialogues in the form of utterances as "dialogue format summarization" and implement this type of summarization as follows. First, we identify the sequence organizations within a dialogue. A sequence organization is a linguistically grounded unit in conversational analysis that consists of initiation toward the other party, response, and subsequent follow-up utterances. Focusing on these units, we rephrased extracted sequence organizations into dialogue exchanges of two to three utterances for summarization. By investigating the effectiveness of these summaries (Fig. 3.11), we found that compared with general text-format summaries, dialogue format summaries are similarly readable and concise while better conveying the realism of dialogues (Yamashita and Higashinaka 2023). In the handover of dialogue, an immediate understanding of the current scenario is desirable. Therefore, the realism achieved through dialogue format summarization is considered to be useful for handover.

The effectiveness of dialogue format summarization was evaluated in an experiment conducted at Nifrel. Operators intervened in dialogues using speech recognition

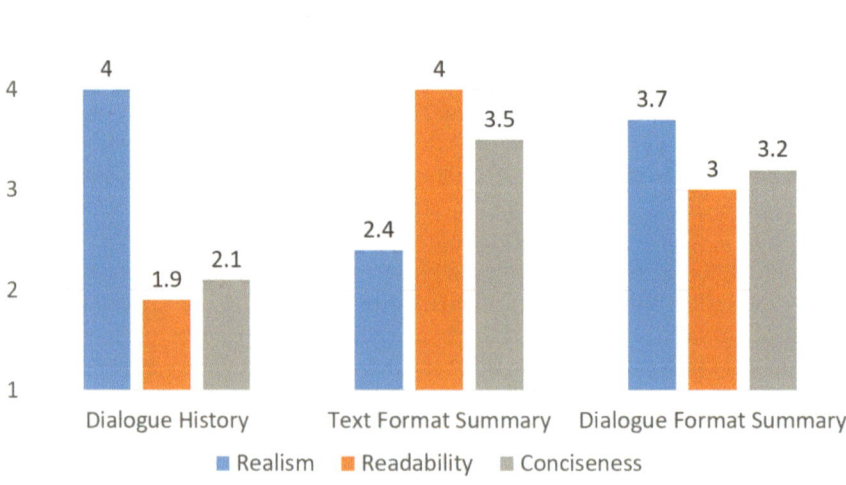

Fig. 3.11 Comparison of dialogue history (no summarization), text-format summarization (conventional summarization), and dialogue format summarization

results and dialogue format summaries. GPT-3 was used to generate dialogue format summaries. Questionnaire results indicated that dialogue format summaries had a smaller number of low usefulness ratings compared with the speech recognition results (Yamashita et al. 2023). This suggests that dialogue format summaries help facilitate the handover of dialogue.

3.5.2.4 Time-Buying Dialogue

Because an operator is not always available, if a situation arises in which an autonomous system cannot solve a dialogue problem, then the system must buy time until an operator becomes available. Here, the dialogue that takes place from the time at which the need for operator intervention is detected (i.e., when the dialogue breaks down) to when an operator successfully intervenes is called a time-buying dialogue. Overall, how well the system continues dialogue when a dialogue breakdown is detected is an important factor of dialogue with CAs.

To allow a CA to perform human-like time buying, findings from the analysis of human–human dialogues are useful. For example, in a previous study, verbal behaviors were examined when clerks were talking to customers on the phone and the information necessary for the clerks to respond was intentionally delayed (López Gambino et al. 2017). The target task was searching for a flight. When the information was displayed with a delay of approximately five seconds for the clerks, whether or not the clerk would speak and what the clerk would say were examined.

The results of this analysis revealed that within the delay time, the duration during which the clerks were speaking, and the duration of silence were approximately equal, and that humans tend to buy time by saying something, rather than simply remaining silent. Clerk utterances were classified into 12 categories and analyzed. The results showed that echoing (e.g., "flight to <place>," "beginning of <month>," which appeared in the customer utterances) accounted for 21% of the utterances, followed by filler utterances (e.g., "uh," "uhm," "mm") at 19% and agent/system state (e.g., "the search for flights is still in progress") at 10%. Simple waiting requests (e.g., "One moment, please") were uncommon, accounting for 6.3% of utterances. Therefore, in human–human dialogue, it is natural to buy time by making some type of utterance. Furthermore, human impressions were also investigated by comparing the human-like behavior described above to a simple waiting strategy (e.g., saying "Please hold on a second" followed by silence) (López Gambino et al. 2019). Utterances were synthesized using two different TTSs and evaluated by crowd workers. The amount of time to be bought was set to 12 s based on the results of preliminary investigations. The results demonstrated the effectiveness of a time-buying strategy when a human-like TTS is used. It should be noted that the analysis was conducted in a task-oriented dialogue with an explicit goal. In such cases, re-purposing the goal is an effective use of such available time, which can be considered as a form of echoing. For non-task-oriented dialogue, providing a summary of the dialogue can be considered as an effective way to buy a small amount of time.

When implementing a time-buying strategy in a real system, we must consider the amount of time that needs to be bought. If its length is less than 5 s, then the system can buy time with fillers. For slightly longer durations (up to approximately 10 s), time can be bought by confirming or recapitulating what was previously discussed during the dialogue without introducing a new topic. The analyses of human–human dialogue described above considered durations of 5 or 12 s. To buy more time, the system must present a new relevant topic and move the dialogue.

Based on these findings, our group has developed a system to realize subdialogues that buy time in a dialogue at the zoo like an aquarium. We assumed that dialogue breakdown detection, which defines the starting point of the time-buying dialogue, is possible. To prevent new dialogue breakdowns from occurring during time-buying subdialogues, the system was designed to engage mainly in system-initiated dialogue based on our previously proposed design guidelines (Komatani et al. 2022). The system was constructed using DialBB (Nakano and Komatani 2023), which is a framework for building dialogue systems, in consideration of its reusability in other domains.

3.5.3 Future Directions

The definition of breakdowns and their detection are important. In particular, when using speech signals, rather than text, as input modalities for CAs operating in the real world, speech-specific scenarios must be considered in addition to textual breakdowns (Higashinaka et al. 2016). Breakdowns that occur in spoken dialogue are not limited to those in linguistic content.

Figure 3.12 presents the four levels of action involved in cooperative dialogue (Paek and Horvitz 2000), inspired by the action levels defined by Clark (1996). This hierarchy represents an action ladder, indicating that lower actions must be completed for higher actions to be completed (upward completion). Furthermore, during a dialogue, the speaker and addressee must engage in their corresponding actions at the same level, as indicated in Fig. 3.9. For example, at the channel level, speaker A executes behavior for addressee B, whereas speaker B must attend to the behavior of speaker A. At the signal level, speaker A presents a signal to addressee B, and addressee B must identify the signal presented by speaker A. In this manner, a joint action is established between the speaker and addressee at each level of action in the dialogue, and the speaker's intention is correctly conveyed to the addressee without any breakdown (Clark 1996).

In spoken dialogue, dialogue breakdown can occur at any level. Specific examples are provided below.

Channel level: B is unaware of the speaker's presence, or that A is talking to B, or that B is not listening to what A says. This can be considered as an issue of engagement in dialogue.

	Speaker A's actions	Addressee B's actions
Conversation	A is proposing to B	B is considering A's proposal
Intention	A is conveying an utterance to B	B is understanding A's utterance
Signal	A is presenting a signal to B	B is identifying A's signal
Channel	A is executing behavior for B	B is attending to behavior from A

Fig. 3.12   Four levels of an action in cooperative dialogue (joint activities)

Signal level: B listens to what A is saying, but A's intentions are not conveyed to B correctly. Speech recognition errors or incorrectly detected speech segments (voice activity detection errors) can occur, resulting in erroneous interpretations. **Intention level**: A's intention is not correctly conveyed to B, even though the linguistic content in the utterance is correctly conveyed. This corresponds to cases in which language understanding is incorrect, even with correct speech recognition results. **Conversation level**: B cannot accept the content of A's utterance. In this scenario, A's intention is correctly conveyed but deemed to be unacceptable by B. This situation corresponds to out-of-topic utterances that are not within the system's functionality. Utterances that do not conform to social conventions in dialogue are also included.

As discussed above, when a speech signal is used as an input to the system, breakdowns beyond those at the intention and conversational levels may occur, and a method to detect such breakdowns accurately is required. Even with the improved accuracy of component technologies such as speech recognition and language understanding, breakdowns are unavoidable given the noise in real-world environments, and the user's lack of fluency and utterance errors. Therefore, CAs operating with a wide variety of users in real environments must be able to detect breakdowns and improve their performance.

Another necessary step is the development of multimodal dialogue using various other sensors such as cameras. In the discussion thus far, we have considered speech signals through the microphone and its language content as input. In contrast, to detect situations in which dialogue leads to a breakdown, information from various sensors can be helpful, including information on facial expressions, posture, and physiological signals. Several studies have attempted to detect user sentiments using such multimodal information (Katada et al. 2023) and the results can be applied to detect various situations, including dialogue breakdown.

3.6   Infrastructure of CG-CA

This section provides an overview of a CG-based CA (CG-CA), covering the design and implementation of a semi-autonomous CA.

3.6.1   CG-Based Cybernetic Avatar

The CG-CA is a type of CA implementation. It is presented as a digital character on a display device and animated during conversations with people. The display device can be a flat-panel display, stereoscopic display, or virtual reality (VR) device. Compared to real-world robots and androids, CG-based avatars have the advantages of low running costs, high portability, and operability (Fox and Gambino 2021). They offer a high degree of expressive freedom, allowing for a range of representations from photorealistic to cartoon-like characters. Autonomous spoken dialogue systems with embodied CG agents (Cassell 2000) have been studied for various tasks. However, CG avatars inevitably lack a sense of perceived reality. Digital characters cannot share physical spaces with users and lack the ability to interact directly, leading to a diminished sense of presence and making it unnatural to perceive them as realistic conversation partners (Powers et al. 2007).

For the practical usage of CG-CAs, their graphical design and behavioral implementation should be considered to enhance the perceived reality of fluent conversations. Specifically, it is essential to pursue an agent design and implementation that fits both human-manipulated and autonomous modes for a semi-autonomous system.

3.6.2   Design Factors of CG Agents

The design framework of a CG agent as an intelligent interface has been discussed for various targets, including healthcare applications (ter Stal et al. 2020) and trust building (Rheu et al. 2021). The design of CG agents for conversational systems is commonly discussed from two perspectives. First, the functions a CG agent should implement, and its appearance should implicitly inform user expectations. Second, to realize CG-CAs as realistic conversation partners, we must focus on design factors that provide the perceived sense of presence and reality required for natural conversation with users. The essential factors are summarized below.

Visual Quality: Because a CG agent is displayed on a screen, the visual quality directly affects user experiences. Low-resolution, low-density, or low-refresh-rate displays spoil the sense of presence and reality and prevent users from perceiving their counterparts as conversational partners.

Human-Likeliness: An embodied appearance that faithfully mimics humans is important for users to perceive a CG agent as a human. Recent CG technologies allow for the rendering of highly human-like expressions in real-time. For animated or cartoon-like characters, users will often reference characters seen on TV or in movies.

Action Capability: During conversations, gestures, and behaviors provide users with a great deal of reality in communication. Non-verbal communication such as eye contact, backchannel responses, gestures, facial expressions, and emotional expressions directly lead to communication realism. Therefore, an effective avatar must perform these functions.

Stereotypes and Gender Bias: As the realism of expression increases, so does the impact of gender biases and stereotypes. There have been many studies actively utilizing gender based on notions such as "male users prefer young female agents," but inappropriate design often becomes a significant issue, overshadowing a system's quality or the usability of the interface. A humanoid design should be carefully tailored to the scope of the application and target users.

Adaptation Gap: CG agents should be designed such that they do not deviate from their actual functions or intellectual capabilities. Humans naturally try to adapt to interlocutors during conversations. Therefore, if there is a large discrepancy (adaptation gap) between the functional model inferred by a user based on the appearance or stereotypical information of the CG agent and the actual functions implemented in the agent, it will become difficult to continue an interaction smoothly (Komatsu et al. 2012).

Transparency: In the context of dialogue systems, particularly those involving AI, the concept of "transparency" has attracted significant attention (Lyons 2013). This is especially relevant in systems that interact with humans in various domains such as customer service, healthcare, education, and personal assistance. When a dialogue system is transparent regarding its decision-making processes or its current state (e.g., being in the process of listening, thinking, or encountering errors), it can enhance user trust. Users are more likely to trust and engage with a system if they understand how to respond or take action. Transparency should preferably be provided non-verbally so that it does not impede the flow of dialogue.

3.6.3 Design Factors of CG Avatars

When users converse with an avatar, the question arises as to whether they feel like they are actually speaking to the avatar itself or simply to an operator who is talking through the avatar. This is a matter of design and purpose, requiring appropriate considerations based on the intended use of a system. In the case of a digital twin modeled after an operator, users naturally perceive the avatar and operator to be identical. Similarly, a unique custom-made avatar is also recognized by a user as a one-of-a-kind entity, leading to the avatar and operator being viewed as the same. In contrast, when using a generic avatar with a subdued personality, the avatar can easily adapt to various roles and contexts; however, it is not suitable for scenarios in which a user needs to identify an operator. We conducted research on the acceptability of CG avatars and identified the following four categories of avatars.

Character Type: This type of avatar focuses solely on the personality of a CG avatar. The primary personality resides in the character and the operator should act accordingly. The operator must act without deviating from the character's defined personality. This type of avatar is suitable for fixed-scenario dialogue tasks and can be challenging to operate in tasks with a high degree of freedom in speech. The use of well-known characters typically falls into this category.

Costume Type: The focus is on the character's personality; however, the personality of the operator is also visible. Using an avatar whose appearance is not overly distinctive reduces the potential for an adaptation gap caused by the personality of the operator. The operator is allowed to deviate while acting as the character. Users perceive an entity that combines the personality of the character with the operator's awareness. CG avatars that are developed starting with character designs typically fall into this category.

Disguise Type: The operator's personality is the main focus and the character's appearance is designed to match the operator's personality. The character's personality is subordinate to the operator and is a form of operator expression. Users perceive an operator acting as a character. The operator behaves almost freely. Digital twins or self-portrait models in which people can easily identify the operator are categorized under this type.

Generic Type: This type does not express the personality of a character. It employs minimal designs, including mannequins or the default avatar designs seen in games, without imparting a personal character. Although generic avatars are versatile and commonly usable by anyone, individuality is recognized only through the voice and behavior of the operator.

3.6.4 Development of a CG-CA System for Semi-autonomous CAs

According to the factors addressed in the preceding subsections, we developed four reference CG-CAs for this project. A widely used avatar should have no strong bias in its appearance, be ambiguous in terms of gender and age, and avoid excessive abstraction to maintain a sense of reality through high-quality 3D modeling. The photorealistic CG-CA "Rubica" (Fig. 3.13) is a generic avatar with minimal personality, but fully detailed expressions, which operates in Unreal Engine. The cartoon-like CG-CA "Gene" (Fig. 3.14) is a generic avatar that aims to provide a truly gender-free and age-free model so that any person can use this avatar. Another cartoon-based CG-CA "Uka" (Fig. 3.15) was also developed as a costume-type avatar with extra body parts representing internal mental states during conversation. The final CG-CA "Nirva" (Fig. 3.16) is a character-type avatar with illustration-based

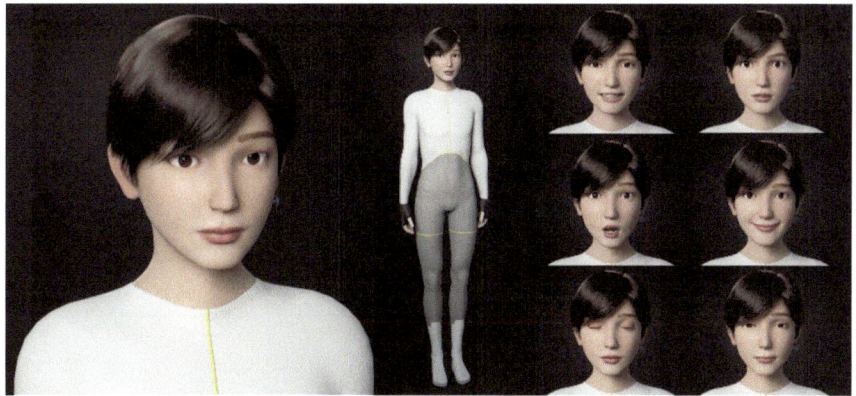

Fig. 3.13 Photorealistic CG-CA (generic) "Rubica"

Fig. 3.14 Cartoon-style CG-CA (generic) "Gene"

high-design characteristics. This avatar was mainly designed for use in exhibitions or on stages with an extra texture animation function to express internal states through surface animation. "Gene" (Lee 2023a) and "Uka" (Lee 2023b) are publicly available on GitHub with a CC-BY 4.0 license.

3.6.5 Architecture of the CG-CA System

The seamless integration of an autonomous dialogue system with an avatar communication system is required to develop a semi-autonomous system. The system should receive not only voice signals but also multimodal signals such as nodding, backchannels, and gestures from both human operators and autonomous systems and respond

Fig. 3.15 Cartoon-style CG-CA (costume) "Uka"

Fig. 3.16 Cartoon-style CG-CA (character) "Nirva"

using a CG-CA. Various types of data transmission are available. A voice signal can be provided either as a continuous audio stream from a live source or in the form of audio files generated by a text-to-speech system. Multimodal actions can be given either as a control sequence stream (e.g., live captured facial expressions) or as a simple text message that invokes predefined actions, which can be commanded by the operator or issued as a result of the action selection scheme of the autonomous system. Specifically, to realize seamless switching between the avatar and autonomous modes in a semi-autonomous system, both modes should be controlled through a common interface.

We designed a common interface foundation that allows for diverse control options. The system is based on an open-source CG agent-based spoken dialogue system toolkit called "MMDAgent-EX" (MMDAgentEX). We added additional

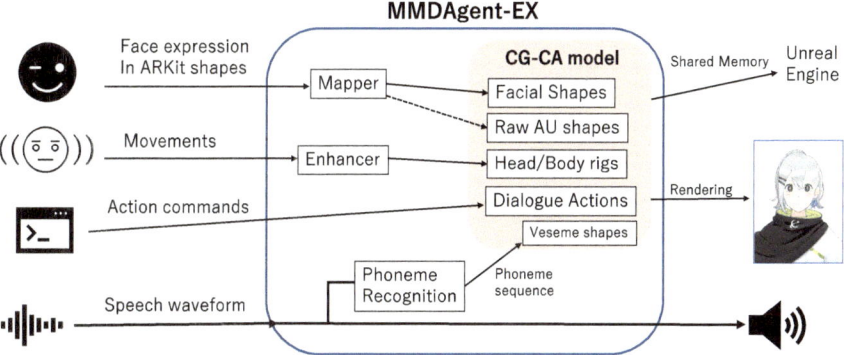

Fig. 3.17 System architecture of MMDAgent-EX for a CG-CA system

features for avatar communication through the remote control of face expressions and movements, as well as a seamless switching mechanism for a semi-autonomous CA system. The system built upon MMDAgent (Lee et al. 2013) improves the performance and capabilities of CG avatars and provides an external application interface for integration with other programming languages for a wide range of interaction designs and external controls.

The system architecture is illustrated in Fig. 3.17. In the avatar mode, the operator's speech audio stream is sent from a remote peer. Concurrently, the facial expressions and head movements of the operator can be streamed in the Apple ARKit facial expression format. Additionally, action commands can be issued by the operator to instruct the CG-CA to perform certain actions. In autonomous mode, speech data are typically given in a single waveform file generated by an autonomous dialogue system. The system can also issue action commands corresponding to utterances. Therefore, the system can handle avatar-based manipulation and system-based response generation through the same interface.

3.7 System Implementations

We have implemented the semi-automated CAs by integrating the components described in the previous sections into several task settings. One is parallel attentive listening (Kawahara et al. 2021), where multiple attentive listening systems with CAs are monitored by a human operator who intervenes in sessions with problems (e.g., few utterances made by users, or few substantial responses made by the system). The system detects problematic sessions and prompts the operator to intervene. The outlook is shown in Fig. 3.18. The interface for the operator shows the dialogue history of each session. It is confirmed that one operator can manage three sessions, and that the semi-autonomous framework improves user evaluations in terms of understanding and empathy (Inoue et al. 2021).

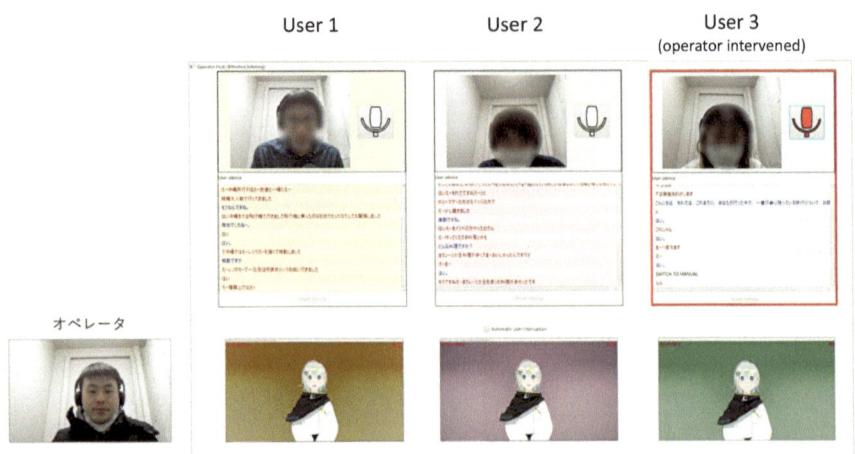

Fig. 3.18 Outlook of parallel dialogue sessions conducted by semi-autonomous CAs

We have also implemented a parallel interview system (Kawai et al. 2022) and a parallel lab guide system (Muraki et al. 2023). A similar framework was implemented for a guide in an aquarium and a trial was conducted for over one month, as described in Sect. 3.5.2.1.

References

Arons B (1992) A review of the cocktail party effect. J Am Voice I/O Soc 12

Baevski A, Zhou H, Mohamed A, Auli M (2020) wav2vec 2.0: a framework for self-supervised learning of speech representations. In: Advances in neural information processing systems

Bartneck C, Belpaeme T, Eyssel F, Kanda T, Keijsers M, Šabanović S (2020) Human-robot interaction. Cambridge University Press, Cambridge, pp 209–245. https://doi.org/10.1017/9781108676649

Brown TB, Mann B, Ryder N, Subbiah M, Kaplan J, Dhariwal P, Neelakantan A, Shyam P, Sastry G, Askell A, Agarwal S, Herbert-Voss A, Krueger G, Henighan T, Child R, Ramesh A, Ziegler DM, Wu J, Winter C, Hesse C, Chen M, Sigler E, Litwin M, Gray S, Chess B, Clark J, Berner C, McCandlish S, Radford A, Sutskever I, Amodei D (2020) Language models are few-shot learners—special version. arXiv.org. https://doi.org/10.48550/arXiv.2005.14165

Cassell J (2000) Embodied conversational interface agents. Commun ACM 43:70–78. https://doi.org/10.1145/332051.332075

Clark HH (1996) Using language. Cambridge University Press, Cambridge

Comon P (1994) Independent component analysis, a new concept? Sig Proc 36:287–314. https://doi.org/10.1016/0165-1684(94)90029-9

Duong NQK, Vincent E, Gribonval R (2010) Under-determined reverberant audio source separation using a full-rank spatial covariance model. IEEE Trans Audio Speech Lang Process 18:1830–1840. https://doi.org/10.1109/TASL.2010.2050716

Fox J, Gambino A (2021) Relationship development with humanoid social robots: applying interpersonal theories to human-robot interaction. Cyberpsychol Behav Soc Netw 24:294–299. https://doi.org/10.1089/cyber.2020.0181

Fujihara Y, Takahashi Y, Miyabe S, Saruwatari H, Shikano K, Tanaka A (2008) Performance improvement of higher-order ICA using learning period detection based on closed-form second-order ICA and kurtosis. In: IWAENC2008: the 11th international workshop on acoustic echo and noise control, Seattle, Washington

Glas DF, Kanda T, Ishiguro H, Hagita N (2008) Simultaneous teleoperation of multiple social robots. In: Proceedings of the 3rd ACM/IEEE international conference on Human robot interaction. ACM, New York, pp 311–318

Glas DF, Kanda T, Ishiguro H, Hagita N (2012) Teleoperation of multiple social robots. IEEE Trans Syst Man Cybern Part A Syst Hum 42:530–544. https://doi.org/10.1109/TSMCA.2011.2164243

Gorin AL, Riccardi G, Wright JH (1997) How may I help you? Speech Commun 23. https://doi.org/10.1016/S0167-6393(97)00040-X

Grais EM, Sen MU, Erdogan H (2014) Deep neural networks for single channel source separation. In: 2014 IEEE international conference on acoustics, speech and signal processing (ICASSP). IEEE, pp 3734–3738

Hasumi T, Nakamura T, Takamune N, Saruwatari H, Kitamura D, Takahashi Y, Kondo K (2023) PoP-IDLMA: product-of-prior independent deeply learned matrix analysis for multichannel music source separation. IEEE/ACM Trans Audio Speech Lang Process 31:2680–2694. https://doi.org/10.1109/TASLP.2023.3293044

Higashinaka R, Funakoshi K, Kobayashi Y, Inaba M (2016) The dialogue breakdown detection challenge: task description, datasets, and evaluation metrics. In: Proceedings of the 10th international conference on language resources and evaluation, LREC 2016, pp 3146–3150

Higashinaka R, D'Haro LF, Abu Shawar B, Banchs RE, Funakoshi K, Inaba M, Tsunomori Y, Takahashi T, Sedoc J (2021) Overview of the dialogue breakdown detection challenge 4. In: Lecture notes in electrical engineering, pp 403–417

Hiroe A (2006) Solution of permutation problem in frequency domain ICA, using multivariate probability density functions. In: Lecture notes in computer science (including subseries Lecture notes in artificial intelligence and Lecture notes in bioinformatics), pp 601–608

Hsu W-N, Bolte B, Tsai Y-HH, Lakhotia K, Salakhutdinov R, Mohamed A (2021) HuBERT: self-supervised speech representation learning by masked prediction of hidden units. IEEE/ACM Trans Audio Speech Lang Process 29:3451–3460. https://doi.org/10.1109/TASLP.2021.3122291

Iizuka S, Mochizuki S, Ohashi A, Yamashita S, Guo A, Higashinaka R (2023) Clarifying the dialogue-level performance of GPT-3.5 and GPT-4 in task-oriented and non-task-oriented dialogue systems. In: The AI-HRI symposium at AAAI fall symposium series

Inoue K, Lala D, Takanashi K, Kawahara T (2018) Engagement recognition by a latent character model based on multimodal listener behaviors in spoken dialogue. APSIPA Trans Signal Inf Process 7:1–16. https://doi.org/10.1017/ATSIP.2018.11

Inoue K, Lala D, Yamamoto K, Nakamura S, Takanashi K, Kawahara T (2020) An attentive listening system with android ERICA: comparison of autonomous and WOZ interactions. In: Proceedings of the 21th annual meeting of the special interest group on discourse and dialogue. Association for Computational Linguistics, Stroudsburg, PA, USA, pp 118–127

Inoue K, Sakamoto H, Yamamoto K, Lala D, Kawahara T (2021) A multi-party attentive listening robot which stimulates involvement from side participants. In: SIGDIAL 2021—22nd annual meeting of the special interest group on discourse and dialogue, proceedings of the conference, pp 261–264

Inoue K, Lala D, Kawahara T (2022) Can a robot laugh with you? Shared laughter generation for empathetic spoken dialogue. Front Robot AI 9. https://doi.org/10.3389/frobt.2022.933261

Ito N, Nakatani T (2019) FastMNMF: joint diagonalization based accelerated algorithms for multichannel nonnegative matrix factorization. In: ICASSP 2019—2019 IEEE international conference on acoustics, speech and signal processing (ICASSP). IEEE, pp 371–375

Kanda T, Shiomi M, Miyashita Z, Ishiguro H, Hagita N (2010) A communication robot in a shopping mall. IEEE Trans Rob 26:897–913. https://doi.org/10.1109/TRO.2010.2062550

Katada S, Okada S, Komatani K (2023) Effects of physiological signals in different types of multi-modal sentiment estimation. IEEE Trans Affect Comput 14:2443–2457. https://doi.org/10.1109/TAFFC.2022.3155604

Kawahara T (2019) Spoken dialogue system for a human-like conversational robot ERICA. In: Lecture notes in electrical engineering, pp 65–75

Kawahara T, Yamaguchi T, Inoue K, Takanashi K, Ward N (2016) Prediction and generation of backchannel form for attentive listening systems. In: Interspeech 2016. ISCA, pp 2890–2894

Kawahara T, Muramatsu N, Yamamoto K, Lala D, Inoue K (2021) Semi-Autonomous avatar enabling unconstrained parallel conversations—seamless hybrid of WOZ and autonomous dialogue systems. Adv Robot 35:657–663. https://doi.org/10.1080/01691864.2021.1928549

Kawai H, Muraki Y, Yamamoto K, Lala D, Inoue K, Kawahara T (2022) Simultaneous job interview system using multiple semi-autonomous agents. In: Proceedings of the 23rd annual meeting of the special interest group on discourse and dialogue. Association for Computational Linguistics, Stroudsburg, PA, USA, pp 107–110

Kim T, Attias HT, Lee S-Y, Lee T-W (2007) Blind source separation exploiting higher-order frequency dependencies. IEEE Trans Audio Speech Lang Process 15:70–79. https://doi.org/10.1109/TASL.2006.872618

Kitamura D, Ono N, Sawada H, Kameoka H, Saruwatari H (2016) Determined blind source separation unifying independent vector analysis and nonnegative matrix factorization. IEEE/ACM Trans Audio Speech Lang Process 24:1626–1641. https://doi.org/10.1109/TASLP.2016.2577880

Komatani K, Takeda R, Nakashima K, Nakano M (2022) Design guidelines for developing systems for dialogue system competitions. In: Lecture notes in electrical engineering, pp 161–177

Komatsu T, Kurosawa R, Yamada S (2012) How does the difference between users' expectations and perceptions about a robotic agent affect their behavior? Int J Soc Robot 4:109–116. https://doi.org/10.1007/s12369-011-0122-y

Kondo Y, Kubo Y, Takamune N, Kitamura D, Saruwatari H (2022) Deficient-basis-complementary rank-constrained spatial covariance matrix estimation based on multivariate generalized Gaussian distribution for blind speech extraction. EURASIP J Adv Signal Process 2022:88. https://doi.org/10.1186/s13634-022-00905-z

Kubo Y, Takamune N, Kitamura D, Saruwatari H (2020) Blind speech extraction based on rank-constrained spatial covariance matrix estimation with multivariate generalized Gaussian distribution. IEEE/ACM Trans Audio Speech Lang Process 28:1948–1963. https://doi.org/10.1109/TASLP.2020.3003165

Lala D, Milhorat P, Inoue K, Ishida M, Takanashi K, Kawahara T (2017) Attentive listening system with backchanneling, response generation and flexible turn-taking. In: Proceedings of the 18th annual SIGdial meeting on discourse and dialogue. Association for Computational Linguistics, Stroudsburg, PA, USA, pp 127–136

Lala D, Inoue K, Kawahara T (2018) Evaluation of real-time deep learning turn-taking models for multiple dialogue scenarios. In: Proceedings of the 20th ACM international conference on multimodal interaction. ACM, New York, pp 78–86

Lala D, Inoue K, Kawahara T (2019a) Smooth Turn-taking by a robot using an online continuous model to generate turn-taking cues. In: 2019 International conference on multimodal interaction. ACM, New York, pp 226–234

Lala D, Nakamura S, Kawahara T (2019b) Analysis of effect and timing of fillers in natural turn-taking. In: Interspeech 2019. ISCA, pp 4175–4179

Lee A (2023a) CG cybernetic avatar "Gene". https://github.com/mmdagent-ex/gene. Accessed 20 Dec 2023

Lee A (2023b) CG cybernetic avatar "Uka". https://github.com/mmdagent-ex/uka. Accessed 20 Dec 2023

Lee DD, Seung HS (1999) Learning the parts of objects by non-negative matrix factorization. Nature 401:788–791. https://doi.org/10.1038/44565

Lee A, Oura K, Tokuda K (2013) MMDAgent—a fully open-source toolkit for voice interaction systems. In: ICASSP2013, pp 8382–8385

Li J (2022) Recent advances in end-to-end automatic speech recognition. APSIPA Trans Signal Inf Process 11. https://doi.org/10.1561/116.00000050

Lim D, Jung S, Kim E (2022) JETS: jointly training FastSpeech2 and HiFi-GAN for end to end text to speech. In: Interspeech 2022. ISCA, pp 21–25

López Gambino S, Zarrieß S, Schlangen D (2017) Beyond on-hold messages: conversational time-buying in task-oriented dialogue. In: Proceedings of the 18th annual SIGdial meeting on discourse and dialogue. Association for Computational Linguistics, Stroudsburg, PA, USA, pp 241–246

López Gambino S, Zarrieß S, Schlangen D (2019) Testing strategies for bridging time-to-content in spoken dialogue systems. In: Lecture notes in electrical engineering, pp 103–109

Lowe R, Noseworthy M, Serban IV, Angelard-Gontier N, Bengio Y, Pineau J (2017) Towards an automatic turing test: learning to evaluate dialogue responses. In: Proceedings of the 55th annual meeting of the Association for Computational Linguistics. Long papers, vol 1. Association for Computational Linguistics, Stroudsburg, PA, USA, pp 1116–1126

Lyons JB (2013) Being transparent about transparency: a model for human-robot interaction. In: AAAI spring symposium—technical report, pp 48–53

Makishima N, Mogami S, Takamune N, Kitamura D, Sumino H, Takamichi S, Saruwatari H, Ono N (2019) Independent deeply learned matrix analysis for determined audio source separation. IEEE/ACM Trans Audio Speech Lang Process 27:1601–1615. https://doi.org/10.1109/TASLP.2019.2925450

Misawa S, Takamune N, Nakamura T, Kitamura D, Saruwatari H, Une M, Makino S (2021) Speech enhancement by noise self-supervised rank-constrained spatial covariance matrix estimation via independent deeply learned matrix analysis. In: 2021 Asia-Pacific signal and information processing association annual summit and conference, APSIPA ASC 2021—proceedings. IEEE

Mitsui Y, Takamune N, Kitamura D, Saruwatari H, Takahashi Y, Kondo K (2018) Vectorwise coordinate descent algorithm for spatially regularized independent low-rank matrix analysis. In: 2018 IEEE international conference on acoustics, speech and signal processing (ICASSP). IEEE, pp 746–750

MMDAgentEX MMDAgent-EX. https://mmdagent-ex.dev/. Accessed 20 Dec 2023

Mochizuki S, Yamashita S, Kawasaki K, Yuasa R, Kubota T, Ogawa K, Baba J, Higashinaka R (2023) Investigating the intervention in parallel conversations. In: International conference on human-agent interaction. ACM, New York, pp 30–38

Mogami S, Takamune N, Kitamura D, Saruwatari H, Takahashi Y, Kondo K, Ono N (2020) Independent low-rank matrix analysis based on time-variant sub-Gaussian source model for determined blind source separation. IEEE/ACM Trans Audio Speech Lang Process 28:503–518. https://doi.org/10.1109/TASLP.2019.2959257

Mori Y, Saruwatari H, Takatani T, Ukai S, Shikano K, Hiekata T, Ikeda Y, Hashimoto H, Morita T (2006) Blind separation of acoustic signals combining SIMO-model-based independent component analysis and binary masking. EURASIP J Adv Signal Process 2006:034970. https://doi.org/10.1155/ASP/2006/34970

Mukai R, Sawada H, Arakt S, Makino S (2004) Blind source separation for moving speech signals using blockwise ICA and residual crosstalk subtraction. IEICE Trans Fundam Electron Commun Comput Sci E87-A:1941–1948

Muraki Y, Kawai H, Yamamoto K, Inoue K, Lala D, Kawahara T (2023) Semi-autonomous guide agents with simultaneous handling of multiple users.

Nakamura T, Kozuka S, Saruwatari H (2021) Time-domain audio source separation with neural networks based on multiresolution analysis. IEEE/ACM Trans Audio Speech Lang Process 29:1687–1701. https://doi.org/10.1109/TASLP.2021.3072496

Nakano M, Komatani K (2023) DialBB: a dialogue system development framework as an information technology educational material. In: The 37th annual conference of the Japanese Society for Artificial Intelligence, Kumamoto, pp 1–4 (in Japanese)

Nishida K, Takamune N, Ikeshita R, Kitamura D, Saruwatari H, Nakatani T (2023) NoisyILRMA: diffuse-noise-aware independent low-rank matrix analysis for fast blind source extraction. In: European signal processing conference 2023 (EUSIPCO 2023)

Nugraha AA, Liutkus A, Vincent E (2016) Multichannel audio source separation with deep neural networks. IEEE/ACM Trans Audio Speech Lang Process 24:1652–1664. https://doi.org/10.1109/TASLP.2016.2580946

Oertel C, Castellano G, Chetouani M, Nasir J, Obaid M, Pelachaud C, Peters C (2020) Engagement in human-agent interaction: an overview. Front Robot AI 7. https://doi.org/10.3389/frobt.2020.00092

Ozerov A, Fevotte C (2010) Multichannel nonnegative matrix factorization in convolutive mixtures for audio source separation. IEEE Trans Audio Speech Lang Process 18:550–563. https://doi.org/10.1109/TASL.2009.2031510

Paek T, Horvitz E (2000) Conversation as action under uncertainty. In: The sixteenth conference on uncertainty in artificial intelligence (UAI'00). Morgan Kaufmann Publishers Inc., San Francisco, pp 455–464

Powers A, Kiesler S, Fussell S, Torrey C (2007) Comparing a computer agent with a humanoid robot. In: Proceedings of the ACM/IEEE international conference on human-robot interaction. ACM, New York, pp 145–152

Qian K, Zhang Y, Chang S, Yang X, Florencio D, Hasegawa-Johnson M (2018) Deep learning based speech beamforming. In: 2018 IEEE international conference on acoustics, speech and signal processing (ICASSP). IEEE, pp 5389–5393

Rheu M, Shin JY, Peng W, Huh-Yoo J (2021) Systematic review: trust-building factors and implications for conversational agent design. Int J Hum Comput Interact 37:81–96. https://doi.org/10.1080/10447318.2020.1807710

Saon G, Kurata G, Sercu T, Audhkhasi K, Thomas S, Dimitriadis D, Cui X, Ramabhadran B, Picheny M, Lim L-L, Roomi B, Hall P (2017) English conversational telephone speech recognition by humans and machines. In: Interspeech 2017. ISCA, pp 132–136

Sawada H, Kameoka H, Araki S, Ueda N (2013) Multichannel extensions of non-negative matrix factorization with complex-valued data. IEEE Trans Audio Speech Lang Process 21:971–982. https://doi.org/10.1109/TASL.2013.2239990

Sawada H, Ono N, Kameoka H, Kitamura D, Saruwatari H (2019) A review of blind source separation methods: two converging routes to ILRMA originating from ICA and NMF. APSIPA Trans Signal Inf Process 8. https://doi.org/10.1017/ATSIP.2019.5

Sekiguchi K, Nugraha AA, Bando Y, Yoshii K (2019) Fast multichannel source separation based on jointly diagonalizable spatial covariance matrices. In: 2019 27th European signal processing conference (EUSIPCO). IEEE, pp 1–5

Shen J, Pang R, Weiss RJ, Schuster M, Jaitly N, Yang Z, Chen Z, Zhang Y, Wang Y, Skerrv-Ryan R, Saurous RA, Agiomvrgiannakis Y, Wu Y (2018) Natural TTS synthesis by conditioning WaveNet on MEL spectrogram predictions. In: 2018 IEEE international conference on acoustics, speech and signal processing (ICASSP). IEEE, pp 4779–4783

Shimada K, Bando Y, Mimura M, Itoyama K, Yoshii K, Kawahara T (2018) Unsupervised beamforming based on multichannel nonnegative matrix factorization for noisy speech recognition. In: 2018 IEEE international conference on acoustics, speech and signal processing (ICASSP). IEEE, pp 5734–5738

Shiomi M, Sakamoto D, Kanda T, Ishi CT, Ishiguro H, Hagita N (2008) A semi-autonomous communication robot. In: Proceedings of the 3rd ACM/IEEE international conference on human robot interaction. ACM, New York, pp 303–310

Stolcke A, Droppo J (2017) Comparing human and machine errors in conversational speech transcription. In: Interspeech 2017. ISCA, pp 137–141

Takahashi Y, Takatani T, Osako K, Saruwatari H, Shikano K (2009) Blind spatial subtraction array for speech enhancement in noisy environment. IEEE Trans Audio Speech Lang Process 17:650–664. https://doi.org/10.1109/TASL.2008.2011517

ter Stal S, Kramer LL, Tabak M, op den Akker H, Hermens H (2020) Design features of embodied conversational agents in eHealth: a literature review. Int J Hum Comput Stud 138:102409. https://doi.org/10.1016/j.ijhcs.2020.102409

Tu Y-H, Du J, Sun L, Lee C-H (2017) LSTM-based iterative mask estimation and post-processing for multi-channel speech enhancement. In: 2017 Asia-Pacific signal and information processing association annual summit and conference (APSIPA ASC). IEEE, pp 488–491

Vaswani A, Shazeer N, Parmar N, Uszkoreit J, Jones L, Gomez AN, Kaiser L, Polosukhin I (2017) Attention is all you need. arXiv.org 1–5. https://doi.org/10.48550/arXiv.1706.03762

Walker MA, Langkilde-Geary I, Hastie HW, Wright J, Gorin A (2002) Automatically training a problematic dialogue predictor for a spoken dialogue system. J Artif Intell Res 16. https://doi.org/10.1613/jair.971

Yamashita S, Higashinaka R (2022) Data collection for empirically determining the necessary information for smooth handover in dialogue. In: 2022 language resources and evaluation conference, LREC 2022, pp 4060–4068

Yamashita S, Higashinaka R (2023) Clarifying characteristics of dialogue summary in dialogue format. In: The 13th international workshop on spoken dialogue systems technology, Los Angeles

Yamashita S, Mochizuki S, Kawasaki K, Kubota T, Ogawa K, Baba J, Higashinaka R (2023) Investigating the effects of dialogue summarization on intervention in human-system collaborative dialogue. In: International conference on human-agent interaction. ACM, New York, pp 316–324

Chapter 4
Human-Level Knowledge and Concept Acquisition

Tatsuya Harada, Lin Gu, Yusuke Mukuta, Jun Suzuki, and Yusuke Kurose

Abstract To increase productivity, it is expected that a single user is able to operate multiple cybernetic avatars (CAs). However, the limited attention span of the user makes it difficult to send direct instructions to all CAs. Therefore, this chapter describes the essential technologies for CAs that solve these problems and behave autonomously according to the user's intentions. First, the realization of spatio-temporal recognition capabilities that enable CAs to move autonomously in an environments that change from moment to moment is described. Following that, methods to implement continuous learning and memory mechanisms to facilitate acquired information reuse in the future are described. In general, the observed data are time series, and future predictions are important to provide appropriate support to users. The time series analysis method is then explained, which is the most important technology. Advanced natural language processing technology is necessary to capture intentions through dialogue with the user and to process large amounts of textual data as prior knowledge and common sense. Examples of the application of these fundamental technologies in the medical field are also presented.

T. Harada (✉) · Y. Mukuta · Y. Kurose
The University of Tokyo, Bunkyo, Tokyo, Japan
e-mail: harada@mi.t.u-tokyo.ac.jp

Y. Mukuta
e-mail: mukuta@mi.t.u-tokyo.ac.jp

Y. Kurose
e-mail: kurose@mi.t.u-tokyo.ac.jp

T. Harada · L. Gu · Y. Mukuta
RIKEN, Wako, Saitama, Japan
e-mail: lin.gu@riken.jp

J. Suzuki
Tohoku University, Sendai, Miyagi, Japan
e-mail: jun.suzuki@tohoku.ac.jp

© The Author(s) 2025   107
H. Ishiguro et al. (eds.), *Cybernetic Avatar*,
https://doi.org/10.1007/978-981-97-3752-9_4

4.1 Introduction

One possible use of cybernetic avatars (CAs) is for a single user to deploy multiple CAs to perform a large amount of work and increase productivity. However, owing to the limited human attention span, a user does not pay attention to the other CAs while operating a CA, and it is not easy to give appropriate instructions to the CAs as a whole. In addition, a remote CA may not be able to receive instructions from users because of the temporary degradation of the communication environment. In other words, for a single user to operate multiple CAs, the CAs must understand the user's intentions and operate continuously, even when the user cannot provide instructions to all the CAs.

The functions that a CA needs to behave autonomously in accordance with the user's intentions are covered in this chapter. Advanced observation skills are essential for CAs to cope with the ever-changing real-world environments and to behave appropriately. In addition, the CA must be able to interact with the user via natural language, act according to the user's instructions, and understand the user's intentions. Therefore, the information obtained from the user and the real-world environment consists of images, speech, natural language, etc., and the CA must process this multimodal information appropriately. The multimodal information obtained from the natural environment is usually time series. Based on this multimodal time series information, the CA makes predictions and assists the user correctly. Because the information received from a real environment contains noise and disturbances, and is only partially available, and it is difficult to fully understand the situation based on the information gathered on the spot. Therefore, with the help of advanced natural language processing, which compares information with a large amount of knowledge and common sense, CAs can gain a proper understanding of the situation. To reuse new experiences, a CA must have a memory mechanism that continuously learns and integrates new experiences with conventional knowledge and common sense. It is expected that by continuing such processes over a long period of time, CAs will acquire concepts comparable to those of humans, and behave appropriately and autonomously in response to ambiguous instructions from the user.

Based on the above functions of CA, the following fundamental technologies are described in this chapter: (1) spatio-temporal understanding of the environment, (2) time series analysis, (3) continuous learning for real-world applications, and (4) language learning, generation, and interpretation. In addition, examples of how these fundamental technologies have been in the medical field are presented.

4.2 Spatio-temporal Understanding of the Environment

4.2.1 Background

Advanced observation skills are essential for CAs to cope with the ever-changing real-world environment and to behave appropriately. Spatio-temporal reconstruction of the surrounding environment is crucial for CAs with a physical body.

By reconstructing the 3D structure of the surrounding environment, it is possible to create a map of the environment and understand its position in space, enabling the CA to move autonomously in the environment. In addition, if the CA can understand temporal changes in the 3D structure, it can recognize human actions and proactively assist humans by predicting their actions.

Spatio-temporal reconstruction is beneficial not only for understanding the geometric structure of the environment and objects, but also for understanding the semantic level of the target object from the visual information. Image data would have several million dimensions, and collecting training images at a scale suitable for this number of dimensions is extremely difficult. However, in the real world there are three-dimensional structures and four-dimensional structures when time is included. Successfully using these low-dimensional structures as prior knowledge can significantly reduce the cost of annotations. For example, a learning a model was considered for recognizing dog breeds such as Akita, Shiba, and Chihuahua from visual information. Dogs are non-rigid objects that change their shape over time. To learn dog breeds with simple supervised learning, a large amount of annotated data must be prepared, in which images of dogs in different poses are paired with dog breeds. If 4D reconstruction of an object from videos is possible, we can generate a large amount of supervised data by generating images from different poses and viewpoints from the reconstructed 4D model, by simply annotating a few video frames with the dog's breed.

The spatio-temporal reconstruction of the environment and the target objects is therefore essential for the autonomous movement of the CA in natural environments and for efficient recognition of the environment at the semantics-level. However, spatio-temporal reconstruction from visual information is still a challenging problem in computer vision. This section presents a method that significantly improves the understanding of 4D geometries based on visual information.

4.2.2 Point Cloud Matching for the Reconstruction of Non-rigid Objects

In this study, reconstructing the shape of an environment or object by acquiring 3D point cloud information from a range sensor is considered. Because an object or a range sensor may move, the information is not necessarily identical even if the same object is observed.

An object or environment is reconstructed by overlaying the point cloud information of partial observations between multiple frames, similar to a patchwork. To stitch the partial point clouds together like a patchwork, a bonding area must be found, i.e., a region where the partial point clouds overlap. Finding the overlapping region in a point cloud is equivalent to determining which points in the observation represent the same part of the object in 3D space. The problem of finding the corresponding points in different frames is called partial point cloud matching. Partial point cloud matching is a central problem in many 3D computer vision applications, including simultaneous localization and mapping (SLAM). Depending on the characteristics of the object, they can be classified into rigid objects, whose shape does not change with time, and non-rigid objects, whose shape changes with time. This study aims to develop a point cloud matching method that is robust to rigid and deformed scenes.

Point cloud matching methods generally consist of two phases: point cloud feature extraction and nearest neighbor search in the feature space. Recent point cloud matching methods often use 3D shape features extracted from 3D convolutional networks (Choy et al. 2019; Thomas et al. 2019). These 3D feature extractors have translation-invariant properties. They are also invariant to rotational transformations to a certain extent, given the max pooling layer commonly used in neural networks and the random rotation-based data augmentation during training.

Although transformation invariance is suitable for the local shape feature representation, it can lead to ambiguity in the partial point cloud matching in situations where similar shape patterns occur frequently. For example, in a scene where the same type of tableware is placed at different positions on a table, local features alone cannot resolve the ambiguity of the matching because multiple similar features are present. In addition, the left and right hands and feet have identical shapes, except that they are symmetrical, which can lead to ambiguities in the shape features.

Humans can discover the correspondence between observed objects and scenes by considering the appearance of objects and the relative positions of local characteristics. This ambiguity can be resolved by leveraging knowledge of the 3D position of point clouds to enhance shape features. To this end Lepard is proposed, which is a novel partial point cloud matching method that takes advantage of 3D position knowledge (Li and Harada 2022a).

The basic structure of Lepard consists of a KPFCN (Thomas et al. 2019), a fully convolution-based feature extractor, a Transformer (Vaswani et al. 2017) that uses self- and cross-attention, along with differentiable matching (Sarlin et al. 2020; Sun et al. 2021). In addition, three methods were introduced to enhance shape features with 3D location information: (1) a framework that completely separates the point cloud representation into feature and position spaces, (2) a position encoding method that explicitly expresses relative 3D distance information by dot products of vectors, and (3) a repositioning module that adjusts the relative positions between point clouds, which is useful for cross-attention and differentiable matching.

Figure 4.1 shows an overview of the proposed Lepard, where $\mathbf{S}$ represents the source point cloud, and $\mathbf{T}$ represents the target point cloud, and the corresponding points are explored between $\mathbf{S}$ and $\mathbf{T}$. Given a set of input points $\mathbf{S}$ and $\mathbf{T}$, the

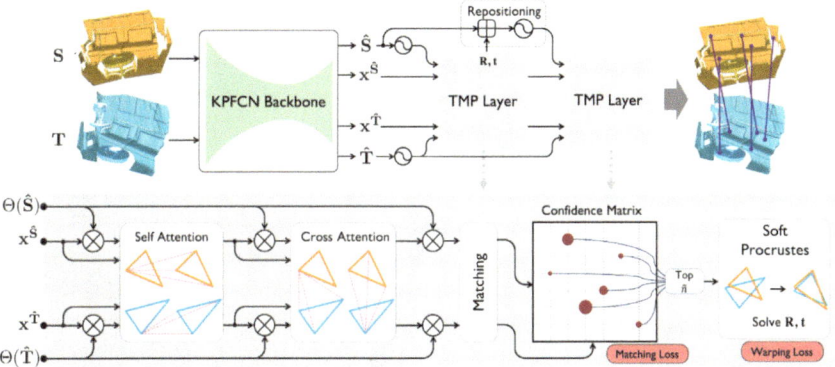

Fig. 4.1 Overview of the proposed partial point cloud matching method: Lepard (Li and Harada 2022a)

KPFCN backbone, a 3D feature extractor, extracts shape features $\mathbf{x}^{\hat{\mathbf{S}}}$ and $\mathbf{x}^{\hat{\mathbf{T}}}$, respectively. Simultaneously, the KPFCN backbone outputs point clouds $\hat{\mathbf{S}}$ and $\hat{\mathbf{T}}$ by grid subsampling of the input point clouds $\mathbf{S}$ and $\mathbf{T}$, respectively. The 3D relative position encoding function $\Theta(\cdot)$ encodes the position information as $\Theta(\hat{\mathbf{S}})$ and $\Theta(\hat{\mathbf{T}})$ from the subsampled point cloud. The encoded position information $(\Theta(\hat{\mathbf{S}}), \Theta(\hat{\mathbf{T}}))$ and shape features $(\mathbf{x}^{\hat{\mathbf{S}}}, \mathbf{x}^{\hat{\mathbf{T}}})$ are then processed in a first transform-matching-procrustes (TMP) layer. This TMP layer consists of a Transformer block with self- and cross-attention, a differentiable matching layer, and a soft procrustes layer for estimating the rotations $\mathbf{R}$ and translations $\mathbf{t}$, which are transformations of the rigid-body fitting. Based on the rotation $\mathbf{R}$ and translation $\mathbf{t}$ estimated by the first TMP layer, the repositioning layer transforms the source point cloud and realigns the position code $\Theta(\hat{\mathbf{S}})$. The second TMP layer uses the updated position code and the transformed features based on the first TMP layer as inputs and outputs the final source and target point cloud matching results.

In addition, a partial point cloud matching benchmark is proposed known as 4DMatch and its low-overlap version, 4DLoMatch. Figure 4.2 shows an example of a 4DMatch/4DLoMatch. Blue represents the source point cloud, and yellow represents the target point cloud. The upper row shows the situation in which the source and the target were superimposed before the deformation. The bottom row shows the ground truth, where the source was deformed to fit the target. The bottom number represents the overlap ratio between source and target. The lower the overlap ratio, the more difficult it is to find the corresponding points between the source and target.

The effectiveness of Lepard was verified by applying it to both rigid-body and deformable point cloud matching problems. For rigid bodies, Lepard in combination with RANSAC and ICP showed a very high registration recall of 93.9% and 71.3% for 3DMatch/3DLoMatch, respectively (Zeng et al. 2017; Huang et al. 2021). For the newly proposed 4Dmatch and 4DLoMatch benchmarks, Lepard achieved a + 27.1% and + 34.8% higher non-rigid matching recall than in their previous work

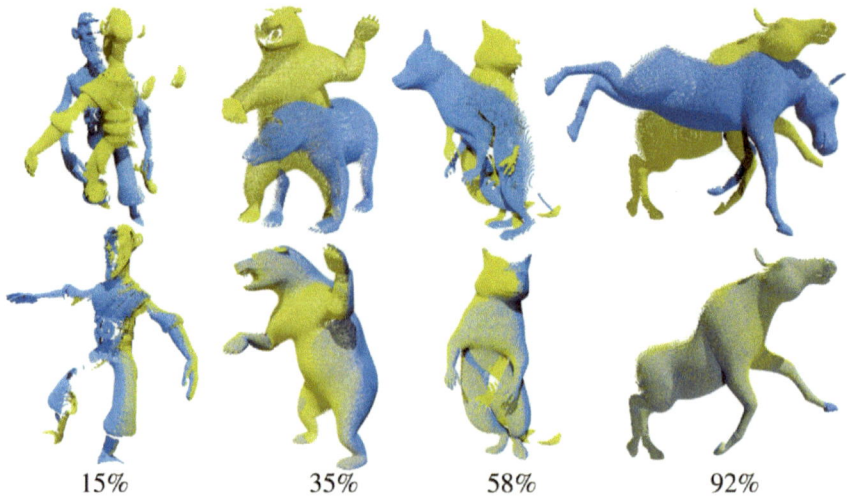

Fig. 4.2 Examples in 4DMatch/4DLoMatch (Li and Harada 2022a)

(Huang et al. 2021). Figure 4.3 shows the qualitative results of point cloud matching on the 4D match benchmark. The "flying dragon" used in this example has bilaterally symmetric shapes. The left side shows the results of the previous method, Predator (Huang et al. 2021), and the right shows the results of the proposed Lepard method. Yellow and blue represent the source and target point clouds, respectively, and the green and red lines connecting the source and target indicate the inliers and outliers, respectively. The number of inliers dominates in the proposed method, whereas the prior method contains many outliers. Visualizing the features with T-SNE also showed that the previous method learned similar features for the two wings of a dragon. In contrast, our method can distinguish between the left and right wings.

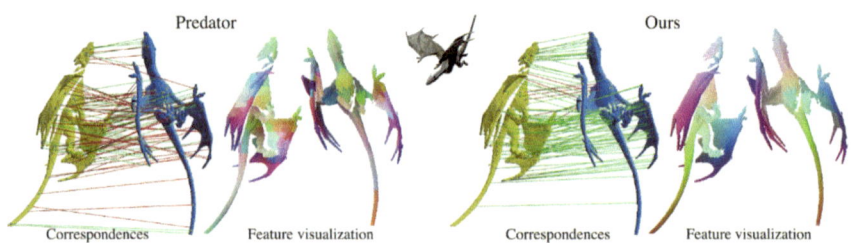

Fig. 4.3 Qualitative results of point cloud matching in the 4Dmatch benchmark (Li and Harada 2022a)

4.2.3   Point Cloud Registration for Reconstruction of Non-rigid Scenes

The previous subsection discussed the point cloud matching problem, whereas this subsection discusses the registration problem of non-rigid point clouds. Non-rigid registration is about finding a transformation that maps one point cloud to another. Once a transformation that aligns the point clouds is identified, it is possible to reconstruct a wide-area object or scene from multiple sets of point clouds, which has many applications, including not only the environment recognition function of the CA, but also autonomous driving, AR/VR, and medical imaging, and others.

Non-rigid registration is a difficult task for three main reasons. The first reason is that range sensor measurements contain noise, outliers, and occlusions. Occlusions often lead to disconnection of the point cloud geometry. Secondly, the overlap between the point clouds may be smaller owing to the deformation of the object or scene and changes in the viewpoint of the range sensor. The third problem is that unlike rigid registration, which only requires the determination of rotation and translation parameters, non-rigid registration requires the estimation of the motion of all points. The third problem is the most complex in non-rigid registration.

Coordinate multilayer perception (MLP) is a popular model for deformation in non-rigid registration using neural networks (Li et al. 2021a; Park et al. 2021). The coordinate MLP takes the 3D coordinates of a point as input and outputs the deformation of that point. However, these are black-box models designed to represent signals on a single scale. They, usually require large networks to fit complex motions, and optimization is usually time consuming.

The focus in this subsection was on the motion decomposition to reduce the complexity of the third problem. Natural non-rigid motions usually form a hierarchical structure, with the upper hierarchy representing the global motion and the lower hierarchy representing the local deformations. For example, walking can be approximated by three levels: (1) global position and posture changes, (2) local joint movements of the arms and legs, and (3) fine fabric deformations due to external forces. Each level represents a different granularity of deformation and has a top-down dependency from global to local deformation.

Based on this insight, a hierarchical motion representation was proposed known as the neural deformation pyramid (NDP) for non-rigid registration (Li and Harada 2022b). Figure 4.4 shows an overview of the NDP: the NDP has a pyramidal structure. It decomposes the motion field by using a sequence of small MLPs to achieve a more interpretable and controllable motion representation and faster optimization. $\mathbf{x}_i^k$ are the 3D coordinates on the pyramid level k. Each pyramid level has a coordinate MLP that takes a sinusoidally encoded 3D point as input and outputs the motion increment $\boldsymbol{\xi}_i^k$ from the previous level and the confidence α_i^k of the motion estimate. The k-th layer transforms the 3D coordinates $\mathbf{x}_i^{k-1}$ using the motion and confidence levels estimated at level k.

$$\mathbf{x}_i^k \leftarrow \mathbf{x}_i^{k-1} + \alpha_i^k \cdot \mathcal{W}\left(\mathbf{x}_i^{k-1}, \boldsymbol{\xi}_i^k\right),$$

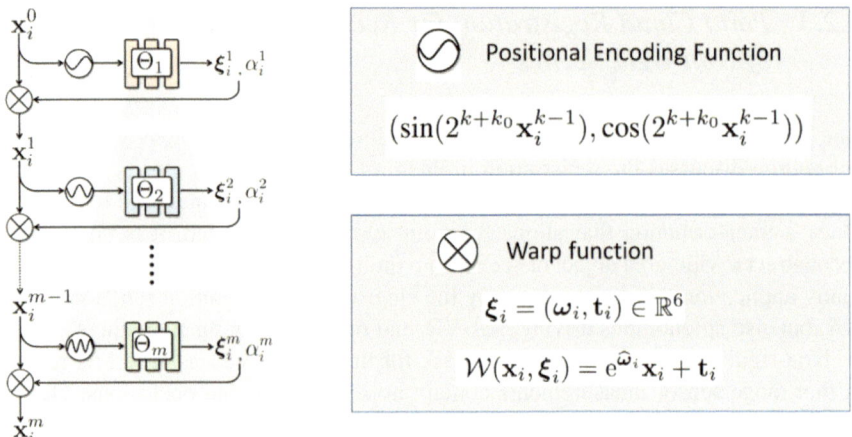

Fig. 4.4 Overview of the neural deformation pyramid (Li and Harada 2022b)

where $\mathcal{W}$ is the warp function.

It was found that the frequency of the sinusoidal function controls the ability of the MLP to represent non-rigidity. That is, lower frequencies produce a smoother signal suitable for conforming to a relatively rigid motion, whereas higher frequencies produce more fluctuations and can represent a highly non-rigid motion. The sinusoidal function starts with lower frequencies in the upper pyramid levels and gradually increases the lower the pyramid level becomes. This allows a multi-level decomposition from rigid to non-rigid motion and leads to faster solutions for training the model than the existing MLP-based approaches. Figure 4.5 shows the result of each hierarchical non-rigid registration pyramid level. The pink color represents the target point cloud and the other colors represent multiple source point clouds. Moving from the upper to the lower levels, the NDP corresponds to rough to fine deformations.

Figure 4.6 shows the quantitative results of non-rigid registration. The source point cloud is shown in pink and the target point cloud in green. The Synorim-pw (Huang et al. 2023) was used for comparison. Because the NDP of the proposed method is a continuous function, all input points can be warped directly. Therefore, it is confirmed that NDP performs better than Synorim, especially on non-overlapping regions, and

Fig. 4.5 Hierarchical non-rigid registration results (Li and Harada 2022b)

that NDP performs better on point cloud registration. NDP also achieved state-of-the-art partial-to-partial non-rigid registration results on the challenging 4DMatch/4DLoMatch (Li and Harada 2022a) benchmarks in unsupervised and supervised settings.

Figure 4.7 shows an example of shape transfer as an application of our non-rigid registration method. The input shapes A, B, C, and D are "Alien Soldier," "Ortiz," "Doozy," and "Jackie" from Mixamo (https://www.mixamo.com/); E and F are "Racoon" and "Bear" from DeformingThings4D (Li et al. 2021b). The arrows indicate the direction of transfer. The dense formulation allows the NDP to reflect changes in other regions of the body at different scales. This allows the transformed shape to nearly match the target shape, while preserving the geometric details of the original shape.

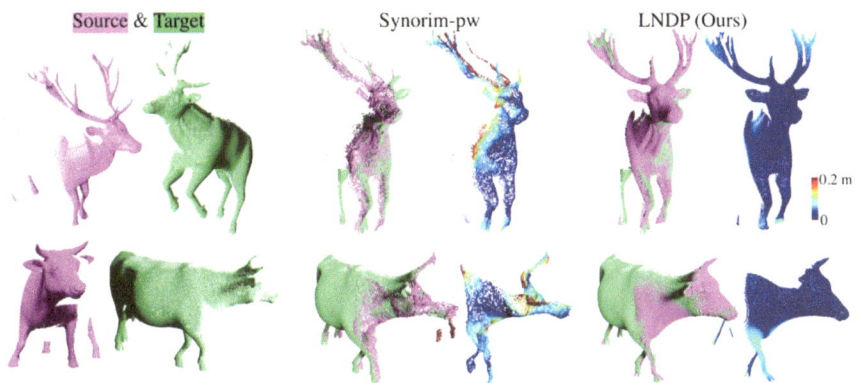

Fig. 4.6 Quantitative results for non-rigid registration (Li and Harada 2022b)

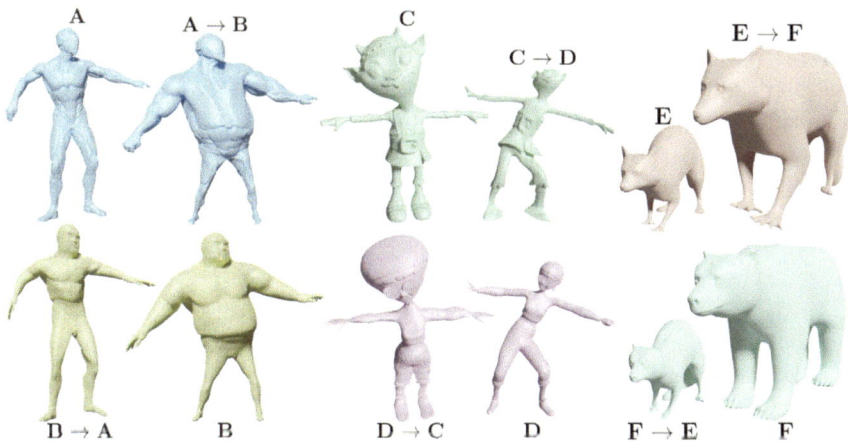

Fig. 4.7 Examples of shape transfer with NDP (Li and Harada 2022b)

4.2.4   Summary

In this section, the methods for spatio-temporal reconstruction from partial point cloud information are explained, which are also essential for autonomous movement of CAs and semantic-level recognition in natural environments. In particular, Lepard, a point cloud matching method was presented that can be used for non-rigid objects, and NDP, a point cloud registration method. Lepard demonstrated the effectiveness of rigid and deformable point cloud matching by leveraging positional knowledge. For NDP, that hierarchical motion decomposition was shown to be is effective for point cloud registration. Future studies will include the implementation of these methods in CAs and the validation of their effectiveness.

4.3   Continuous Learning and Memory Mechanism

4.3.1   Artificial Representation for Continuous Learning

The environment is constantly changing, and the relationships between the CA, operators, and users are also changing. Therefore, the knowledge and concepts within the environment also change. Continuous learning is essential for the constant acquisition of knowledge and concepts. Continuous learning provides a human-like memory mechanism for the CA. When a human-like memory mechanism is realized, CAs become highly compatible with both the operators they serve and the interactors involved in the conversation. This section discusses different artificial intelligence frameworks that enable CAs to continuously learn from and adapt to their environment.

This section consists of two parts: Sect. 4.3.2 Artificial constant representation of the visual signal and Sect. 4.3.3 Encoding and decoding knowledge from language.

When the human brain stores the input information received from the sensory organs as short- or long-term memory, the encoded concepts are perceptions, and information such as shape, size, and brightness is constant. Therefore, when a specific object is perceived, the visual signal sent to the CA can change from moment to moment. However, despite these fluctuations in stimulus effects, the CA should perceive the signal with almost identical characteristics. This section therefore proposes three artificial solutions for encoding visual information in a constant perceptual form.

As the environment changes, knowledge related to real-world data is also constantly changing. In the course of evolution, this constant change has been naturally encoded in the human natural language. By encoding and decoding natural language, CAs can cope with the emergence of new ambiguous concepts or words through continuous learning from communication, media, and other data sources. In addition, doctors rely on previous records rather than just the patient's current condition when making a diagnosis. Using the techniques described in this section, CAs in the medical-related field can facilitate the treatment of patients based on knowledge of their medical history.

4.3.2 Artificial Constant Representation for Visual Signals

Manifold Learning for Dark Images

Dark environments pose a major challenge for CAs. The computational photography community has proposed several human vision-oriented algorithms restoring normally illuminated images (Lv et al. 2021). Unfortunately, the restored image is not necessarily beneficial for the high-level visual understanding tasks of the CA. Because enhancement/restoration approaches are optimized for human visual perception, they may produce artifacts that are misleading for subsequent tasks. For example, if an elderly person falls at night, it is important for the CA to recognize the identity and contact relatives or friends before providing appropriate assistance according to the elderly person's medical history.

Nevertheless, the prevailing approaches encounter two main problems: target inconsistency and data inconsistency. Data inconsistency leads to complications when it is assumed that the training data should reflect the conditions used in the evaluation. For example, the object detection models of CAs are often trained on well-lit and clear images. When faced with challenging lighting conditions, these models resort to fine-tuning with augmented dark images, but are unable to examine the intrinsic structure under illumination variations. Echoing Tolstoy's adage that "Happy families are all alike; every unhappy family is unhappy in its own way," the existing datasets (Deng et al. 2009; Everingham et al. 2010; Lin et al. 2014), despite their comprehensiveness, have difficulty in comprehensively capturing the diverse distribution of real-world conditions encountered during training.

In this context, our objective was to bridge the above two gaps in a unified framework. As shown in Fig. 4.8, a normally illuminated image can be subjected to a parametric transformation (t_{deg}) to generate its degraded low-illumination counterpart. Building on this transformation, a novel multitask autoencoding transformation (MAET) is proposed to extract transformation-equivariant convolutional features for object detection in low-light images. Training of MAET involves two tasks: (1) learning the intrinsic representation by decoding the low-illumination-degrading transformation based on unlabeled data and (2) decoding object positions and categories based on labeled data. As illustrated in Fig. 4.8, MAET was trained to encode pairs of normally lit and low-light images with a Siamese encoder E and decode their degradation parameters, such as noise level, gamma correction, and white balance gains, with the decoder D_{deg}. This approach allows our model to capture the intrinsic visual structure corresponding to the illumination variance. In contrast to previous approaches (Lore et al. 2017; Lv et al. 2021), which used an oversimplified synthesis, our degradation model was developed taking into account the physical noise model of the sensors and image signal processing (ISP). Subsequently, the object detection task was performed by decoding the bounding box coordinates and classes using the decoder D_{obj} based on the representation encoded by E.

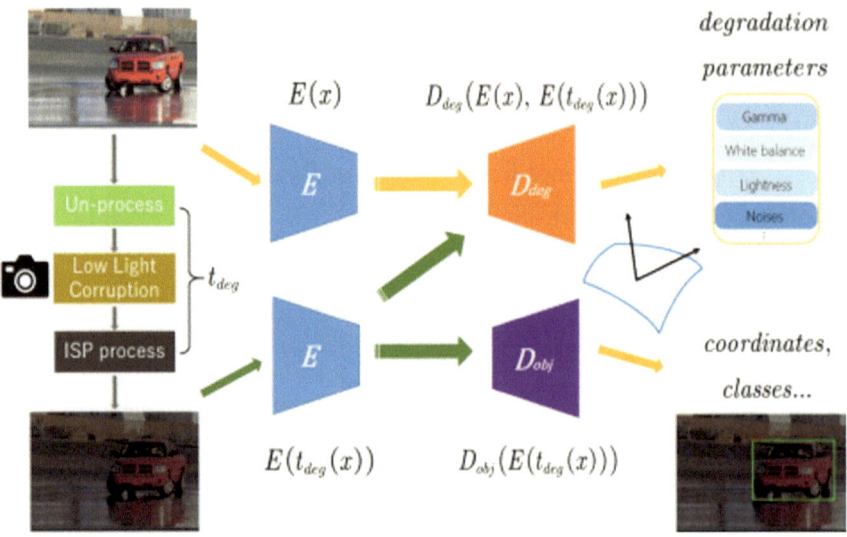

Fig. 4.8 The illustration of the system (Cui et al. 2021)

Whereas MAET effectively regulates network training by predicting low-light-degrading parameters, the joint training of object detection and transformation decoding becomes overly intertwined through a shared backbone network. Although this improves the detection of dark objects by MAET regularity, there is a potential risk of overfitting the object-level representation to the self-supervisory imaging signals. To solve this problem, disentangling the tasks of object detection and transformation decoding by introducing an orthogonal tangent regularity is proposed. This regularization method assumes that the multivariate outputs of two tasks form a parametric manifold. The disentanglement of the multitask outputs along this manifold can be formulated geometrically by maximizing the orthogonality between the tangents along the outputs of the different tasks.

The proposed framework can be directly trained end-to-end using standard target detection datasets such as COCO (Lin et al. 2014) and VOC (Everingham et al. 2010) to enable the detection of low-light images. Although YOLOv3 (Redmon et al. 2016) was used for illustration purposes, it is important to note that the proposed MAET is a general framework that is easily applicable to other mainstream object detectors.

Manifold Learning for Low-Resolution Images

High-level vision tasks have achieved great success mainly because of the availability of large-scale datasets (Deng et al. 2009; Everingham et al. 2010; Lin et al. 2014). These datasets mainly consist of images captured by commercial cameras and are characterized by higher resolution and better signal-to-noise ratio (SNR). However, models trained and optimized on these high-quality images may lose performance when a CA encounters a low-resolution signal (Dai et al. 2016).

To improve vision algorithms for degraded low-resolution images, image prepro-cessing with super-resolution (SR) algorithms is a solution. Many existing enhance-ment methods, especially super-resolution (SR) algorithms (Bell-Kligler et al. 2019), operate under the assumption that the target images follow a known and fixed degra-dation model, denoted by $t(x)$ for the degraded low-resolution (LR) image and x for the original high-resolution (HR) input. However, the efficacy of these enhancement algorithms can decrease significantly if the actual degradation deviates from this assumption (Gu et al. 2019). Thus, in machine perception tasks, a higher resolution does not necessarily guarantee better performance in high-level tasks.

Instead of explicitly improving an input image with a fixed restoration module, an intrinsic equivariant representation for various resolutions and degradations is used. Drawing inspiration from "Alice in Wonderland," where Alice remarks, "I know who I was when I got up this morning, but I think I must have been changed several times since then," our approach is akin to encoding Alice's ability to adapt to diverse sizes and transformations, much like the changes she undergoes in the story, such as being small enough to squeeze through a door, or large enough to shed a pool of tears. This concept emphasizes the need for an equivariant representation to effectively capture Alice's identity in the world.

Building on the encoded representation shown in Fig. 4.9, an end-to-end frame-work is proposed for object detection in low-quality images. To capture the intri-cate patterns of the visual structures, groups of downsampling degradation transfor-mations are used under varying downsampling rates, noise levels, and degradation kernels as self-supervised signals.

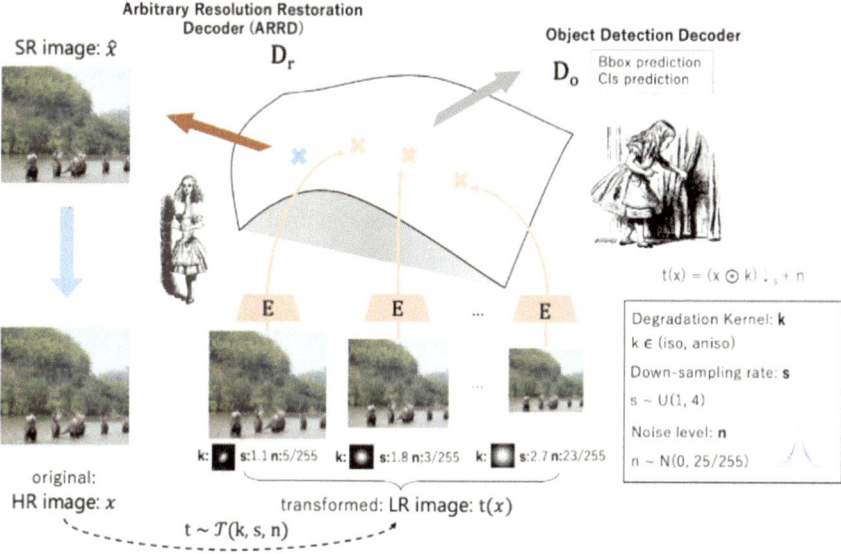

Fig. 4.9 Illustration of the proposed framework (Cui et al. 2022b)

During training, a degraded LR image $t(x)$ is generated from the original HR image x using a random degradation transformation t. As shown in Fig. 4.9, an arbitrary-resolution restoration decoder (ARRD) is introduced to train Encoder E for the degradation equivariant representation $E(t(x))$. The ARRD implicitly decodes t to reconstruct the original HR data x from the representation $E(t(x))$ of different degraded LR images $t(x)$. When a self-supervised signal is reconstructed, the representation should capture the dynamics of its changes at different resolutions and other degradations as well as possible. The nature of reconstructing HR data allows us to capitalize on advances in rapidly-growing SR research by directly using successful architectures.

For the encoded representation $E(t(x))$, an object detection decoder D_o is also introduced, which guides and supervises the encoder E in encoding the image structures relevant to the subsequent tasks. The object detection decoder D_o is responsible for the detection task, in which information about the location and class of the object is extracted. During inference, the target image undergoes direct processing by the encoder E and the object detection decoder D_o, as depicted in Fig. 4.9, which facilitates the detection process.

In contrast to methods based on preprocessing modules, our inference pipeline demonstrates higher computational efficiency. This efficiency results from our approach, which avoids the explicit reconstruction of image details and streamlines the detection process. To cover the diverse degradations and resolutions in a real-world scenario, degraded $t(x)$ was generated by randomly sampling a transformation t according to a practical downsampling degradation model (Liu and Sun 2014).

Extensive experiments show that the proposed framework achieves SOTA results for two mainstream public datasets, MS COCO (Lin et al. 2014) and KITTI (Geiger et al. 2012), under different degradation conditions (resolution, noise, and blur).

Lightness Correction

The different lighting conditions encountered in the real world pose a challenge to both the visual appearance of the operator and the automatic recognition tasks of the CA, such as cognition recognition (MacDorman and Ishiguro 2006) and stereo matching (Ishiguro et al. 1992). Images captured under inadequate illumination (Fig. 4.10, top-left) exhibit limited photon counts and unwanted in-camera noise. Conversely, outdoor scenes often face intense lighting conditions such as direct sunlight (Fig. 4.10, bottom left), resulting in image saturation owing to the limited dynamic range of the sensors and non-linearities in the camera's image pipeline. Both underexposure and overexposure can coexist, which complicates matters. Spatially varying illuminations cast by shadows may lead to contrast ratios of 1000:1 or more.

Various techniques have been proposed to improve these difficult lighting conditions, including low-light enhancement (Lv et al. 2021; Cui et al. 2021) and exposure correction (Afifi et al. 2021). Low-light enhancement methods aim to restore details while suppressing the accompanying noise. Exposure correction methods focus on adjusting underexposed or overexposed images to reconstruct clear images with short or long exposure times. As illustrated in Fig. 4.8, our goal is to develop a unified

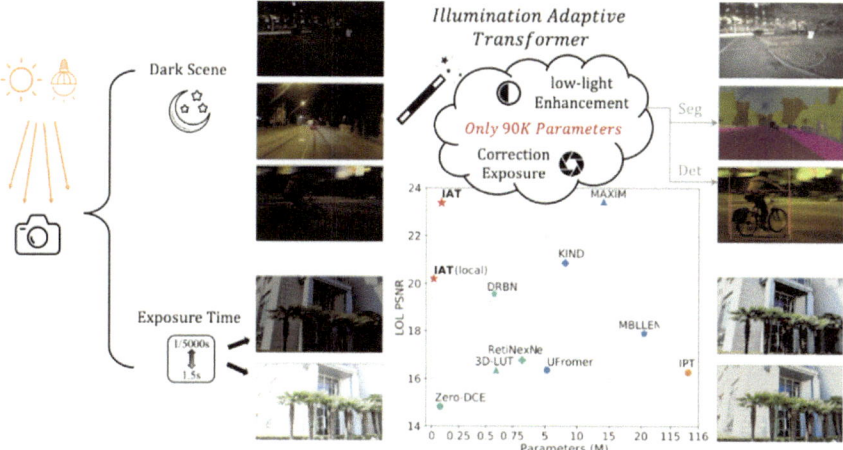

Fig. 4.10 Illustration of our method (Cui et al. 2022a)

lightweight framework that improves both visual appearance and subsequent recognition tasks under the challenging real-world illumination conditions encountered by CA systems.

To address this challenge, a novel two-branch transformer-based model is proposed that consists of a pixel-wise local branch, f, coupled with a global image processing (ISP) branch.

In the local branch, the input image is mapped to a latent feature space and replaced the attention block of the transformer with a depth-wise convolution for a lightweight design. In the global branch, the transformer's attention queries is used to control and adjust global ISP-related parameters, such as the color transform matrix and gamma value. In addition, the learned queries change dynamically under different lighting conditions, such as over-and under-exposure, simultaneously. Let us take an input sRGB image $I_i \in R\,H \times W \times 3$ under the lighting condition L_i, where $H \times W$ denotes the size dimension and 3 denotes the channel dimension ({r, g, b}). Our illumination adaptive transformer (IAT) is proposed to transfer the input RGB image I_i to a target RGB $I_t \in R\,H \times W \times 3$ under the proper uniform light L_t.

Instead of using a U-Net style structure (Ronneberger et al. 2015) that downsamples the images before upsampling, our intention is to maintain the input resolution through a local branch to preserve the informative details. Therefore, a transformer-style architecture for the local branches is proposed. Compared to the popular U-Net style structures (Ronneberger et al. 2015; Wei et al. 2021; Afifi et al. 2021), our structure process images with arbitrary-resolution images without resizing.

Extensive experiments were conducted with various real-world and synthetic datasets. The results show that the proposed IAT (Cui et al. 2022a, b) achieves state-of-the-art performance across a spectrum of low- and high-level tasks. Notably, our IAT model involves only 0.09 million parameters, which is significantly smaller than the current state-of-the-art transformer-based models (Chen et al. 2021; Tu et al.

2022). In addition, our average inference speed was 0.004 s per image, outperforming the state-of-the-art methods, which typically require approximately 1 s per image.

4.3.3 Encoding and Decoding Knowledge from Language

Knowledge in Medical Reports

Medical informatics has actively used hospital databases to feed data-hungry deep learning algorithms, focusing in particular on Chest X-ray datasets such as MIMIC (Johnson et al. 2016) and Chexpert (Irvin et al. 2019). These datasets have attracted considerable attention in the field of vision-language (VL) modality (Miraglia et al. 2023). Researchers have used natural language processing (NLP) to mine per-image common disease labels or to investigate report generation and answering specific predefined questions.

Despite significant progress, technical challenges remain due to the heterogeneity, systemic biases, and subjective nature of medical reports. Automatically mining labels from reports can be problematic because rule-based approaches do not handle uncertainty and negation well. Training an automated radiology report generation system to match the report can compensate for bias, but may struggle with radiologists' abstract logic.

Therefore, a novel medical image difference VQA system is suggested, which aligns the CA closely with practice of radiologists. MIMIC-Diff-VQA is the first comprehensive and large-scale dataset for visually answering questions about medical image differences. It includes 164,324 pairs of medical images, with 700, and 703 question–answer pairs. The questions were related to various attributes, including abnormalities, presence, location, level, type, view, and differences. This dataset provides a valuable resource to advance research in the field of medical image analysis and, in particular, to solve the difficult task of answering visual questions related to medical image differences. Radiologists often compare current and previous images to assess disease progression. Therefore, the MIMIC-Diff-VQA dataset is presented, which contains pairs of "main" (present) and "reference" (past) images, along with question and answer pairs derived from MIMIC reports. This dataset meets the needs of medical image difference tasks and supports the inherently interactive nature of radiology reports in clinical practice.

To deal with the challenging task of comparing medical image differences, an expert knowledge-aware image difference graph representation learning model is proposed. This model leverages the features of different anatomical structures and constructs a graph that encodes spatial, semantic, and implicit relationships. The resulting graph difference features were processed using LSTM networks with attention modules for answer generation.

To summarize, this subsection presents a unique dataset and an advanced system for improving the understanding and interpretation of medical images. It enables the CA to directly answer questions raised by radiologists in clinical settings.

Color Knowledge in Language

The question of whether a CA can have the same color perception mechanism as humans is a complex and intriguing problem that can be examined from the perspective of machine learning. Human color perception involves the visual reception and neural processing of light stimuli when the light spectrum interacts with the cone cells located in the eyes. In addition, the physical properties of objects, geometry, incident illumination, and other factors contribute to the overall perception of color.

Ongoing research (Berlin and Kay 1969) suggests that human language continually evolving to acquire new color names, resulting in an increasingly refined color-naming system. This evolutionary process is hypothesized to have been driven by pressures of communication efficiency and perceptual structures.

The application of these principles to the CA involves the development of a color perception system that mimics or approximates the human perceptual mechanism. This may involve incorporating mechanisms for recognizing and categorizing colors based on similar perceptual features. The challenge is to create a system that not only accurately identifies colors, but also understands the nuances of color naming, given the evolving nature of language and the diverse cultural and anthropological influences on color perception.

Similarly, Zaslavsky et al. (2022) measured communication efficiency in color naming by analyzing information complexity according to the information bottleneck (IB) principle (Zhou et al. 2023). It has been argued that the network recognition accuracy reflects the communication efficiency when the number of colors is limited. Because human color naming is influenced by both perceptual structure and communication efficiency, it is necessary to integrate these aspects into both perception and machines. Consequently, a novel end-to-end color quantization transformer, CQFormer (Su et al. 2023), is proposed in this subsection to discover artificial color-naming systems. This model considers the dual requirement of preserving the perceptual structure and optimizing the communication efficiency in the quantization process.

The CQFormer pipeline, depicted in Fig. 4.11, consists of two branches: annotation and palette. The annotation branch annotates each pixel in the input RGB image with an suitable quantized color index. The index map was painted using the corresponding color from a color palette. The palette branch localizes the color palette within the entire RGB color space by applying a novel approach that detects key points using explicit attention queries of the transformer.

During the training stage, represented by the red and black lines in Fig. 4.11a, the palette branch interacts with the input image and reference palette queries to maintain the perceptual structure by minimizing the loss of perceptual structure loss. This perception-centric design groups similar colors, and ensures that the color palette adequately represents the color-naming system defined by the world color survey (WCS), color naming stimulus grids. Finally, the quantized image is forwarded to a high-level recognition module for machine-accuracy tasks, such as classification and detection. The joint optimization of CQFormer and the subsequent high-level module enabled a balance between perception and machine considerations. This approach

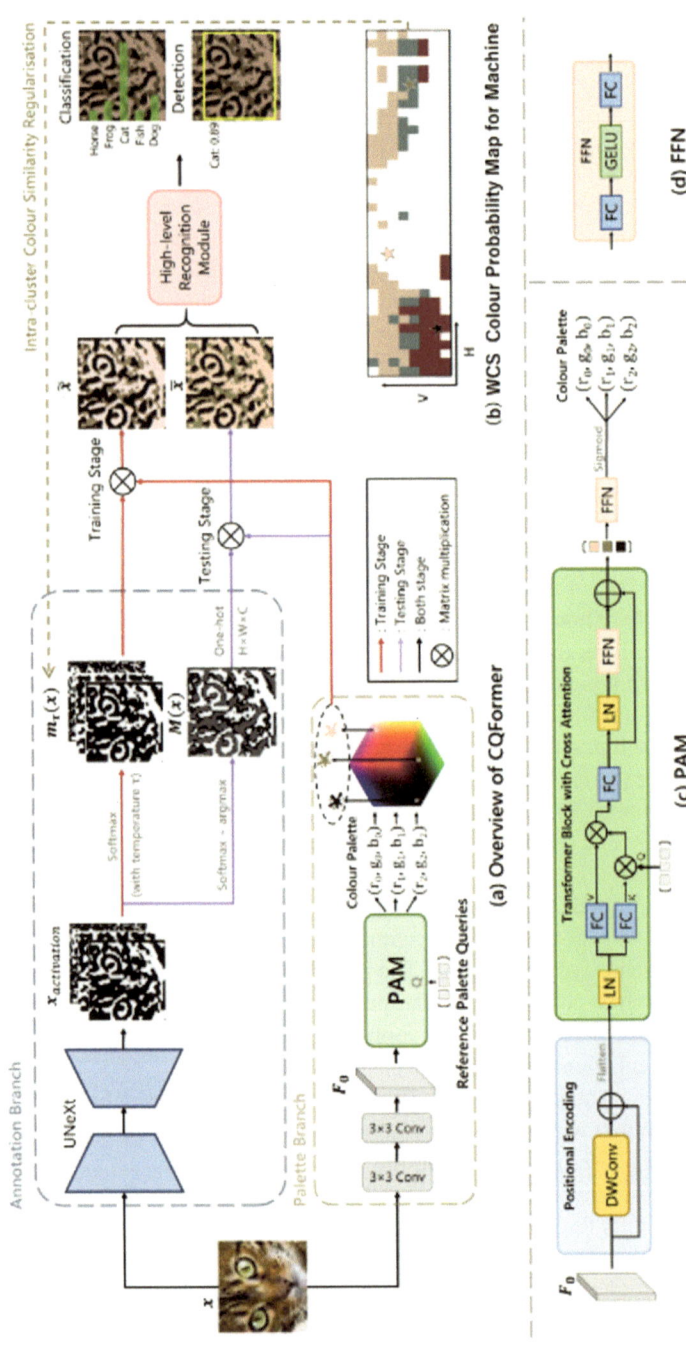

Fig. 4.11 Architecture of the system (Su et al. 2023)

ensures that the color quantization process not only aligns with human perceptual structures but also effectively meets the requirements of machine learning tasks.

In addition to automatic discovery of the color-naming system, CQFormer provides an effective solution for extreme compression of image storage while maintaining high performance in high-level recognition tasks. For example, CQFormer achieved a remarkable top-1 accuracy of 50.6% on the CIFAR100 (Krizhevsky 2009) dataset, even with a 1-bit color space (i.e., two colors). The ability to achieve high accuracy with extremely low-bit quantization shows its potential for integration in quantization network research.

4.4 Research on Time Series Analysis

When a CA cooperates with a person, it is expected to support the person's actions smoothly and without stress if the CA can understand the person's intentions in advance and act before the person does. Our focus was on the time series prediction of human behavior as a system for anticipating human intentions. If the system can predict how a person will behave based on the movements observed thus far, the CA can perform supporting actions according to the predicted actions. In this section, a machine learning-based approach to general time series prediction is presented, including time series prediction of human motion.

In machine learning-based methods, a future predictor is obtained by learning patterns from training data. In other words, some model structures have been prepared to generate predicted future time series from the observed series as inputs. Pairs of predicted future time and input series data were prepared as training data. The loss function was then designed according to the deviation between the predicted time series and the true future time series in the training data, and the optimal model parameters in the given model structure were calculated to minimize the loss function. Therefore, for an accurate prediction, it is necessary to (1) use good training data, (2) use a model structure that reflects the nature of the task, and (3) successfully design a metric between the predicted data and training data as a loss function. In Sect. 4.4.1, the design of the loss function between time series is explained. In Sect. 4.4.2, the model structure of time series estimation is described, and Sect. 4.4.3 provides some prospective avenues for future research.

4.4.1 Loss Function

To train a time series forecasting model, the distance metric between series as a loss function must be formulated so that the forecasted series is close to the true series and the model is trained to reduce loss. Therefore, the performance of the forecasting model depends on the nature of its loss function.

What properties of the loss function should be satisfied to predict human behavior? For example, if the actions of walking toward a chair are considered, stopping in front of the chair and sitting down on the chair, it is relatively unimportant how long it takes to walk until one finally reaches the chair. However, if you attempt to sit down before reaching the chair, you will fall.

Ideno et al. (2021) focused on the characteristics that the time spent on each action is variable and diverse and that the order in which the actions occur is important in predicting the time series of such actions, they extended the dynamic time warping to the distance between variable-length time series as a distance measure for the prediction model.

Dynamic time warping is a metric that measures the similarity of time series by matching between the elements of two time series. Dynamic time warping between a time series A of length M and a time series B of length N is obtained by calculating the distance between $A[m](1 \leq m \leq M)$ and $B[n](1 \leq n \leq N)$ at each point in time, and then calculating the path starting from $A[1]$ and $B[1]$ and ending with $A[M]$ and $B[N]$ by increasing m or n by 1 so that the sum of the distances of corresponding $A[m]$ and $B[n]$ is minimized.

Updating the prediction model so that the matching loss between the model's predicted time series A and the ground-truth-supervised future time series B is reduced would be desirable. Although dynamic time warping is suitable for comparing time series with fixed lengths, from this perspective it is not sensible to directly compare the values of dynamic time warping when predicted time series A with different lengths are input, because the length of the path on which the sum changes is formed when the length of the input series changes. Therefore, it is not appropriate to compare the values of dynamic time warping directly if different lengths of the predicted time series A are used as input. In addition, because the focus is on the order of actions and it is not desired that the length depends on the order of actions, dynamic time warping that sums over paths, is not suitable because the value increases as the length of the path increases.

Therefore, the loss function proposed by Ideno et al. (2021) enables loss function comparisons of time series of different lengths in two steps: Specifically, the generative model generates a set of time series end probabilities along with predicted future actions, and the outer step computes the optimal substring end position r based on the summation of the matching loss of substring $A(r)$ with ground truth B, the probability that r is the true end position of B and the loss owing to the difference between the lengths of B and r. To account for the variable-length behavior in the calculation of internal dynamic time warping, an extension is made to allow for a del operation that removes series elements when matching between series.

To evaluate the proposed method put forth by Ideno et al. (2021), variable-length time series data was created that consisted of behavioral categories and the performance of the prediction model was then evaluated. The results showed a 30% reduction in series length prediction error and a 70% reduction in dynamic time warping loss for the model using ordinary least squares error, and a similar but faster performance for the other methods using dynamic time warping.

4.4.2 Model Structure

First, the self-attention layer used in the transformer is described, which is the standard for time series modeling. Let $\{x_t\}_{t=1}^{T}$ denote the input time series, $Q_t = W_Q x_t$, $K_t = W_K x_t$, $V_t = W_V x_t$ are calculated using learnable matrices W_Q, W_K, W_V. The output time series $\{y_t\}_{t=1}^{T}$ is then calculated as $y_t = \frac{\sum_{n=1}^{T} \exp(Q_t, K_n) V_n}{\sum_{n=1}^{T} \exp(Q_t, K_n)}$. Q is the query, K is called the key, and V is the value. Each time, the V that corresponds to the K with a higher similarity to the input query Q affects the output. The most important thing is that only the values of Q, K, V are referenced at each time point, and the information at time t is not used. Therefore, in ordinary time series forecasting, a positional embedding reflecting t is used to concatenate the time information with input x.

In Umagami et al. (2023), this self-attention layer, which focuses only on element values and does not use element location information, was used for time series prediction by exploiting its symmetry for multi-variate time series.

For example, consider a situation in which multiple individuals cooperate with each other or multiple organizations conduct commercial activities and collaborate. In such cases, if the information of multiple people and companies is simply arranged and processed as a vector, the information of the time series of the movements of multiple homogeneous entities is lost. To make predictions while preserving the multi-agent nature, Umagami et al. (2023) proposed a prediction task that takes into account the symmetry of order switching among agents. In other words, if we want to predict the behavior of individuals A, B, C, and D, it is assumed that A, B, C, and D are homogeneous agents and build a model that gives the same prediction results when each individual's information is treated as an input as $[A, B, C, D]$ or as $[D, C, A, B]$.

To create a model with symmetry against such reordering, Umagami et al. (2023) focused on the property in which only the elements of the self-attention layer described earlier were used, and the position information of the elements was not used. In other words, recognition is performed by applying the self-attention layer not only in the temporal direction, but also in the inter-agent direction. In the above example, this implies that a self-attention layer is applied between the four entities A, B, C, and D. As a result the recognition is independent of the order in which A, B, C, and D are entered.

In addition to global symmetry, which allows all subjects to switch their order, this study also considered the symmetry of hierarchical reordering, which takes into account the subjects' characteristics. For example, a person may have information about the group to which they belongs, such as the organization, rank, and type of industry in the case of a company. In the example above, let us assume A_1 and A_2 belong to group A and B_1 and B_2 belonging to group B. For the input $[[A_1, A_2], [B_1, B_2]]$, it is allowed to swap the order within the same group such as $[[A_2, A_1], [B_1, B_2]]$, or between groups such as $[[B_1, B_2], [A_1, A_2]]$, but it is not allowed to swap the order across groups such as $[[A_1, B_2], [A_2, B_1]]$ because the group structure is broken. In other words, the number of allowed order transformations is reduced, and richer information can be obtained. A model has also been developed that exploits the group-aware symmetry in such reordering.

To create a symmetric prediction model for such hierarchical order transformations, the calculation was performed hierarchically by dividing it into self-attention mechanisms within and between groups. In the above example, self-attention is first calculated within $[A_1, A_2]$ and within $[B_1, B_2]$. Then, an A vector is created that summarizes the information in $[A_1, A_2]$ and a B vector that summarizes the information in $[B_1, B_2]$, and self-attention is calculated within A and B. By integrating this information, discriminative information can be obtained with various hierarchical characteristics while allowing for order transformation.

When applied to high-dimensional time series, such as time series prediction of player behavior in an NBA game or a dataset of economic indicators, the proposed model exhibits superior prediction performance compared to methods that disregard covariance to ordinal transformations, Moreover, it is confirmed that the prediction accuracy is further improved by considering group information.

Forecasting performance and forecasting speed are important for the application of time series forecasting. For example, if a system that predicts one second into the future takes two seconds to compute, the future will arrive while the computation is in progress, rendering the system useless. This section explains how time series prediction can be accelerated.

First, the computational complexity of the self-attention layer is considered. In the above self-attention layer, the similarity between all pairs of elements must be calculated to obtain the output. In other words, for each time series length T, $O(T^2)$ is required. For example, when viewing a time series with a finer resolution or use the distant past as input, T becomes larger, and the computation of $O(T^2)$ becomes more laborious. Several methods have been proposed to reduce the computational burden on the self-attention layer.

Jung et al. (2022a) introduced a grouped self-attention model that accounts for sparsity in the temporal direction, where the influence of nearby time is modeled in detail, whereas the influence of distant time is modeled roughly, thereby reducing computational complexity.

The grouped self-attention layer consists of a global and a local attention layer. The layer is calculated in this space. The global attention layer projects Q, K, and V onto a low-dimensional space and calculates the self-attention layer in that space. This enables the acquisition of global correlation information, while reducing the computational effort by reducing the dimensionality. In contrast, the local attention layer obtains detailed relationships at the local level. Specifically, the input time series was divided into a shorter time domain, and self-attention was calculated within this time domain. In this way, it is possible to obtain locally detailed information while reducing the number of computations required.

The proposed model was applied to a standard time series forecast dataset, and it was confirmed that the accuracy of the proposed method is equivalent to that of ordinary self-attention, and that the increase in computational complexity is slower than that of ordinary self-attention when the series length increases.

Kamata et al. (2022) proposed a highly efficient time series generation method for time series prediction in spiking neural networks to reduce the computational complexity. That is, if the continuous input signal at time t is x_t, the current continuous membrane potential is u_t, and the binary output signal is o_t, then the leaky integral firing model used in the spiking neural network calculates the output by $u_t = \tau_{\text{decay}} u_{t-1}(1 - o_{t-1}) + x_t, o_t = H(u_t - V_{\text{th}})$, where τ_{decay} is the decay rate, H is a Heaviside function that outputs 1 if the input is positive and 0 if the input is negative, and V_{th} is the threshold. In other words, u_t decays with τ_{decay} until it fires, but gradually increases with x_t. When u_t exceeds V_{th}, it fires and u_t resets to zero. This model mimics the motion of neurons in the human brain, and because the computation is performed using binary signals, it has the advantage of being faster and less expensive than GPU-based neural network computations that use ordinary continuous signals.

In Kamata et al. (2022), the stochastic generative model, the variational autoencoder, which forms the basis for probabilistic time series forecasting, was extended using spiking neural networks.

Variational autoencoder is a method to obtain a stochastic generative model by the maximum likelihood estimation method to find the parameter theta that maximizes the likelihood $p_\theta(x)$ of the observed data x. Usually $p_\theta(x)$ is expressed as $\int p_\theta(x|z)p(z)\mathrm{d}z$, and using another encoder model $q_\phi(z|x)$, the variational lower bound $E_{q_\phi(z|x)}\big[\log p_\theta(x|z)\big] - \text{kl}\big[q_\phi(z|x), p(z)\big]$ is used to train the generative model. Here, $p_\theta(x|z), p(z), q_\phi(z|x)$ are usually expressed as Gaussian distributions, where kl is the kl divergence, which is a continuous distribution. The actual data generation is sampled by first sampling z from $p(z)$ and then calculating $p_\theta(x|z)$.

The difficulty of applying spiking neural networks to a variational autoencoder is that the sampling distribution is expressed as a Gaussian distribution. In other words, the problem is how to perform probabilistic inferences using spiking neural networks that can only handle binary data.

In Kamata et al. (2022), a method for probabilistic sampling of the spiking neural network output was proposed, in which C layers were prepared, and a sample from the Bernoulli distribution was approximated by randomly selecting one of the outputs. For example, suppose $C = 3$ and the layer outputs are $[0, 1, 0]$. By outputting one of layer outputs at random, it can be considered a sample from a distribution, where 0 is the output with a probability of 2/3, and 1 is the output with a probability of 1/3. Therefore, by replacing $p_\theta(x|z), p(z), q_\phi(z|x)$ by the proposed sampling method, a generative model based on a spiking neural networks can be constructed. In particular, $p(z)$ itself is also trained. In addition, it is shown that the performance of the generative model is improved by measuring the maximum mean discrepancy of $(z), q_\phi(z|x)$ as a distance measure instead of the kl distance between the binomial distributions.

In Kamata et al. (2022), a prediction experiment using time series encoded MNIST image data showed a 30% reduction in prediction error and a 3% improvement in inception score compared to a conventional neural network, and the number of product calculations was reduced by approximately one order of magnitude in terms of computational complexity.

4.4.3   Future Directions

At the beginning of Sect. 4.4, the three elements required for machine learning-based time series forecasting were mentioned: training data, model structure, and loss function. Because the loss function is discussed in Sect. 4.4.1, and the model structure in Sect. 4.4.2, future directions from the perspective of the data are now discussed. First, the datasets used for human motion prediction are discussed. Datasets such as NTU RGB+D and Human3.6M, which are currently mainly used for action recognition, are based on simple actions such as "walking" and "weaving hand" taken by subjects, and the joint positions in each frame are captured by the motion capture system. Future prediction is performed by predicting the time series of joint positions in several frames and seconds in the future based on the time series of joint positions in the past few seconds. This method, which uses only a simple time series of joint positions has the advantage of allowing a wide range of system applications in the sense that predictions can be made when only the joint positions are known; however, it also has some problems.

For example, there is the problem that semantic information cannot be captured. For example, even when the joint motion is the same, the long-term future behavior and supportive actions to be taken by the CA may vary greatly depending on whether the right hand reaches forward to pick up an object or to open a door. However, it is difficult to obtain this information from the joint position time series alone. Furthermore, the same action can be performed multiple times. In other words, when considering the operation of picking up an object, it is important that the hand is in a position where it can eventually grasp the object, and which of the many possible paths it should take on the way there is relatively unimportant. If only the joint positions are known, it is difficult to design a prediction loss that takes into account the degrees of freedom of these intermediate trajectories, because it is not known which time of the trajectory is important.

To solve this problem, it is useful to include not only the person, but also the surrounding environmental information in the input and recognition. In other words, by integrating and recognizing the environment and behavior, i.e., where the person is currently located, what is around them, what kind of relationship exists between them, what kind of behavior the person performs in that environment, and what kind of series the joint position has in that environment–semantics can be generated in the time series and future predictions can be made considering the meaning. Constructing a model that can predict the future based on multimodal inputs that go beyond a mere time series, and creating a dataset with various sensor data and teacher labels for this purpose are future research directions.

4.5 Language Learning and Generation and Its Interpretation

Natural language is the most essential and straightforward medium of human-to-human communication. In fact, throughout human history, we have always used natural language to exchange and record information, such as ideas, feelings, facts, with each other in daily life. Natural language is also considered essential in the inter-action between CAs, which are intended to act autonomously in physical or virtual environments, and their operators or users. To this end, CAs are expected to possess not only rudimentary linguistic abilities, but also linguistic abilities comparable to those of humans.

CAs primarily require two linguistic skills. The first is the ability to understand human language. The second is the ability to form coherent sentences that are appropriate for a given context. Ideally, a CA should be able to handle a wide variety of languages to adapt to the native language of the operator or user. In addition to these features, it is desirable for CAs to demonstrate their language interpretation processes to provide feedback to the operator.

This section therefore discusses the language processing component of the CA, focusing on the communication between the developer (human) and the CA. Figure 4.12 shows abstract diagrams of the practical applications of (1) language understanding, (2) language generation, and (3) human-understandable methods in real-world CA use cases in the real world.

Many potential research topics can drastically improve the usability of CAs for language-handling capabilities. However, it is impossible to present all improve-ments in a limited space. This section focuses on two specific topics: cross-lingual communication support (CSS) and the stability of the training process in large language models (LLMs). CSS is comparable to the application of NLP technolo-gies. In contrast, the stability issue in LLM training is a fundamental technology for improving basic resources for many NLP applications. The remainder of this section describes these two research topics.

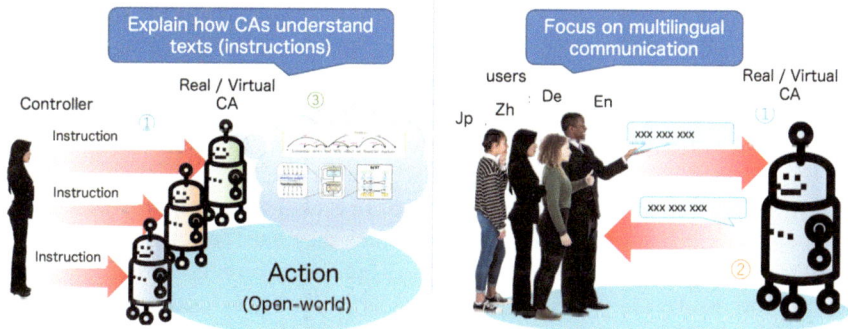

Fig. 4.12 Illustration of our target research topics, namely, language understanding and generation

4.5.1 Cross-Lingual Communication Support

In the increasingly interconnected world of globalization, there are more opportunities to engage with others across language and cultural barriers. For example, we try to connect in online virtual reality (VR) spaces or attend international gatherings, such as the World Expo. While English is often the lingua franca in these cross-cultural interactions, the uncomfortable reality is that not everyone is fluent in English and those with limited ability often face significant barriers. Under these circumstances, it could be beneficial to converse in one's native language. This is where machine translation systems have proven to be an important tool in promoting cross-lingual communication.

Machine translation, based on state-of-the-art neural networks and deep learning, has made great advances, especially in the translation of factual content such as news articles. These advances have demonstrated that computers can translate sentences from one language to another while retaining the intended message. However, the challenge becomes even greater when colloquial language is translated into everyday conversations such as those found in chats and negotiations. Unlike written documents, where the facts are clear, conversations are typically shrouded in ambiguity, and there is no one-to-one correspondence of sentences between different languages. Unique idiomatic expressions, cultural nuances, differences in social norms, and various emotional expressions typical of human communication further complicate this challenge. In real-life interactions, precise expressions can take on different meanings depending on the context, surrounding emotional atmosphere, or implicit cues. Consequently, machine translation systems are currently unable to deal with the subtleties and complexity of interlingual communication. This shows that there is an urgent need for more sophisticated solutions that can deal with the nuances of human interaction.

Investigation of Cross-Lingual Communication Support Requirements
Achieving a perfect error-free chat translation system is a challenge due to linguistic peculiarities that make it impossible to achieve perfection. Therefore, a viable alternative approach is considered. The suggested approach is to improve translation software by warning of possible mistranslations to avoid confusion (note that we categorize all translations that are unsuitable for any reason, including cultural reasons and nuances, as mistranslations). Although this approach is promising, the perception and effect of such warning messages have not yet been comprehensively revealed.

To investigate this, a survey was conducted to examine the effectiveness of warnings about potential mistranslations in chats as an alternative approach to improve the experience of cross-lingual communication (Li et al. 2023). Participants take part in a simulated cross-lingual chat scenario in which they must select the most meaningful response from three options. Each time a translation error occurred, a warning message was displayed. At the end of the chat, participants answered questions about their perception of the warning messages. A survey was conducted and the responses were collected using crowdsourcing.

The participants' decisions regarding the warning messages and divided them into three cases: (1) they entered the same scenario in both the round with warnings and the round without warnings and did not change their choices; (2) they entered the same scenario in both rounds and changed their choices, and (3) they did not change their choices because they entered other branches in advance. The first case demonstrated that the warnings did not influence the participants, whereas the second demonstrated that they were influenced. In the third case, although it is impossible to compare whether participants changed their choice in the same scenario since they changed earlier, it is still considered as an indirect influence due to the equivalence between having no warning messages and having no erroneous translations. In fact, 103 participants stated that they changed their choice because they made sure that there were no incorrect translations, which is about 75% who changed their choice, either directly or indirectly, because of the warning messages. Therefore, it can be concluded that the warning messages encouraged participants to change their behavior. We also observed that the participants expected the warning message to (1) point out the specific error in the translation, (2) indicate the correctness rate of the translation, and (3) make alternative translation suggestions. These observations show that alerts are helpful in cross-lingual chats and indicate what features participants want for the warning messages.

This survey is the first to investigate the effect of mistranslation warnings in cross-lingual chats. It provides valuable insights for the development of an assistant function that recognizes incorrect chat translations and warns users about them. The results are thus useful the development of an assistant function that recognizes and warns users of translation errors in cross-lingual conversations. Finally, it was concluded that these warning messages were helpful. In addition, the limitations of this survey were clarified. Careful measures were taken in the survey design phase to minimize potential leading effects on participants' judgment by randomizing the order and neutralizing the question style. Despite diligent efforts, it must be recognized that it is difficult to eliminate all influences on the individuals who participated in the survey. Therefore, the need for further optimization is recognized to ensure the fairness and validity of the responses. Refinement is warranted to further minimize this bias.

Translation Error Detection in Cross-Lingual Conversations

Based on the survey above, translation error detection in machine translation is practical to a certain extent when communicating in different languages. Next, a detection system was constructed and then its performance was evaluated (Li et al. 2022). It was assumed that the translation system would use a system provided by a web-based API. This assumption leads to the idea that a translation error detection system can be constructed in such a way that it does not depend on the features of a particular translation system. This is because it is very likely that when creating a support system for cross-language communication, the translation part will use a translation system provided on the web via an API. Therefore, a more realistic setting was assumed for our experiments.

A chat is defined as a two-utterance colloquial dialog between two people using different languages to detect incorrect translations. The focus was on predicting whether the second utterance (i.e., the response) was translated correctly. The preceding context, the translation of the context, the response, and the translation were fed into the error detector. The detector then predicted the translated response using other utterances as reference data. The detector then determined whether the translated response was incorrect. Consider the following example task for evaluating the Japanese translation of an English utterance: Here, the Japanese speaker's initial utterance JA1 is translated into EN1, and the English speaker's response EN2 is translated into JA2. For example, the detector assesses the utterance "Thanks (in Japanese)," which is not an accurate translation of the utterance "I agree." The detector was provided with the previous context (JA1, EN1, and EN2) as reference data to predict whether the translation was accurate and coherent. When the detector indicates the translation EN2 of the response JA2, the reference data include EN1, JA1, and JA2 in the opposite direction.

The results of our experiments show that the error detector can classify incorrect translations in chats to a certain degree. Based on the accuracy values, it can be concluded that the error detector performs better than simple baseline methods such as majority and minority classifiers. The results suggest that the proposed method can solve tasks without relying on luck. However, although the detector was able to distinguish between poor translations or coherence problems, it failed to detect non-obvious errors. The error detector did not perform well in predicting the translations generated by the high-quality neural machine translation model. One possible reason for this is that the detector was trained on a dataset in which incorrect examples were more obvious. As can be seen from the results, even with current technology it is still a challenge to achieve near-zero errors. It may be necessary to drastically improve the methods used in current systems.

4.5.2 Improving the Usability of Large Language Models (LLMs)

Research in the field of natural language processing (NLP) has focused on the development and analysis of LLMs. The reason for this is because LLMs have become essential and fundamental research resources for text data processing. LLMs are indispensable components of NLP-related systems. We also take advantage of LLMs. In this subsection, how to improve the stability of LLM training is discussed. This is an important research topic for building LLMs, as their training process is often unstable and difficult to manage.

Background: Stability of Language Learning
Language models play a central role in today's natural language processing. With the public availability of various learning tools and the verification of appropriate settings, the training of language models has become relatively stable. In addition,

the problem of the amount of training data, which used to be a major challenge, is becoming less of a concern as large amounts of training data are available online. Consequently, it is relatively easy to create language models of a specific size if sufficient computational resources are available and no time constraints need to be considered. However, the situation differs when attempting to train models of the order of GPT-3 with 175 billion parameters, Llama2 with 70 billion parameters, or when incorporating original elements in the models (Brown et al. 2020; Touvron et al. 2023). In such cases, the training process becomes more unstable and there is a high potential for loss divergence or learning failure. In such situations, it is essential to ensure the stability of the learning process.

Stability of Training Transformers
In the paper "Attention is All You Need," Vaswani et al. (2017) presented a model structure of neural networks that is suitable for language models and is called Transformers. Transformers have become the base model for language models, and almost all known language models use Transformers. However, many improvements have been made to enhance performance in various aspects. In this case, as an improvement from the perspective of stability during learning, it has been clarified from both theoretical and experimental perspectives that the stability of learning differs significantly depending on the location of layer normalization (Takase et al. 2023). More specifically, it has been shown that there are two major methods to date, depending on the location of layer normalization, and that there is a trade-off between performance and stability in each. At the same time, it has also been shown that the vanishing gradient problem, which was widely verified in early deep learning studies, is one of the reasons for the lack of stability, especially when the number of layers is significantly increased.

4.5.3 Summary

Section 4.5 discussed language understanding and generation, which can be used as fundamental functions in automated CAs. Several detailed functions are related to language understanding and generation. Two main topics are presented. The first is a cross-lingual communication support system. The other is the trade-off between performance and stability based on the position of layer normalization when training Transformers. As for the first part of the cross lingual communication support system, it is essential to use CAs both globally and internationally. The second part of the trade-off becomes more critical for deeper Transformers, as deeper Transformers lead to a more unstable training process. Thus, both research topics are crucial for improving the capabilities of CAs.

4.6 Cybernetic Avatars for Medical Applications

4.6.1 *Background*

Since the advent of deep learning, machine learning techniques have developed considerably in many fields. Medical applications are no exception. Many medical applications using machine learning algorithms have been reported to achieve accuracy comparable to that of doctors (Gulshan et al. 2016; Ehteshami Bejnordi et al. 2017). In particular, with the recent emergence of generative artificial intelligence (AI) represented by ChatGPT, much research has been conducted on the application of these technologies in medicine (Yang et al. 2022; Singhal et al. 2023; Thirunavukarasu et al. 2023).

Medicine is an important application of CAs. One reason for this is the shortage of doctors in Japan. According to OECD Health Statistics in 2020, the number of professionally active physicians per 1000 people was 2.67. This figure is far below the average for OECD countries (3.72). If CAs can help physicians solve problems, that is a big step, and it is our belief that CAs have the potential to do so.

A scenario in which the CA assists physicians with diagnosis is considered. If CAs could interview and diagnose patients instead of physicians, physicians would only need to confirm their diagnoses, significantly reducing the amount of time it takes to diagnose each patient. This also applies when doctors are far away from the patient. Currently, remote medical care is based on screen-based interviews where the patient must make a diagnosis based on limited information. However, a more reliable diagnosis can be made if the CA intervenes and collects information.

If we consider developing a framework in which CAs conduct interviews and make diagnose instead of physicians, one of the solutions is to develop a large language model that is applicable to this situation. Large-language models, as represented by ChatGPT, have the potential to interview and diagnose patients using the CA. However, the development of a large language model requires a large amount of data. The data of general purpose large-scale language models, such as ChatGPT, are mainly created from data available on the Internet.

However, it is difficult to collect large amounts of medical data because it is not available on the Internet. Therefore, most current research on large-scale medical language models does not use real medical data, but data extracted from medical textbooks or papers on PubMed (Han et al. 2023; Moor et al. 2023).

In addition, a medical interview should not only capture the information from the dialogue with the patient, but also the patient's condition at the time of the interview. Furthermore, a diagnosis requires not only text information, but also information from other modalities, such as medical images. In other words, physicians integrate information from various modalities to make diagnoses. The same is true for CAs. Above all, a diagnosis based on medical images is important.

With this in mind, our research focus was on learning algorithms based on limited data. In this section, some of the results of our research are presented.

4.6.2 Weakly Supervised Learning Method for Chest X-Ray Images

As previously mentioned, deep learning requires a large amount of data for training. In addition, the same number of annotations is required to indicate locations or labels of the diseases. However, the annotation cost for medical images is high and time consuming because annotations for medical images requires expert knowledge, and only physicians can annotate them. Weakly supervised learning methods have been studied to reduce annotation costs. In this subsection, a weakly supervised method for detecting lesions in chest X-ray images is considered. Given a system that predicts the locations of the lesions and the class label (lesion name), our weakly supervised method requires only the class labels compared to the general supervised learning method that requires a bounding box to indicate the location of the lesion and the class label of the lesion for training.

A class activation map (CAM) (Zhou et al. 2016) is one of the most popular methods for weakly supervised lesion localization to localize lesion sites using only image-level supervision. Most methods using CAM or its variations apply CAM as a postprocessing technique to generate a heatmap and lesion bounding boxes after training their models (Wang et al. 2017; Sedai et al. 2018). In contrast, probabilistic-CAM (PCAM) pooling leverages the localization capability of CAM during training (Ye et al. 2020).

When analyzing chest radiographs, it is important to consider the relationships and dependencies between diseases and lesions when radiologists make a diagnosis. For example, pulmonary consolidation refers to a region of lung tissue that is filled with fluid rather than air. This fluid is often introduced as part of pulmonary edema, where there is an accumulation of fluid in the tissues and air spaces of the lungs. Another example is that consolidation should always be present on the chest X-ray when diagnosing pneumonia. Therefore, a method is proposed that uses the relationship between lesions and their locations to improve the accuracy of weakly supervised lesion detection using chest X-ray images.

The proposed method is shown in Fig. 4.13. A feature map graph representational probabilistic class activation map (FGR-PCAM) framework is proposed, that not only leverages images and labels for weakly supervised learning, but also incorporates the dependencies between diseases and lesions. FGR-PCAM efficiently captures the dependencies between symptoms using the localization capability of PCAM. First, the localized feature maps of each thoracic disease were extracted using the PCAM framework as the backbone. The extracted feature maps representing each disease were fed to the proposed visual interaction (VI) module as nodal features of a graph. The VI module uses the propagation of a gated graph neural network (GNN) with node features as the output of the PCAM and the adjacency matrix as the label co-occurrence. Graph propagation was used to capture the mutual interactions between lesion-specific feature maps. The VI module incorporates lesion-specific feature maps instead of semantic-specific feature vectors as node features; thus, it contains position information compared to the node features of vectors. The output of the

VI module, which captures the interactions between the lesions, is fed into the fully connected layer to classify the diseases.

To demonstrate the efficiency of the proposed method, the models were tested with the CheXpert (Irvin et al. 2019) and ChestX-ray14 datasets (Wang et al. 2017). The CheXpert dataset contains 224,316 chest radiographs from 65,240 patients, with 14 observations extracted from medical reports. The validation set of CheXpert4 consists of 200 chest radiographs that were manually annotated by three board-certified radiologists. In the experiments, models were also trained with those five selected disease labels of "Cardiomegaly," "Edema," "Consolidation," "Atelectasis," and "Pleural Effusion" to compare with the baseline data. The ChestX-ray14 dataset contains 112,120 frontal-view X-ray images, of which 51,708 contain one or more pathologies. The performance of all 14 labels was reported. ChestX-ray14 provides hand-labeled bounding boxes for a small number of images to validate the localization accuracy of weakly supervised learning. Eight of the 14 diseases were annotated using 984 bounding boxes. These bounding boxes were used to compare the performance of the proposed method. The localization accuracy was tested with eight disease labels as follows: "Atelectasis," "Cardiomegaly," "Effusion," "Infiltration," "Mass," "Nodule," "Pneumonia," and "Pneumothorax."

Tables 4.1 and 4.2 show the experimental results. Table 4.1 shows the mean AUC for the CheXpert and ChestX-ray14 datasets. Table 4.2 shows the mean IOU scores for the ChestX-ray14 dataset. Our proposed method outperformed the PCAM. A qualitative analysis was also performed. As shown in Fig. 4.14. Our proposed method, FGR-PCAM, correctly predicted the localization of lesions compared to PCAM.

4.6.3 Domain Adaptive Multiple Instance Learning for Pathological Images

In the field of pathological diagnosis, physicians use microscopy to examine tissue slide images to detect cancer and other diseases. Several studies have been conducted to integrate image recognition technologies to automate diagnoses and reduce the burden on physicians. Because a detailed examination at the cellular level is necessary for diagnosis, the size of a whole-slide (WSI) can reach $10^5 \times 10^5$ pixels. Owing to memory constraints, WSIs are often broken down into smaller patch images for input into the classification models. Patch-level annotation is very expensive because it requires the expertise of a physician and considerable time to annotate large WSIs. However, per-slide labels, which indicate whether a WSI has an abnormality adds little cost to annotation. Only per-slide labels can improve the patch-level classification performance of WSI.

Multiple instance learning (MIL) is a form of weakly supervised learning in which a single label is assigned to a bag of instances. In the context of pathological image analysis, MIL methods treat the patch image as an "instance" and the entire slide as a "bag." (Hou et al. 2016; Hashimoto et al. 2020). Although MIL proves to be

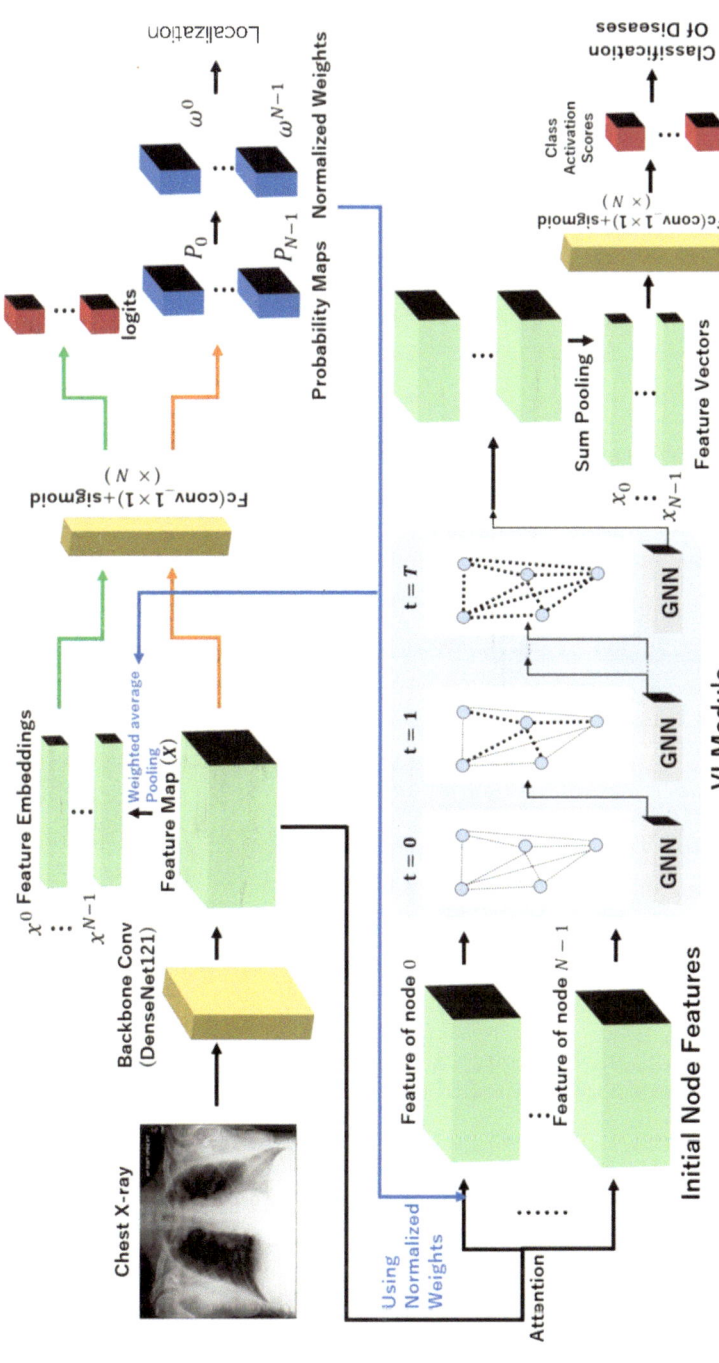

Fig. 4.13 FGR-PCAM (Jurg et al. 2022a)

Table 4.1 Mean AUC score for the CheXpert and ChestX-ray14 datasets: The bold values indicate the results of our proposed method

Dataset/Model	PCAM	FGR-PCAM
CheXpert	89.96 ± 0.07	**90.11 ± 0.48**
ChestX-ray14	81.73 ± 0.21	**82.10 ± 0.10**

Table 4.2 Mean IOU for the ChestX-ray14 datasets: The bold values indicate the results of our proposed method

Dataset/Model	PCAM	FGR-PCAM
ChestX-ray14	37.39 ± 1.128	**38.80 ± 1.187**

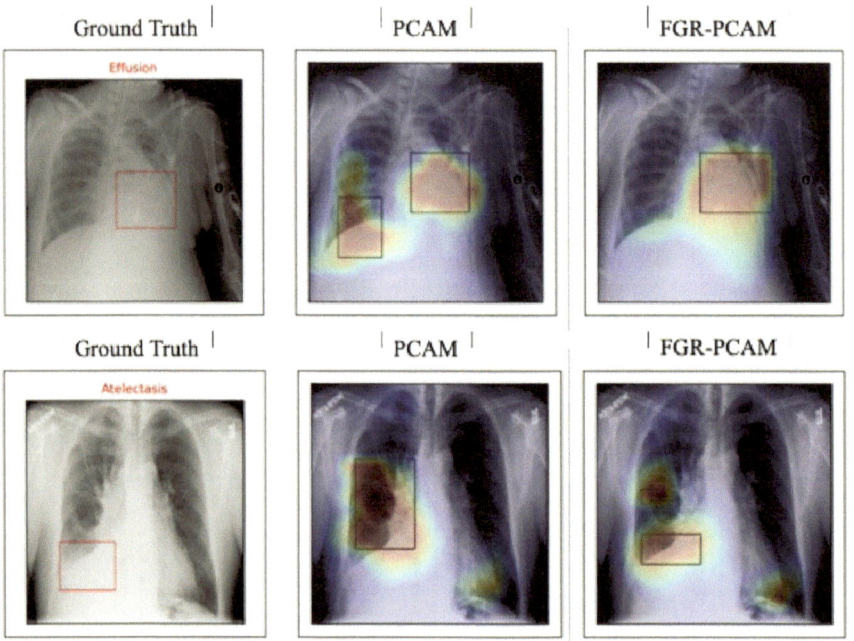

Fig. 4.14 Qualitative analysis of FGR-PCAM: red bounding boxes indicate the ground truth and black bounding boxes indicate the predicted bounding boxes (Jung et al. 2022b)

a valuable approach to reduce annotation costs, its accuracy tends to be lower than that of models trained with patch-level labels.

Alternatively, leveraging information from other datasets can improve classification performance without incurring additional annotation costs. Domain adaptation (DA) is a method that uses different domains to improve the performance of target data (Ganin and Lempitsky 2015; Saito et al. 2018).

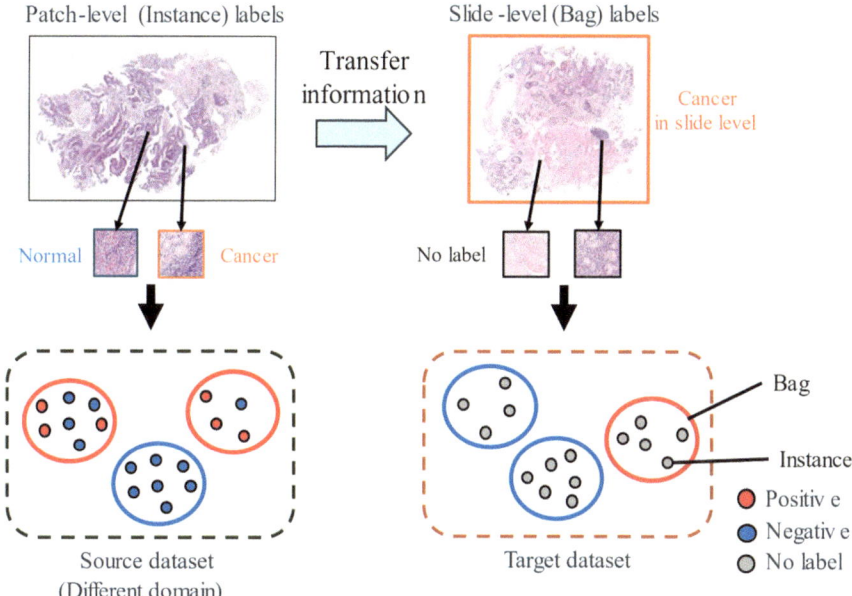

Fig. 4.15 Our task setting (Takahama et al. 2023)

In this context, a novel task setting is presented that is aimed at improving the patch-level classification performance of a target dataset by using only slide labels. This was achieved by incorporating information from a labeled source dataset originating from another domain (Fig. 4.15). However, owing to the qualitative differences in the supervised information between the source and target datasets, there is no guarantee that simply merging the two datasets will lead to an improvement in performance. Consequently, a new training pipeline is proposed that includes the generation of pseudo-labels and ensures high reliability by integrating information from both the source and target datasets.

An overview of the proposed method is shown in Fig. 4.16. Our pipeline consists of three main components: an encoder (G), a bag classifier (F_B), and an instance classifier (F_I). Each individual instance in the bags from both the target and source datasets was fed into the encoder (G) to extract the corresponding feature vectors. In contrast, a feature vector from each instance is input into the instance classifier F_I to obtain the binary prediction score of the instance label. Because the target has no instance label, the source instances with instance labels and the target instance with pseudo-labels were used to train F_I. At the time of inference, the target instance features are entered into F_I to obtain the prediction scores of the target instances. AttentionDeepMIL (Ilse et al. 2018) was used as F_B. In this method, the bag feature is the weighted sum of the instance features, and its weight is learnable.

To improve the prediction performance of the target instances through an instance classifier (F_I), a domain adaptation loss designed for the distribution matching of the

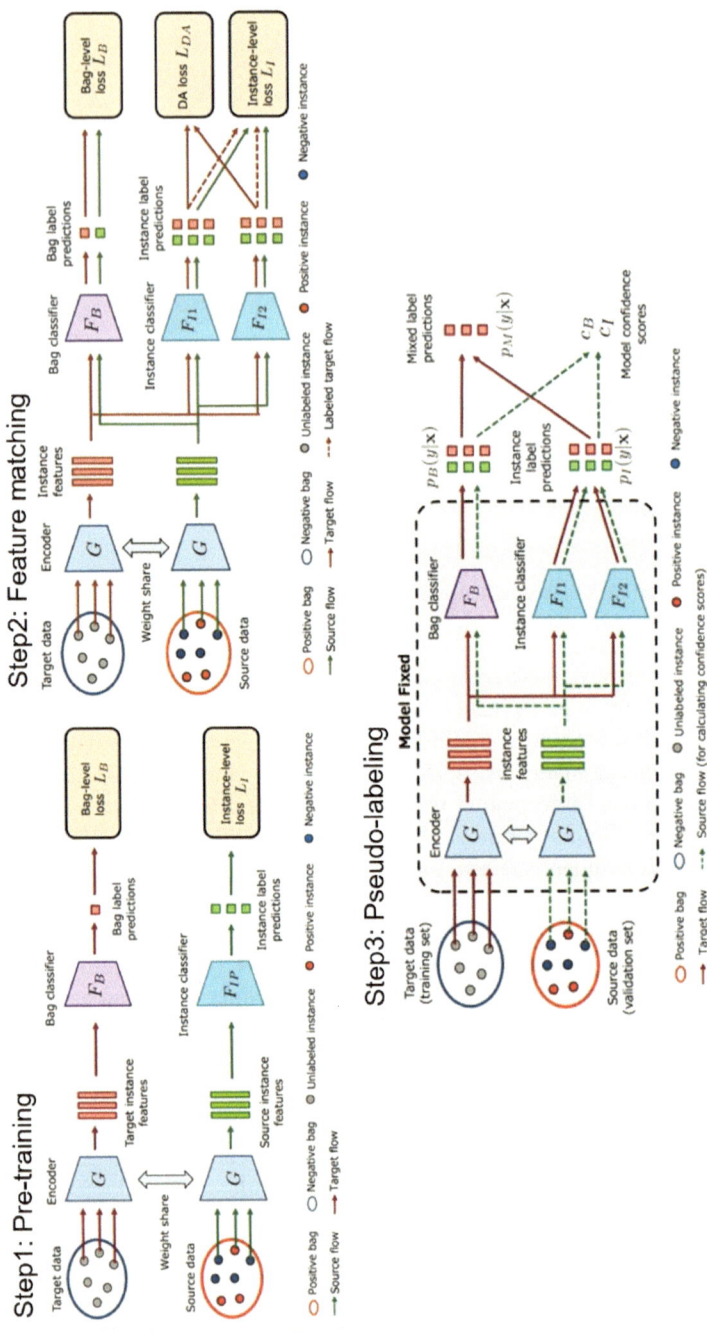

Fig. 4.16 Training process of the proposed method (Takahama et al. 2023)

intermediate features was integrated. Specifically, MCD (Saito et al. 2018) was used as the domain adaptation (DA) loss that facilitates feature distribution matching while taking category information into account. MCD achieves this by training features that are distant from the class boundary. Concurrently, the encoder G is trained to minimize the discrepancy loss for the target domain, while two classifiers are trained to alternately maximize the discrepancy loss.

Despite achieving matching feature distributions, numerous misclassified instances may occur if the decision boundaries of the source and target are not aligned. To address this challenge, our model was directly optimized for target instance prediction. This entails assigning pseudo-labels to the target instances and their integration into the training process of an instance classifier (F_I). Given that all instances in the negative bag are negative, our primary focus is on the instances in the positive bags. To derive dependable pseudo-labels, the outputs of two classifiers, namely, F_B and F_I were used. Because these classifiers were trained with distinct supervisory information, they have different properties. By integrating the information from both classifiers, pseudo-labels with higher reliability can be generated.

Figure 4.16 shows the complete training process for the proposed method. The method under consideration consists of the following three steps: In Step 1, supervised learning is performed for F_B using the target bag labels and F_I using the source instance labels. In this phase, a single-instance classifier called F_{IP} is used. The inclusion of Step 1 improves training stability, and lays the groundwork for obtaining reliable pseudo-labels at the beginning of Step 3. Once Step 1 converges, two instance classifiers, F_{I_1} and F_{I_2}, are initialized and followed by the alternating execution Steps 2 and 3. In Step 2, the model is trained using the feature matching loss of the DA. F_B is trained using both source and target data. In addition, F_{I_1} and F_{I_2} are trained with data containing the source instances, the target instances with pseudo-labels from positive bags, and the sampled target instances from negative bags. In Step 3, the model parameters are set and pseudo-labels are assigned to the target instances from the positive bags.

Experimental results are presented to confirm the effectiveness of the proposed method. A detailed evaluation is also performed. To showcase the effectiveness of the proposed method, a novel dataset of pathological images was compiled. This dataset consists of WSI obtained from two distinct body parts, namely "Stomach" and "Colon," encompassing 997 and 1368 WSIs, respectively. In contrast to many previous studies that used WSIs from two different datasets of a single organ as the source and target (Huang et al. 2017; Ren et al. 2018), our approach introduces a more challenging setting. Two different organs were intentionally selected, the stomach and the colon, to create a larger domain gap, which makes our experimental setup more demanding and suitable to illustrate the effectiveness of our method. For the experiments, the colon was designated as the source domain and the stomach as the target domain.

Table 4.3 lists the results of each method, and makes it clear that our proposed approach outperforms the other methods and achieves scores comparable to the "ideal case". Figure 4.17 shows heat maps illustrating the estimated patch labels in the WSIs of the target test set for each trained model. The heatmaps of "Source only" and

Table 4.3 The classification performance of each method for the pathology dataset: The bold values indicate the results of our proposed method

	Accuracy	PR-AUC
Attention MIL	72.4 ±5.62	66.0 ±1.24
Source only	82.5 ±1.20	66.9 ±1.00
MCDDA	76.0 ±3.09	51.8 ±2.78
Ours	**86.0 ±4.11**	**83.4 ±3.48**
Ideal case	91.1 ±0.79	87.1 ±2.05

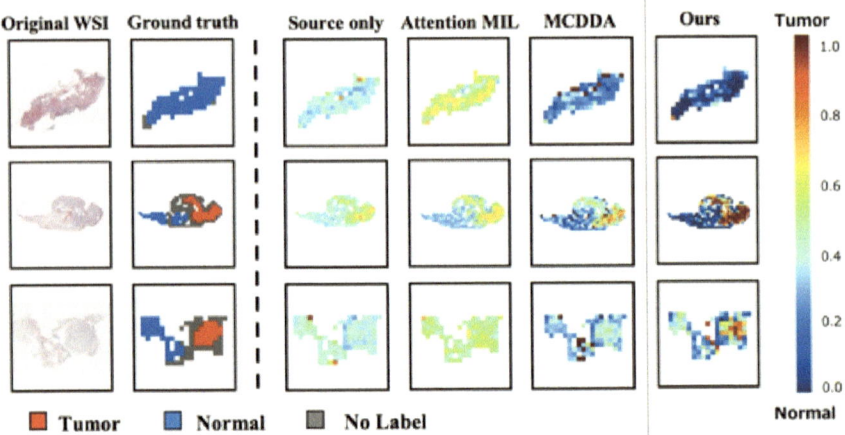

Fig. 4.17 Prediction heatmaps of the positive prediction scores for the target stomach dataset (Takahama et al. 2023)

"Attention MIL" show little difference between the scores of the normal and abnormal areas, which implies that the predictions appear relatively vague. The heatmaps of "MCDDA" appear to be relatively reasonable, but there are some regions with high abnormality scores in the normal region. There are instances of high abnormality scores in normal regions, which are unfavorable for practical use because even small abnormality areas require medical examination by doctors. Additionally, "MCDDA" does not effectively detect abnormal regions in the example below.

In contrast, our proposed method generated prediction maps with clear differences in prediction scores between the normal and abnormal regions, demonstrating qualitative validity. These findings highlight the effectiveness of our method, even in real-world applications such as pathological imaging.

4.6.4 Mental Health

As mentioned above, medical interviews are one of the areas in medicine where a CA is particularly useful. In this subsection, the functions that CAs should provide during medical interviews are discussed. During a medical interview, the doctor not only asks the patient questions but also makes a diagnosis based on information such as facial color, movement, and tone of voice. Therefore, CAs must be able to make diagnoses based on multimodal information such as images and sounds, in addition to textual information. Psychiatry is one of the most common departments to use this type of information to make diagnoses. Therefore, we are working on the diagnosis of psychiatric disorders through medical interviews as an application of CA.

A well-known study that examined the diagnosis of mental illness through medical interviews was the AVEC2019 (Ringeval et al. 2019). This competition predicts the severity of depression using machine learning based on facial expressions, tone of voice, and dialogue with a depressed patient during a medical interview.

Some studies have used large-scale language models to assess psychiatric disorders. The MentalLLaMA (Yang et al. 2023) is a large language model based on the LLaMA developed by Meta and constructed to assess mental disorders from SNS posts. However, these models are based on textual information actively posted by patients and do not consider information not contained in the text.

Therefore, we are currently working on a way for the diagnosis of psychiatric disorders to be performed by a CA. We are also working on developing a diagnostic algorithm that takes into account not only textual, but also multimodal information. Since interviewing to diagnose psychiatric disorders is time consuming, we expect that the development of this algorithm will be of great help to psychiatrists by enabling a CA to diagnose mental disorders that many people suffer from.

References

Afifi M, Derpanis KG, Ommer B, Brown MS (2021) Learning multi-scale photo exposure correction. In: 2021 IEEE/CVF conference on computer vision and pattern recognition (CVPR). IEEE, pp 9153–9163

Bell-Kligler S, Shocher A, Irani M (2019) Blind super-resolution kernel estimation using an internal-GAN. In: Proceedings of the 33rd international conference on neural information processing systems, pp 284–293

Berlin B, Kay P (1969) Basic color terms: their universality and evolution. University of California Press, Berkeley

Brown TB, Mann B, Ryder N et al (2020) Language models are few-shot learners. In: Advances in neural information processing systems

Chen H, Wang Y, Guo T et al (2021) Pre-trained image processing transformer. In: 2021 IEEE/CVF conference on computer vision and pattern recognition (CVPR). IEEE, pp 12294–12305

Choy C, Gwak J, Savarese S (2019) 4D Spatio-temporal ConvNets: Minkowski convolutional neural networks. In: 2019 IEEE/CVF conference on computer vision and pattern recognition (CVPR). IEEE, pp 3070–3079

Cui Z, Qi G-J, Gu L et al (2021) Multitask AET with orthogonal tangent regularity for dark object detection. In: 2021 IEEE/CVF international conference on computer vision (ICCV). IEEE, pp 2533–2542

Cui Z, Li K, Gu L et al (2022a) You only need 90K parameters to adapt light: a light weight transformer for image enhancement and exposure correction. In: 2022 British machine vision conference (BMVC)

Cui Z, Zhu Y, Gu L et al (2022b). Exploring resolution and degradation clues as self-supervised signal for low quality object detection. In: 2022 The European conference on computer vision ECCV 2022, vol 13669. Springer, Cham

Dai D, Wang Y, Chen Y, Van Gool L (2016) Is image super-resolution helpful for other vision tasks? In: 2016 IEEE winter conference on applications of computer vision (WACV). IEEE, pp 1–9

Deng J, Dong W, Socher R et al (2009) ImageNet: a large-scale hierarchical image database. In: 2009 IEEE conference on computer vision and pattern recognition. IEEE, pp 248–255

Ehteshami Bejnordi B, Veta M, Johannes van Diest P et al (2017) Diagnostic assessment of deep learning algorithms for detection of lymph node metastases in women with breast cancer. JAMA 318:2199. https://doi.org/10.1001/jama.2017.14585

Everingham M, Van Gool L, Williams CKI et al (2010) The pascal visual object classes (VOC) challenge. Int J Comput vis 88:303–338. https://doi.org/10.1007/s11263-009-0275-4

Ganin Y, Lempitsky V (2015) Unsupervised domain adaptation by backpropagation. In: Proceedings of the 32nd international conference on machine learning, pp 1180–1189

Geiger A, Lenz P, Urtasun R (2012) Are we ready for autonomous driving? The KITTI vision benchmark suite. In: 2012 IEEE conference on computer vision and pattern recognition. IEEE, pp 3354–3361

Gu J, Lu H, Zuo W, Dong C (2019) Blind super-resolution with iterative kernel correction. In: 2019 IEEE/CVF conference on computer vision and pattern recognition (CVPR). IEEE, pp 1604–1613

Gulshan V, Peng L, Coram M et al (2016) Development and validation of a deep learning algorithm for detection of diabetic retinopathy in retinal fundus photographs. JAMA 316:2402. https://doi.org/10.1001/jama.2016.17216

Han T, Adams LC, Papaioannou J-M et al (2023) MedAlpaca—an open-source collection of medical conversational AI models and training data

Hashimoto N, Fukushima D, Koga R et al (2020) Multi-scale domain-adversarial multiple-instance CNN for cancer subtype classification with unannotated histopathological images. In: 2020 IEEE/CVF conference on computer vision and pattern recognition (CVPR). IEEE, pp 3851–3860

Hou L, Samaras D, Kurc TM et al (2016) Patch-based convolutional neural network for whole slide tissue image classification. In: 2016 IEEE conference on computer vision and pattern recognition (CVPR). IEEE, pp 2424–2433

Huang Y, Zheng H, Liu C et al (2017) Epithelium-stroma classification via convolutional neural networks and unsupervised domain adaptation in histopathological images. IEEE J Biomed Health Inform 21:1625–1632. https://doi.org/10.1109/JBHI.2017.2691738

Huang S, Gojcic Z, Usvyatsov M et al (2021) PREDATOR: registration of 3D point clouds with low overlap. In: 2021 IEEE/CVF conference on computer vision and pattern recognition (CVPR). IEEE, pp 4265–4274

Huang J, Birdal T, Gojcic Z et al (2023) Multiway non-rigid point cloud registration via learned functional map synchronization. IEEE Trans Pattern Anal Mach Intell 45:2038–2053. https://doi.org/10.1109/TPAMI.2022.3164653

Ideno A, Mukuta Y, Harada T (2021) Generation of Variable-length time series from text using dynamic time warping-based method. In: ACM multimedia Asia. ACM, New York, pp 1–7

Ilse M, Tomczak JM, Welling M (2018) Attention-based deep multiple instance learning. In: Proceedings of the 35th international conference on machine learning, pp 2127–2136

Irvin J, Rajpurkar P, Ko M et al (2019) CheXpert: a large chest radiograph dataset with uncertainty labels and expert comparison. Proc AAAI Conf Artif Intell 33:590–597. https://doi.org/10.1609/aaai.v33i01.3301590

Ishiguro H, Yamamoto M, Tsuji S (1992) Omni-directional stereo. IEEE Trans Pattern Anal Mach Intell 14:257–262. https://doi.org/10.1109/34.121792

Johnson AEW, Pollard TJ, Shen L et al (2016) MIMIC-III, a freely accessible critical care database. Sci Data 3:160035. https://doi.org/10.1038/sdata.2016.35

Jung B, Mukuta Y, Harada T (2022a) Grouped self-attention mechanism for a memory-efficient transformer

Jung B, Gu L, Harada T (2022b) Graph interaction for automated diagnosis of thoracic disease using x-ray images. In: Medical imaging 2022: image processing. SPIE, pp 135–147

Kamata H, Mukuta Y, Harada T (2022) Fully spiking variational autoencoder. Proc AAAI Conf Artif Intell 36:7059–7067. https://doi.org/10.1609/aaai.v36i6.20665

Krizhevsky A (2009) Learning multiple layers of features from tiny images

Li Y, Harada T (2022a) Lepard: learning partial point cloud matching in rigid and deformable scenes. In: 2022 IEEE/CVF conference on computer vision and pattern recognition (CVPR). IEEE, pp 5544–5554

Li Y, Harada T (2022b) Non-rigid point cloud registration with neural deformation pyramid. In: Advances in neural information processing systems

Li X, Pontes JK, Lucey S (2021a) Neural scene flow prior. In: Advances in neural information processing systems

Li Y, Takehara H, Taketomi T et al (2021b) 4DComplete: non-rigid motion estimation beyond the observable surface. In: 2021 IEEE/CVF international conference on computer vision (ICCV). IEEE, pp 12686–12696

Li Y, Suzuki J, Morishita M et al (2022) Chat translation error detection for assisting cross-lingual communications. In: Proceedings of the 3rd workshop on evaluation and comparison of NLP systems, pp 88–95

Li Y, Suzuki J, Morishita M et al (2023) An investigation of warning erroneous chat translations in cross-lingual communication. In: Proceedings of the IJCNLP-AACL 2023 student research workshop

Lin T-Y, Maire M, Belongie S et al (2014) Microsoft COCO: common objects in context. In: Fleet D, Pajdla T, Schiele B, Tuytelaars T (eds) Computer vision—ECCV 2014: 13th European conference, Zurich, Switzerland, 6–12 Sept 2014, proceedings, part V. Springer, Cham, pp 740–755

Liu C, Sun D (2014) On Bayesian adaptive video super resolution. IEEE Trans Pattern Anal Mach Intell 36:346–360. https://doi.org/10.1109/TPAMI.2013.127

Lore KG, Akintayo A, Sarkar S (2017) LLNet: a deep autoencoder approach to natural low-light image enhancement. Pattern Recognit 61:650–662. https://doi.org/10.1016/j.patcog.2016.06.008

Lv F, Li Y, Lu F (2021) Attention guided low-light image enhancement with a large scale low-light simulation dataset. Int J Comput Vis 129:2175–2193. https://doi.org/10.1007/s11263-021-01466-8

MacDorman KF, Ishiguro H (2006) The uncanny advantage of using androids in cognitive and social science research. Interact Stud 7:297–337. https://doi.org/10.1075/is.7.3.03mac

Miraglia L, Di Dio C, Manzi F et al (2023) Shared knowledge in human-robot interaction (HRI). Int J Soc Robot. https://doi.org/10.1007/s12369-023-01034-9

Moor M, Huang Q, Wu S et al (2023) Med-Flamingo: a multimodal medical few-shot learner

Park K, Sinha U, Barron JT et al (2021) Nerfies: deformable neural radiance fields. In: 2021 IEEE/CVF international conference on computer vision (ICCV). IEEE, pp 5845–5854

Redmon J, Divvala S, Girshick R, Farhadi A (2016) You only look once: unified, real-time object detection. In: Proceedings of the IEEE computer society conference on computer vision and pattern recognition

Ren J, Hacihaliloglu I, Singer EA et al (2018) Adversarial domain adaptation for classification of prostate histopathology whole-slide images. In: Frangi AF, Schnabel JA, Davatzikos C et al (eds) Medical image computing and computer assisted intervention—MICCAI 2018: 21st international conference, Granada, Spain, 16–20 Sept 2018, proceedings, part II. Springer, Cham, pp 201–209

Ringeval F, Schuller B, Valstar M et al (2019) AVEC 2019 workshop and challenge: state-of-mind, detecting depression with AI, and cross-cultural affect recognition. In: Proceedings of the 9th international on audio/visual emotion challenge and workshop. ACM, New York, pp 3–12

Ronneberger O, Fischer P, Brox T (2015) U-Net: convolutional networks for biomedical image segmentation. In: Navab N, Hornegger J, Wells WM, Frangi AF (eds) Medical image computing and computer-assisted intervention—MICCAI 2015. Springer International Publishing, Cham, pp 234–241

Saito K, Watanabe K, Ushiku Y, Harada T (2018) Maximum classifier discrepancy for unsupervised domain adaptation. In: 2018 IEEE/CVF conference on computer vision and pattern recognition. IEEE, pp 3723–3732

Sarlin P-E, DeTone D, Malisiewicz T, Rabinovich A (2020) SuperGlue: learning feature matching with graph neural networks. In: 2020 IEEE/CVF conference on computer vision and pattern recognition (CVPR). IEEE, pp 4937–4946

Sedai S, Mahapatra D, Ge Z et al (2018) Deep multiscale convolutional feature learning for weakly supervised localization of chest pathologies in X-ray images. In: Shi Y, Suk H-I, Liu M (eds) Machine learning in medical imaging: 9th international workshop, MLMI 2018, held in conjunction with MICCAI 2018, Granada, Spain, 16 Sept 2018, proceedings. Springer, Cham, pp 267–275

Singhal K, Azizi S, Tu T et al (2023) Large language models encode clinical knowledge. Nature 620:172–180. https://doi.org/10.1038/s41586-023-06291-2

Su S, Gu L, Yang Y et al (2023) Name your colour for the task: artificially discover colour naming via colour quantisation transformer. In: Proceedings of the IEEE/CVF international conference on computer vision

Sun J, Shen Z, Wang Y et al (2021) LoFTR: detector-free local feature matching with transformers. In: 2021 IEEE/CVF conference on computer vision and pattern recognition (CVPR). IEEE, pp 8918–8927

Takahama S, Kurose Y, Mukuta Y, et al (2023) Domain adaptive multiple instance learning for instance-level prediction of Pathological Images. In: 2023 IEEE 20th international symposium on biomedical imaging (ISBI). IEEE, pp 1–5

Takase S, Kiyono S, Kobayashi S, Suzuki J (2023) B2T connection: serving stability and performance in deep transformers. In: Findings of the association for computational linguistics: ACL 2023. Association for Computational Linguistics, Stroudsburg, PA, USA, pp 3078–3095

Thirunavukarasu AJ, Ting DSJ, Elangovan K et al (2023) Large language models in medicine. Nat Med 29:1930–1940. https://doi.org/10.1038/s41591-023-02448-8

Thomas H, Qi CR, Deschaud J-E et al (2019) KPConv: flexible and deformable convolution for point clouds. In: 2019 IEEE/CVF international conference on computer vision (ICCV). IEEE, pp 6410–6419

Touvron H, Martin L, Stone K et al (2023) Llama 2: open foundation and fine-tuned chat models

Tu Z, Talebi H, Zhang H et al (2022) MAXIM: multi-axis MLP for image processing. In: 2022 IEEE/CVF conference on computer vision and pattern recognition (CVPR). IEEE, pp 5759–5770

Umagami R, Ono Y, Mukuta Y, Harada T (2023) HiPerformer: hierarchically permutation-equivariant transformer for time series forecasting

Vaswani A, Shazeer N, Parmar N et al (2017) Attention is all you need. In: Advances in neural information processing systems

Wang X, Peng Y, Lu L et al (2017) ChestX-Ray8: hospital-scale chest X-ray database and benchmarks on weakly-supervised classification and localization of common thorax diseases. In: 2017 IEEE conference on computer vision and pattern recognition (CVPR). IEEE, pp 3462–3471

Wei K, Fu Y, Zheng Y, Yang J (2021) Physics-based noise modeling for extreme low-light photography. IEEE Trans Pattern Anal Mach Intell 44:1–1. https://doi.org/10.1109/TPAMI.2021.3103114

Yang X, Chen A, PourNejatian N et al (2022) A large language model for electronic health records. NPJ Digit Med 5:194. https://doi.org/10.1038/s41746-022-00742-2

Yang K, Zhang T, Kuang Z et al (2023) MentaLLaMA: interpretable mental health analysis on social media with large language models

Ye W, Yao J, Xue H, Li Y (2020) Weakly supervised lesion localization with probabilistic-CAM pooling

Zaslavsky N, Garvin K, Kemp C et al (2022) The evolution of color naming reflects pressure for efficiency: evidence from the recent past. J Lang Evol 7:184–199. https://doi.org/10.1093/jole/lzac001

Zeng A, Song S, Niessner M et al (2017) 3DMatch: learning local geometric descriptors from RGB-D reconstructions. In: 2017 IEEE conference on computer vision and pattern recognition (CVPR). IEEE, pp 199–208

Zhou B, Khosla A, Lapedriza A et al (2016) Learning deep features for discriminative localization. In: 2016 IEEE conference on computer vision and pattern recognition (CVPR). IEEE, pp 2921–2929

Zhou L, Liu Y, Zhang P et al (2023) Information bottleneck and selective noise supervision for zero-shot learning. Mach Learn 112:2239–2261. https://doi.org/10.1007/s10994-022-06196-7

Chapter 5
Cooperative Control of Multiple CAs

Takayuki Nagai, Tomoaki Nakamura, Komei Sugiura, Tadahiro Taniguchi, Yosuke Suzuki, and Masayuki Hirata

Abstract In a world where Cybernetic Avatars (CAs) are active in real society, it is expected that one person will control multiple CAs or multiple CAs will cooperate with each other to perform a task. For one operator to control multiple CAs simultaneously, technologies with which one person can operate multiple CAs are required. CAs should work while understanding the intentions of the operator according to the task and environment. In addition, it is assumed that not only able-bodied people but also people with disabilities, such as amyotrophic lateral sclerosis (ALS) patients, will control CAs. This chapter outlines new technologies for realizing the simultaneous remote and coordinated control of multiple CAs (flexible CA control) from various perspectives.

T. Nagai
Osaka University, Toyonaka, Osaka, Japan
e-mail: takato@sys.es.osaka-u.ac.jp

T. Nakamura
The University of Electro-Communications, Chofu, Tokyo, Japan
e-mail: tnakamura@uec.ac.jp

K. Sugiura
Keio University, Yokohama, Kanagawa, Japan
e-mail: komei.sugiura@keio.jp

T. Taniguchi (✉)
Ritsumeikan University, Kusatsu, Shiga, Japan
e-mail: taniguchi@em.ci.ritsumei.ac.jp

Y. Suzuki
Kanazawa University, Kanazawa, Ishikawa, Japan
e-mail: suzuki@se.kanazawa-u.ac.jp

M. Hirata
Osaka University, Suita, Osaka, Japan
e-mail: mhirata@ndr.med.osaka-u.ac.jp

© The Author(s) 2025
H. Ishiguro et al. (eds.), *Cybernetic Avatar*,
https://doi.org/10.1007/978-981-97-3752-9_5

5.1   Introduction

In a world where CAs are active in real society, it is assumed that not only does one person necessarily control a single CA, but also that one person controls multiple CAs, or that multiple CAs cooperate with each other to accomplish tasks. For example, in a hospital room where coordinated tasks such as preparation for treatment, examination, treatment, and explanation occur frequently, CAs can accomplish their tasks more efficiently by coordinating with multiple CAs rather than acting alone. Cooperation between CAs can provide various services that cannot be performed by a single CA. For one operator to operate multiple CAs simultaneously, it is necessary to have the technology to use multiple CAs that work while understanding the operator's intentions depending on the task and environment. It is also necessary to consider the relationships and coordination with people in the environment. Furthermore, it is naturally assumed that not only able-bodied people but also people with disabilities, such as patients with ALS, will control CAs (Takeuchi et al. 2020).

This chapter discusses multiple-CA control techniques, but what is the problem with multiple-CA controls in the first place? This issue can be organized in terms of a probabilistic generative model (generative artificial intelligence (AI)). Figure 5.1 shows the elemental technologies related to the multiple-CA controls described in this section using a generative model. Note, however, that the model in the figure is not strictly a generative model; rather, ease of understanding is prioritized. The figure implies that given a set of random variables and an inference method, and with appropriate training data, it is possible to realize the control of multiple CAs, taking into account a variety of factors. However, to control multiple CAs, inference is expected to become very complex, and although various means have been proposed to solve such implementation problems (Nakamura et al. 2018; Taniguchi et al. 2020a, b), it is not an easy task. However, the intent of the discussion here with the generative model is to provide a bird's-eye view of all elemental technologies, and this model clarifies the position of each technology described in this chapter.

This chapter outlines new technologies for establishing simultaneous remote control and coordinated control of multiple CAs (flexible CA control) from the following perspectives. Note that the numbers in Fig. 5.1 correspond to the numbers below.

(1) Language-based control of multiple CAs and a shared control method based on intention estimation: A method using large language models (LLMs) as one of the methods for the centralized control of multiple CAs and a method for integrating human CA control and autonomous CA control based on intention estimation are described.

(2) Autonomous control model for CAs: For multiple CAs to act autonomously, they must make action decisions based on the current environment and the state of others. This section outlines the model that serves as the basis for such decisions.

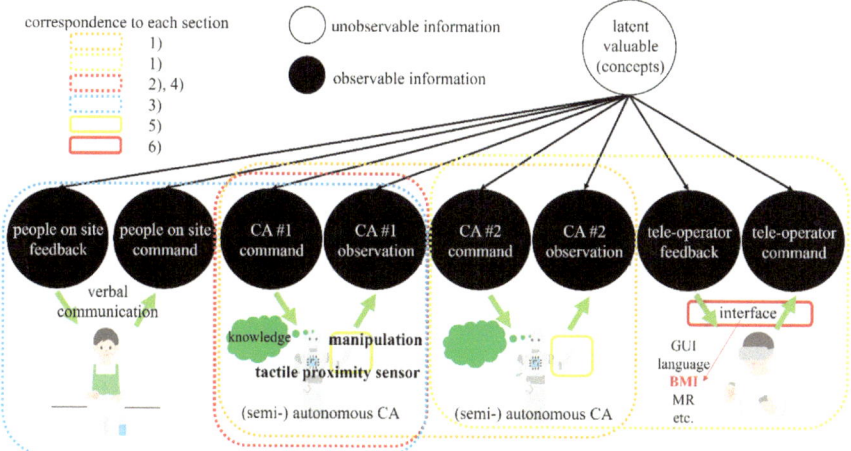

Fig. 5.1 Probabilistic generative model representation of multiple-CA control. From (1) to (6) corresponds to each section

(3) Dialogue technology that responds to changes in the environment and situation: This section outlines a flexible dialogue that responds to dynamic changes in the environment and situations that occur when multiple CAs work together to manipulate objects in a living environment.

(4) Common sense and local knowledge for autonomous physical CAs: When a CA performs daily physical support in a field environment, including offices and hospitals, it is necessary to efficiently acquire and support not only common sense linguistic knowledge such as that of LLMs but also new local knowledge in that environment. Therefore, it is necessary to efficiently acquire and support new local knowledge in the environment. This section outlines the research related to these issues.

(5) Manipulation techniques for physical CAs: The development of tactile sensing capabilities for hands and their integration into autonomous/semi-autonomous systems to enable CAs to skillfully perform physical tasks are discussed.

(6) Harmonized control of autonomous CAs based on a voluntary brain–machine interface (BMI): This section describes efforts to realize CAs that can be controlled as desired by harmonizing control by BMI, which excels in voluntariness, and autonomous systems, which excels in autonomy.

5.2   Development of Flexible CA Control Technologies

5.2.1   Multiple-CA Control Using LLMs

5.2.1.1   Multiple-SayCan

SayCan (Ahn et al. 2023), proposed by Google, uses a large-scale language model to interpret the purpose of ambiguous human language instructions, convert them into robot commands, and execute tasks. In this study, we applied the SayCan mechanism to control multiple CAs. Although a simple approach would be to assign the action plan output by the LLM to each CA, it is necessary to make action decisions and task assignments that consider the constraints that arise from the use of multiple CAs, such as collisions during simultaneous operations and the characteristics of each CA. The authors proposed a method in which an LLM handles the dependencies among tasks and assigns tasks using mathematical optimization (the subtask partitioning method) (Obata et al. 2023a). We also considered the use of the Chain of Thought (CoT) method (Wei et al. 2022), which is an LLM method that has shown excellent results in many problem forms. In this subsection, we describe these methods.

First, we consider constraints in the execution of multiple CA tasks. The complexity of the task varies depending on the language instructions given by the user. The definition of dependency here is that when a dependency exists between two or more tasks and task AB has a dependency, task B cannot be executed unless a particular task A has been completed. For example, the sequence of actions to realize the instruction "pile up the red, blue, and yellow blocks in the order blue, yellow, red from the bottom" would be (A) place the yellow block on top of the blue block, (B) place the red block on top of the yellow block, and so on. Because of the risk of collision when task A and task B are performed simultaneously, task B cannot be executed until task A is completed. Thus, AB is a task with a dependency relationship.

Some constraints depend on the characteristics of the CAs; each CA has its own characteristics, and there are restrictions on the tasks that can be executed according to the characteristics. Even with the same language instruction, if the CA changes, the task assignment must be adjusted according to the CA's characteristics.

5.2.1.2   Subtask Partitioning Method

To execute multiple-CA tasks, it is necessary to infer their dependencies. For this purpose, LLM was used. Simultaneously, when selecting the tasks necessary to realize the language instructions, it infers whether there are dependencies between each task and divides them into subtasks. The CA-dependent constraints are then addressed by setting the skill affordances of each CA. Tasks were assigned to each CA using a mathematical optimization method based on the results of subtask partitioning and skill affordances. An overview of the subtask partitioning method is

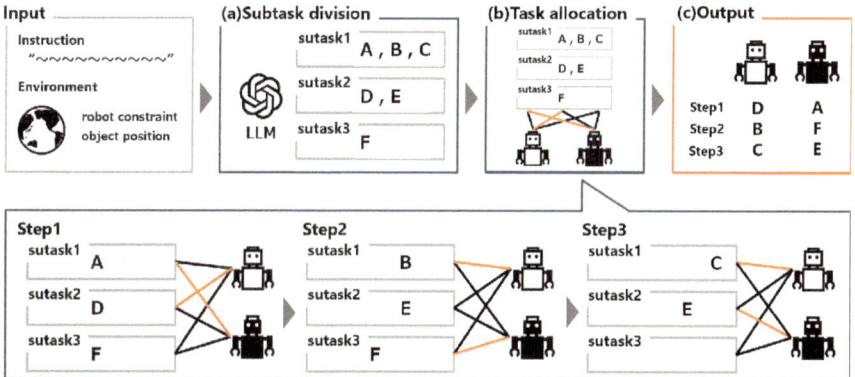

Fig. 5.2 Overview of the subtask partitioning method

shown in Fig. 5.2. The subtask partitioning method consists of two processes: subtask partitioning using an LLM and task assignment using mathematical optimization.

Subtask Partitioning: Using LLM, tasks corresponding to language instructions are selected from predefined behaviors. It then divides the tasks into one or more subtasks based on the dependencies among the tasks (Fig. 5.2a). The following two preconditions were used for subtask partitioning:

1. Tasks within the same subtask have a dependency
2. Tasks in different subtasks do not have dependencies.

A five-shot prompt was created according to the preconditions, and the dependencies between tasks were handled by an LLM (GPT3 Davinci was used in the experiment).

Set Up Skill Affordances: CAs must maintain a set of skills to perform physical tasks. It does not matter how the skills are implemented; for example, they can be acquired through reinforcement learning or manually designed behaviors. Importantly, the feasibility score (skill affordance) of each skill can be obtained from environmental information such as camera images. The Q-values can be used for reinforcement learning. In subsequent experiments, the distance from the CA to the grasping object was used.

Task Assignment: Tasks are assigned to multiple CAs using a linear programming assignment problem based on skill affordances (Fig. 5.2b). In the first of the subtasks generated in the previous step, an allocation problem was generated with the sum of the skill affordances in each CA as the objective function. The allocation problem is solved, and the solved task is assigned as the first task in the first step of each CA. Subsequently, a new allocation problem is generated by moving one task forward in the subtask to which the selected task belongs. This operation is repeated until all tasks are assigned, and the final solution obtained at each step is the output (Fig. 5.2c).

5.2.1.3   CoT Method

The CoT method approaches each of these issues using stepwise reasoning. The proposed method comprises four steps and uses a one-shot prompt for reasoning.

[Step 1: Reasoning about execution tasks] Reasons for tasks required to realize language instructions.

[Step 2: Processing CA constraints] Process constraints based on the characteristics of each CA in the tasks enumerated in Step 1. Confirm the blocks that can be grasped based on the constraint set and infer tasks that can be executed accordingly.

[Step 3: Process dependencies among tasks] Infer tasks that can be executed by each CA while satisfying the dependencies between tasks. Check for nonexecuted tasks, infer dependencies among tasks, and infer tasks to be executed by each CA. This process was repeated until all tasks were assigned.

[Step 4: Output execution tasks] The results of the inference in Step 3 are output as commands that can be recognized by the CAs.

5.2.1.4   Experiments

The usefulness of the proposed method is evaluated by performing task assignments for various language instructions in a simulation environment. An example of the experimental environment is shown in Fig. 5.3. Two-arm robots were placed on a table in the environment, and four types of red, blue, yellow, and green blocks were placed (additional objects were placed according to the task). We evaluated the performance of each method from the object recognition and positional acquisition of each object in the simulator environment to the task assignment stage.

Fig. 5.3 An example of the experimental environment

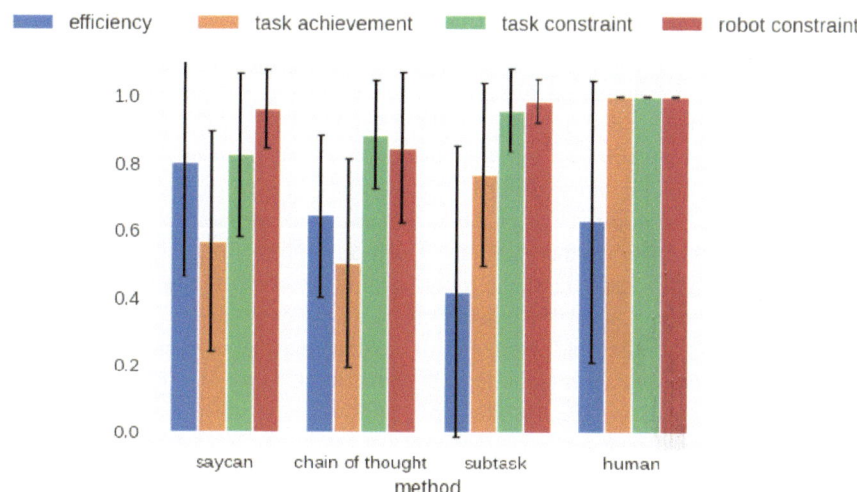

Fig. 5.4 Results of the experiment

The four performance evaluations are efficiency, task achievement, task-dependent constraint compliance, and robot-dependent constraint compliance. However, it should be noted that efficiency can be increased without constraint compliance, as the evaluation here depends on planning.

The experimental results are shown in Fig. 5.4. Overall, the existing and CoT methods showed high scores for the efficiency item, whereas the subtask partitioning method showed high scores for the other items. Some results show that the existing and CoT methods score higher than the human task assignment, which is the correct response in this experiment, in terms of efficiency, indicating that the existing method ignores constraints and reduces the robot's waiting actions to increase efficiency. This also has a significant impact on the success rate of the task because it ignores the constraints and the task accomplishment item shows a lower score. In any case, the subtask partitioning method resulted in the highest compliance with constraints and the highest task accomplishment.

5.2.2 Shared CA Control Based on User-Intention Estimation

In the previous subsection, we described an example in which two CAs cooperated to execute commands using a remote operator that provided verbal commands. In this subsection, we introduce a shared control approach in which the remote operator controls the CA in coordination with autonomous control. This assumption is particularly important for CA manipulation based on BMI (Parvizi and Kastner 2018) using intracranial EEG, which is measured using electrodes implanted intracranially. Intracranial EEG is expected to have high decoding performance (Parvizi and Kastner

2018) and is suitable as an input for BMI. However, high concentration is required for the user to intentionally generate EEG signals. In addition, the command generation frequency of intracranial EEG is not sufficiently high compared to joysticks and keyboards, which are commonly used for robot operations today.

Shared control, in which an autonomous system assists the user, has been widely applied to robot manipulation using biometric signals to reduce the user load and enable more precise control. For example, Zhuang et al. introduced shared control in manipulating a robotic hand using myoelectricity, where the user controls when fine manipulation is required and the robot's autonomy controls when robustness is required, thereby enabling stable object grasping(Zeng et al. 2020).

This subsection outlines a shared control method based on user-intention estimation for CA base movement, assuming a CA operation using an intracranial EEG-based BMI (Obata et al. 2023b). However, although this subsection assumes the remote control of a mobile CA using BMI, the same assumption applies to CA operations in poor communication environments.

5.2.2.1   Shared Control Based on Intention Estimation

Overview of Shared Control: Figure 5.5 presents an overview of the shared control to be introduced. This method comprises the following four modules:

(1) User Module: This module receives inputs from the user, defines them as speed commands for the robot, and outputs them to each module.
(2) Goal Module: Receives the position of obstacles and the initial position of the robot at the start of a task, and then receives the current position from the robot at regular intervals. Upon receiving a user command from (1), the system estimates the user's goal intention based on this information and calculates the confidence level for that estimate.

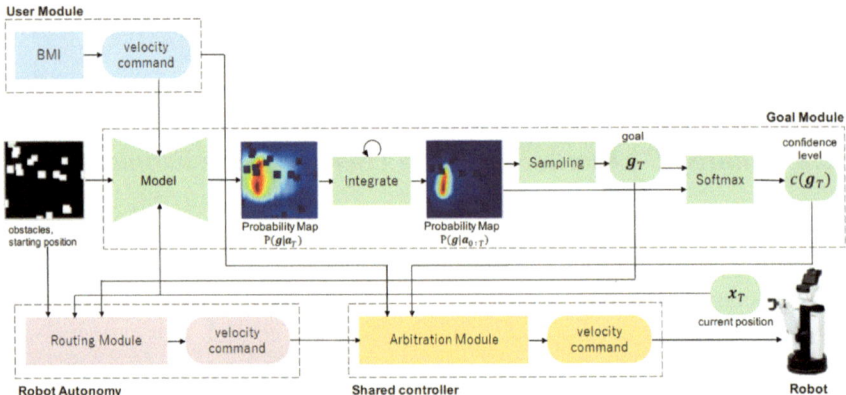

Fig. 5.5 Overview of the shared control method

(3) Robot Autonomy Module: This module generates autonomous commands by autonomously generating a path to the goal position estimated in (2) and calculating the speed at the current position.
(4) Shared Controller: This controller combines user commands received from (1) and autonomous commands received from (3) based on the estimation in (2) to generate control commands for the robot.

Estimation of the User's Goal Intention: The user's goal intention in the Goal Module is estimated using the following two steps:

[Step 1] Estimate the probability distribution of the user's intended goal position based on user commands at a certain time.
[Step 2] Sample goal positions from the above probability distribution and calculate their confidence levels.

An important problem is the estimation of the distribution of user intentions. Here, we focus on the fact that the artificial potential field can be calculated by providing an environmental map, current position of the CA, and goal position from which the path to the goal can be calculated. Because the environmental map and CA's position can be assumed to be known, given the goal position appropriately, ideally, the current user's operation commands should match the CA's movement commands obtained from the computed path. Furthermore, if we define the difference between the operation command and calculated movement command of the CA as the estimation error of the goal position, we can realize a neural network that infers the goal position from the current position of the CA and the user operation command. In this case, it is important to note that all training data can be generated automatically as long as an environmental map is provided. The architecture of the model is shown in Fig. 5.6. The goal position (user intention) was estimated by sampling the distribution estimated by the model. In addition, the probability values were normalized to provide a confidence level for the estimation.

Generation of Shared Control Commands: In the Shared Controller, the user and autonomous commands are combined to generate shared control commands, which are determined using the goal (user intention) estimated by the Goal Module and its confidence level. Finally, the calculated shared control commands are inputted to the robot.

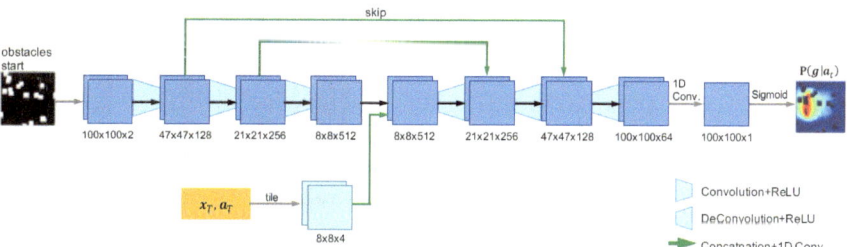

Fig. 5.6 The architecture of the model for estimating user's goal position

5.2.2.2 Experiment

Assuming that the CA will be operated using intracranial EEG in the future, we conducted an experiment assuming that the user can generate commands at 0.5 Hz. The CA, on the other hand, was assumed to be able to receive commands at 5 Hz. In this experiment, the user inputs commands using a joystick. For this experiment, several rectangular obstacles were placed in a two-dimensional simulation environment, as shown in Fig. 5.7. Two environments, A (Fig. 5.7a) and B (Fig. 5.7b), were also prepared, with black x in the figure as the start and red x as the goal. The user uses a joystick to move the CA indicated by the red circle in Fig. 5.7 from the designated start to the goal while avoiding obstacles. The trajectory of the CA and its collision with obstacles were compared when the CA was moved using only user commands and when shared control was applied.

The trajectories of the CAs when executing tasks in Environments A and B are shown in Fig. 5.7c–f, respectively. Black x indicates the start position, red x indicates the goal position, and the blue dots are the trajectories of the CAs. Figure 5.7c, e show the results of the shared control, and Fig. 5.7d, f show the results of the user command. The position of the CA when the user command was received, the probability distribution of the estimated user-goal intention, and the estimated goal are shown in Fig. 5.8 (Environment A). In Fig. 5.8, the blue dots in the upper row indicate the CA's position, the red x indicates the sampled goal, and the lower row shows the corresponding probability distribution. When visualizing the probability distributions, the probability values are scaled to facilitate relative color discrimination.

These figures show that the trajectories generated with the aid of shared control were smoother and more direct than those generated without assistance. Furthermore, when the task was executed using only user commands, collisions with obstacles were observed. However, these collisions were avoided using shared control. Regarding the estimation of the user's goal intention, Fig. 5.8 shows that the position with the highest probability value converged to the goal position over time.

5.3 Hierarchical Control for Autonomous CAs

This study presents a learning model that enables multiple CAs to act autonomously. For CAs to act autonomously, they need a model that hierarchically structures the environment and can predict the environmental changes caused by their actions. Furthermore, for multiple CAs to collaborate, they must communicate with each other and select the appropriate actions. This study first introduces HVGH (Nagano et al. 2022), which is a model that enables CAs to learn the spatiotemporal structures of their environments. It then introduces a model that enables CAs to produce messages for communication and exhibit cooperative behavior using these messages.

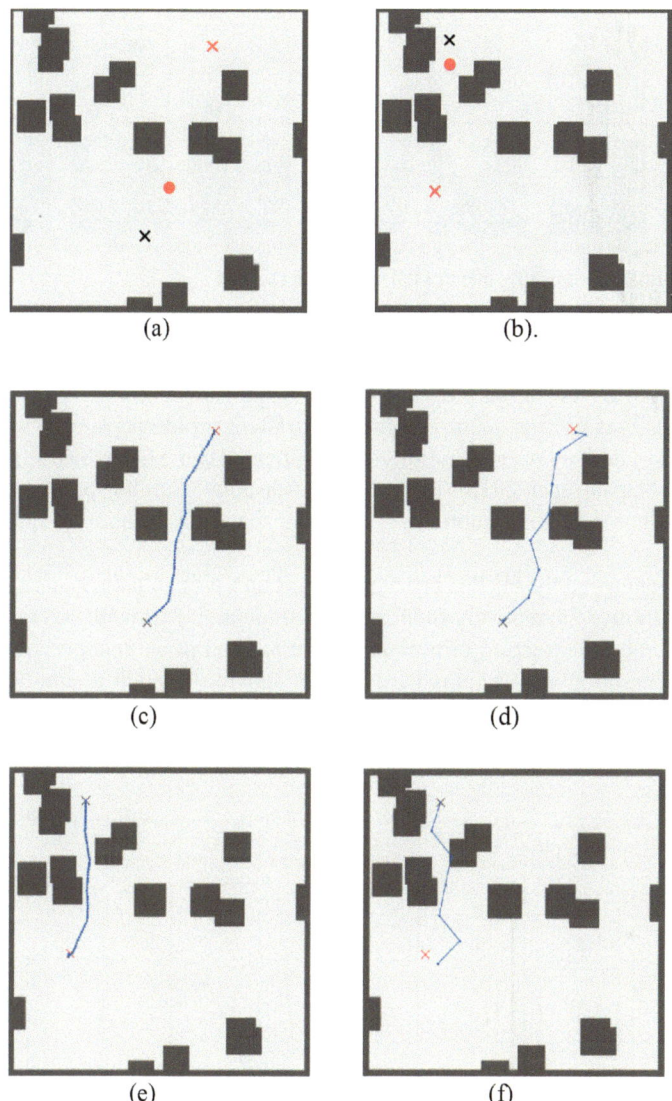

Fig. 5.7   Two-dimensional simulation environment

5.3.1   Acquisition of Spatiotemporal Structure of Environment by CAs

For a CA to act autonomously, a model that structures high-dimensional sensor information and predicts environmental changes is required. In a previous study (Nagano et al. 2022), we proposed a model that enables a mobile CA to learn the

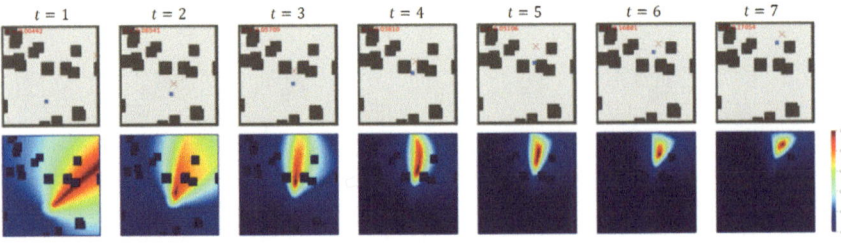

Fig. 5.8 Estimated probability maps of the user's goal position

spatiotemporal structure of an environment from first-person images captured by the CA. An overview of the proposed model is shown in Fig. 5.9. A convolutional autoencoder was used to learn a compressed latent representation of first-person images. This latent representation was then segmented using GP-HSMM (Nakamura et al. 2017; Nagano et al. 2018). This segmentation allows similar spatial information to be represented in a common state, such that temporal changes in states can be learned.

One advantage of the HVGH is that it is based on a probabilistic generative model and can be trained on relatively small amounts of data. Experiments have shown that the spatiotemporal structure of a maze can be learned in an unsupervised manner from first-person images acquired from a CA moving through a simulated maze. Furthermore, we demonstrated that the performance was better with a small amount of data than with models that used deep neural networks.

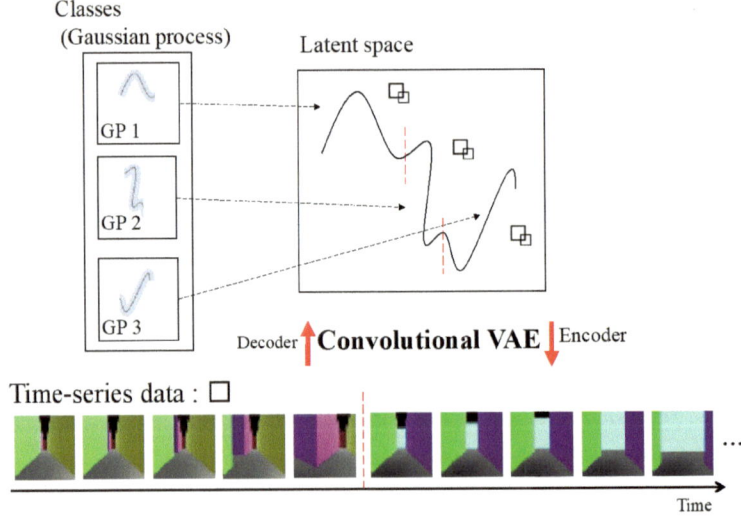

Fig. 5.9 Overview of the HVGH

5.3.1.1 HVGH

Figure 5.10 presents a graphical model of the HVGH. In this model, c_j is a class index, and the number of classes is assumed to be infinite; π_c represents the probability of transition from class c and is generated from the GEM distribution (Sethuraman 1994) parameterized by β and η, which is generated by the Dirichlet process:

$$\beta \sim \text{GEM}(\gamma),$$
$$\pi_c \sim \text{DP}(\eta, \beta).$$

The j-th class c_j is determined by the $j-1$-th class c_{j-1} and transition probability π_c. Furthermore, the sequence of latent variables Z_j is generated by the Gaussian process parameterized by ϕ_c:

$$c_j \sim P(c|c_{j-1}, \pi_c),$$
$$Z_j \sim \text{GP}(Z|\phi_{c_j}),$$

where ϕ_c represents the set of sequences of latent variables classified into class c. The segments are generated by the decoder of the VAE p_{dec}:

$$X_j \sim p_{\text{dec}}(X|Z_j).$$

By connecting the segments X_j, the observed sequence S can be obtained by connecting segments X_j. The model parameters were learned in an unsupervised manner from only the observed sequence S using a blocked Gibbs sampler and stochastic gradient descent.

Fig. 5.10 Graphical model of HVGH

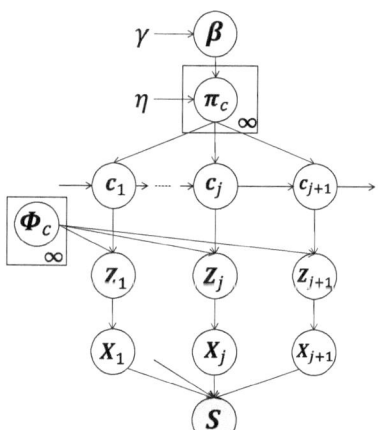

5.3.1.2 Experimental Results

We conducted unsupervised segmentation of video images from a CA first-person viewpoint, and the images were obtained by simulating a three-dimensional maze. Figure 5.11 depicts the results of the segmentation, which show that appropriate segmentation can be obtained using image similarity. Moreover, the structure of the maze can be learned as transition probability $P(c|c_{j-1})$. This result indicates that HVGH enables unsupervised learning of spatiotemporal structures.

Moreover, HVGH achieved higher accuracy with less data than a previously proposed segmentation method based on recurrent neural networks (Kim et al. 2019) that introduced binary variables to represent segment switching.

5.3.2 Emergent Communications Based on Gaussian Process Latent Variable Model (GPLVM)

For multiple CAs to cooperate autonomously, symbols must be used for communication with each other. However, such symbols vary from task to task, and must emerge unsupervised according to the task. In this study, we propose a symbol emergence model based on a Gaussian process.

Symbol emergence is formulated as the inference of a shared symbol s that can represent the observations of the two CAs, as shown in Fig. 5.12.

$$s \sim p(s|o_a, o_b) \tag{5.1}$$

However, because we assume that the two CAs are separate individuals whose internal states o_a, o_b, cannot be observed by the other CA, the symbols cannot be inferred using only this formulation. Therefore, we use the Metropolis-Hastings naming game (MHNG) (Hagiwara et al. 2022; Taniguchi et al. 2023b) proposed by Taniguchi et al. to infer s on the assumption that the internal state of one CA cannot be directly observed by another. Furthermore, s is a categorical variable in the conventional method, whereas in this study, we consider the inference to be a

Fig. 5.11 Segmentation result for HVGH

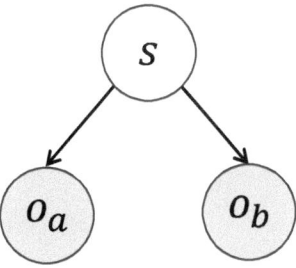

continuous latent variable. Thus, we use the GPLVM, in which the models of $p(o_a|s)$
and $p(o_b|s)$ are Gaussian processes.

5.3.2.1   MHNG

MHNG is an inference method based on the Metropolis-Hastings (MH) algorithm,
where the following equation is used as the target distribution to generate sample s
from the posterior distribution (Eq. 5.1).

$$p(s|o_a, o_b) \approx Cp(s|o_a)p(s|o_b), \qquad (5.2)$$

where C is the normalization term and the product of the expert's approximation
is used to transform the equation. The proposed distribution $Q(s^*)$ of the CA-A-
generating message S^* is described as

$$Q(s^*) = p(s^*|o_a) \qquad (5.3)$$

Based on Eqs. (5.2) and (5.3), the acceptance rate r of a new sample s^* in the MH
algorithm can be obtained as:

$$r = \frac{Q(s)p(s^*|o_a, o_b)}{Q(s^*)p(s|o_a, o_b)} = \frac{p(s^*|o_b)}{p(s|o_b)} \qquad (5.4)$$

That is, for proposal s^* by CA-A, CA-B can determine whether to accept the
proposal using only its own information, and vice versa. Thus, both CAs send a
message sampled from their own internal states to the other and update their parameters if the other accepts the message they send. A shared symbol is a sample according
to Eq. (5.1) is generated by iterating this process.

5.3.2.2   GPLVM-based MHNG

In a GPLVM, given N symbols, $S = [s_1, s_2, \ldots, s_N]^{\mathrm{T}}$, the probability that the corresponding training data $O_b = \left[o_{b_1}, f_{b_2}, \ldots, f_{b_N}\right]^{\mathrm{T}}$ is generated can be calculated as:

$$p(O_b|S) \propto \frac{1}{|K|^{\frac{D}{2}}} \exp\left(-\frac{1}{2}\mathrm{tr}\left(K_b^{-1}F_bF_b^{\mathrm{T}}\right)\right), \tag{5.5}$$

where D is the dimensionality of o_b, K is the Gram matrix whose elements correspond to the i-th row, the j-th column is $k(s_i, s_j)$, and k is the kernel function. Thus, if a new symbol s_{n^*} is proposed for the n-th observation, $S^* = \left[s_1, s_2, \ldots, s_n^*, \ldots, s_N\right]^{\mathrm{T}}$, and the acceptance rate (from Eq. 5.4) is obtained as:

$$r = \frac{p(O_b|S^*)p(S^*)}{p(O_b|S)p(S)} = \frac{p(O_b|S^*)}{p(O_b|S)},$$

where $p(S)$ is assumed to have a uniform distribution. Thus, the acceptance rates calculated using Eq. (5.4) can be computed for GPLVM.

However, the shape of the distribution represented by Eq. (5.5), when s is a variable, is complex, and samples from the proposed distribution $Q(S^*) = P(S^*|O_a)$ cannot be simply generated. Therefore, the samples from the proposed distribution were generated using the MH algorithm. Let $p(S)$ be a uniform distribution; let the target distribution be $Q(S^*) = p(S^*|O_a) \propto p(O_a|S^*)$. Using the Gaussian distribution $N(s^*|s, \sigma^2)$, which is symmetric with respect to the current sample S and new sample S^*, as the proposal distribution, the acceptance rate is calculated as follows:

$$r' = \frac{p(O_a|S^*)N(s|s^*, \sigma^2)}{p(O_a|S)N(s^*|s, \sigma^2)} = \frac{p(O_a|S^*)}{p(O_a|S)} \tag{5.6}$$

Using these equations, the MH method generates a sample from one CA and proposes it for the other CA. Subsequently, by accepting or rejecting the sample using Eq. (5.6) with the information of the other CA, symbol emergence can be achieved using a GPLVM-based MHNG.

5.3.2.3   Experiments

In the experiment, 67 objects from 11 categories, as shown in Fig. 5.13, were used. As shown in Fig. 5.13, CA-A observed the frontal images of these objects, and CA-B observed the rear images. The features o_a, o_b are extracted from the trained CLIP using a vision encoder (Radford et al. 2021). The s was set to two dimensions.

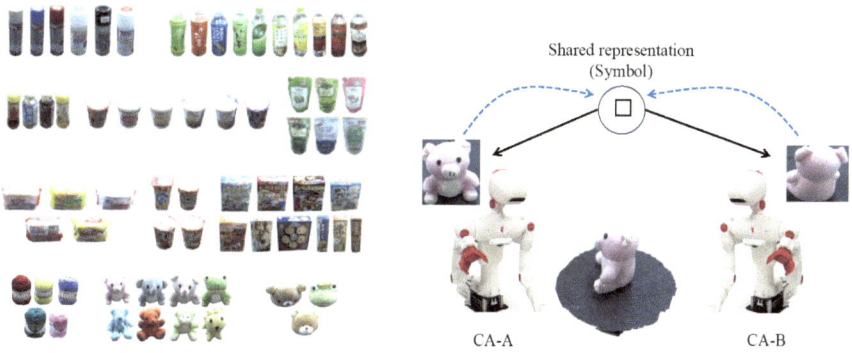

Fig. 5.13 Left: objects used in the experiment. Right: overview of the experiment

Fig. 5.14 Emergent symbols from two CAs

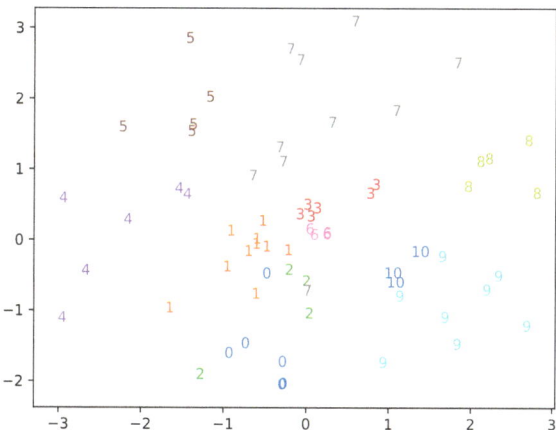

Figure 5.14 shows the two-dimensional shared latent variables trained using GPLVM. The numbers represent the category indices of the objects, and their positions are symbols represented as continuous two-dimensional vectors. Symbols representing each category are separated in space. This indicates that the CAs can generate shared symbols that represent the object categories without directly observing the internal states of other CA.

5.3.3 Cooperative Control Based on Emergent Communication for Autonomous CAs

As described in the previous subsection, CAs generate symbols representing their observations. Using these symbols, the CA communicates its observations or states to others. In this study, we propose a method that enables CAs to learn and generate cooperative behavior by utilizing symbols as messages.

5.3.3.1 Action Decision Based on the States and Messages

Figure 5.15 presents an overview of the proposed method. First, the two CAs infer messages about their own states s_A^t and s_B^t. Figure 5.16 depicts a graphical model of the message emergence. The message emerges depending on the state and reward for the cooperative action r_m^t. Using the MHNG, messages can be generated without directly observing the other states.

Each CA $i \in \{A, B\}$ generates actions based on their own state s_i^t and the message inferred by MHNG:

$$a_i^t \sim \pi_i(\cdot | s_i^t, m^t)$$

Here π_i represents the CA policy. The CAs can discover each other's states indirectly through an inferred message; therefore, they can generate actions depending on their own state and that of the other CA. Thus, the message has the ability to change

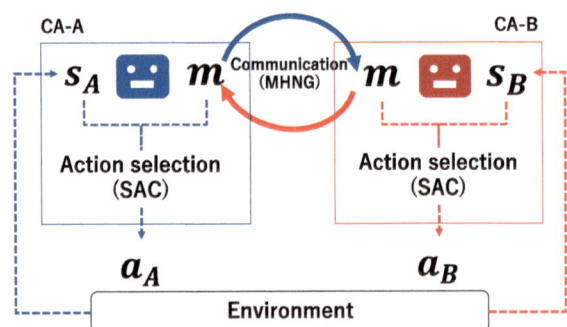

Fig. 5.15 Overview of the cooperative controls for autonomous CAs

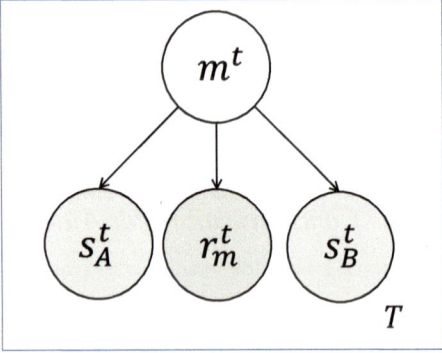

Fig. 5.16 Graphical model of message emergence

the actions of the CAs, and the actions of the CAs are adjusted so as to generate cooperative actions through communication. Value v_i^t of action a_i^t is evaluated using the following action value function Q_i:

$$v_i^t = Q_i(s_i^t, a_i^t, m^t)$$

Functions π_i and Q_i were implemented using neural networks and trained using the soft actor–critic (SAC) algorithm. The policy and action value functions are represented as π_{φ_i} and Q_{θ_i}, respectively, using the network parameters φ_i and θ_i, and the objective function to be minimized is obtained as follows:

$$J_\pi(\varphi_i) = E\left[\alpha \log \pi_{\varphi_i}(a_i^t | s_i^t, m^t) - Q_{\theta_i}(s_i^t, a_i^t, m^t)\right],$$

$$J_Q(\theta_i) = E\left[\frac{1}{2}\left(Q_{\theta_i}(s_i^t, a_i^t, m^t) - \widehat{Q}(s_i^t, a_i^t, m^t)\right)^2\right],$$

where

$$\widehat{Q}(s_i^t, a_i^t, m^t) = r_i^t + \beta r_m^t + \gamma E\left[V(s_i^{t+1}, m^{t+1})\right],$$

$$V(s_i^t, m^t) = E\left[\alpha \log \pi_{\varphi_i}(a_i^t | s_i^t, m^t) - Q_{\theta_i}(s_i^t, a_i^t, m^t)\right],$$

where α is the weight of the entropy regularization term, β is the weight of the reward for cooperative actions, and γ is the discount rate. The parameters were estimated using a stochastic gradient descent to minimize the objective functions:

$$\varphi_i \leftarrow \varphi_i - \lambda_{\varphi_i} \nabla_{\varphi_i} J_\pi(\varphi_i)$$

$$\theta_i \leftarrow \theta_i - \lambda_{\theta_i} \nabla_{\theta_i} J_Q(\theta_i)$$

Here λ_{φ_i} and λ_{θ_i} represent the learning rates.

5.3.3.2 Experiments

For our evaluation, we used a simple spread task in a multiagent particle environment (Mordatch and Abbeel 2018). In this task, the two CAs attempt to achieve their goals without colliding. The environment and examples of the generated actions are presented in Fig. 5.17. First, the policies of the CAs and messages were trained for over 1200 episodes. In each episode, the goals were randomly changed and the CAs randomly moved 25 steps in the environment. Next, for evaluation, the CAs were subjected to another 200 episodes using trained policies and messages. When the proposed method was compared with the method without communication, the total number of collisions was found to be 289 and 500, respectively. This result indicates that messages that are effective in solving cooperative tasks performed by the CAs can emerge.

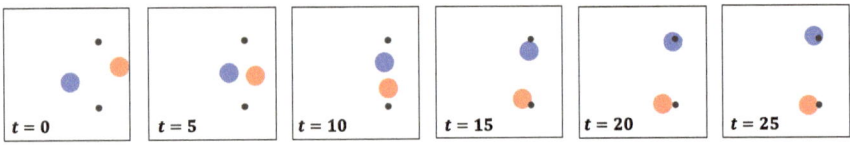

Fig. 5.17 Cooperative behavior generated by the two CAs

5.4   Research and Development of Embodied Dialogue CAs

Embodied CAs hold significant promise for assisting older individuals and those with disabilities. There is a growing trend toward the standardization of CAs to offer a diverse range of support functions (Yamamoto et al. 2019). Within this array of capabilities, the skill of navigating within indoor settings is of utmost importance, as it serves as a fundamental requirement for numerous everyday activities, including tasks such as retrieving a glass of water from the kitchen. However, for most CAs, the capacity for language interaction, designed to be accessible to non-expert users, remains restricted because of the intricacies involved in comprehending natural language.

The instructions provided by humans often contain ambiguities, posing a significant challenge for CAs when it comes to identifying both the target object and destination. According to Shridhar et al. (2020), humans achieved a 91.0% success rate on the ALFRED dataset, a widely recognized benchmark in Vision-and-Language Navigation (VLN) (Anderson et al. 2018b), that involves tasks related to object manipulation. By contrast, state-of-the-art methods (e.g., Murray and Cakmak 2022) have achieved success rates of approximately 46%.

Extensive research has been conducted in the field of Embodied AI, including competitive benchmarking events for CAs operating in standardized domestic environments (e.g., RoboCup@Home (Iocchi et al. 2015)), which closely aligns with the task addressed in this study. Most existing approaches in the realm of natural language understanding for Embodied AI focus on navigation (e.g., Shah et al. 2023), object manipulation (e.g., Shridhar et al. 2021), or a combination of the two (e.g., Khandelwal et al. 2022). Representative Embodied AI tasks include VLN and Object Goal Navigation (ObjNav) (Anderson et al. 2018a). VLN tasks involve instructing an agent to navigate and/or manipulate objects in 3D environments based on natural language instructions.

To address the multimodal language understanding of CAs, it is essential to establish clear objectives and milestones. We developed objectives based on service dog tasks because they are well defined. We categorized the hardware-feasible tasks from the 108 service dog tasks specified in (IAADP 2023). Table 5.1 lists the task categories.

Table 5.1 Domestic service tasks that can be performed using standard hardware

Category	# Tasks	Descriptions
Retrieve	12	Bring portable phone to any room in house; Fetch a beverage from a refrigerator or cupboard; Fetch food bowl(s); Pick up dropped items like coins, keys etc., in any location; Bring clothes, shoes, or slippers laid out to assist with dressing; Assist to tidy house or yard—pickup, carry, deposit designated items; Retrieve purse from hall, desk, dresser or back of van; Fetch basket with medication and/or beverage from cupboard; Seek & find teamwork direct the dog with hand signals, vocal cues to: retrieve an unfamiliar object out of partner's reach, locate TV remote control, select one of several VCR tapes atop TV cabinet; Put trash, junk mail into a wastebasket or garbage can; Deposit empty soda pop can or plastic bottle into recycling bin; Dirty food bowl [dog's]—put into kitchen sink; Put silverware, non-breakable dishes, plastic glasses in sink
Carry	8	Move bucket from one location to another, indoors and outdoors; Carry item(s) from the partner to a care-giver or family member in another room; Send the dog to obtain food or other item from a care-giver and return with it; Deliver items to "closet" [use a floor marker to indicate drop location]; Put prescription bag, mail, other items on counter top; Transport textbooks, business supplies or other items up to 50 lbs in a wagon or collapsible cart, weight limit depends on dog's size, physical fitness, type of cart, kind of terrain; Backpacking—customary weight limit is 15% of the dog's total body weight; 10% if a dog performing another task, such as wheelchair pulling in addition to backpacking; Total weight includes harness (average 3–4 lbs.). Load must be evenly distributed to prevent chafing
OpenClose	20	Open cupboard doors with attached strap; Open drawers via strap; Open refrigerator door with a strap or suction cup device; Open interior doors via a strap with device to turn knob; Answer doorbell and open front door with strap attached to lever handle; Open or close sliding glass door with a strap or other tug devices; Shut restroom door that opens outward via a leash tied to doorknob; Close stall door that opens outward in restroom by delivering end of the leash to partner; Shut interior home, office doors that open outward; Pull a drapery cord to open or close drapes; Cupboard door or drawers—nudge shut; Dryer door—hard nudge; Stove drawer—push it shut; Dishwasher door—put muzzle under open door, flip to shut; Refrigerator and freezer door—close with nudge; Cupboard door—shut it with one paw; Dryer door— shut it with one paw; Refrigerator and freezer door—one forepaw or both; Close heavy front door, other doors—jump up, use both forepaws, Haul open heavy door, holding it ajar using six foot lead attached to back of harness, other end of lead attached to door handle or to a suction cup device on a glass door

(continued)

Table 5.1 (continued)

Category	# Tasks	Descriptions
Follow	2	Carry items following a partner using a walker, other mobility aids; Find the care-giver on command, lead back to location of disabled partner
SoftObjManip	7	Bring in groceries—up to ten canvas bags; Unload suitable grocery items from canvas sacks; Unload towels, other items from dryer; Transfer merchandise in bag from a clerk to a wheelchair user's lap; Assist partner to load clothing into top loading washing machine; Operate rope device that lifts blanket and sheet or re-covers disabled person when he or she becomes too hot or cold; Alternatively, take edge of a blanket and move backwards, tugging to remove it or assist someone to pull the blanket up to their chin if cold

The tasks were extracted from service dog tasks

5.4.1 Multimodal Language Understanding in "Retrieve" Tasks

We define multimodal language understanding for fetching instructions (MLU-FI) as the task of identifying the target object based on an instruction and contextual regions. Figure 5.18 illustrates a typical scenario for this task. Within this scene, we consider the specific instruction, "Go to the hallway and bring me the photo above the table." In this context, the objective of the model is to pinpoint a photo of the wall. The MLU-FI can be characterized by the following aspects:

- Input: An image and instruction sentence.
- Output: A bounding box or segmentation mask.

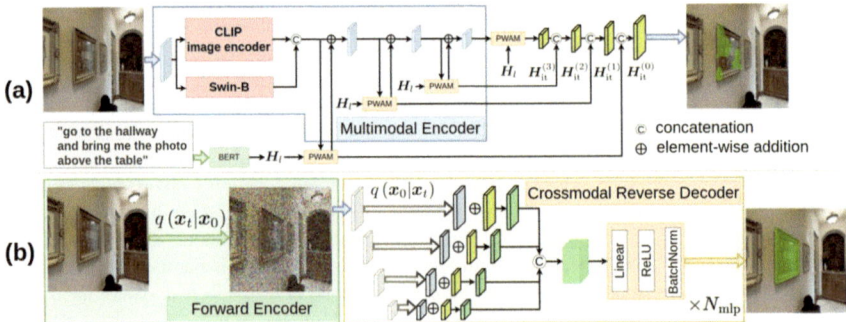

Fig. 5.18 Structure of our approach. Our approach is divided into two steps: **a** the intermediate training step and **b** the diffusion step. The model in step **a** is composed of two primary components, the crossmodal encoder and the pixel-word attention module (PWAM) (Anderson et al. 2018b). In step **b**, the model comprises two key modules, the forward encoder and the crossmodal reverse decoder

We developed several methods to handle the MLU-FI tasks (Magassouba et al. 2018, 2019, 2020; Ishikawa and Sugiura 2021; Otsuki et al. 2023; Iioka et al. 2023). Figure 5.18 illustrates the structure of the multimodal diffusion segmentation model (MDSM) (Iioka et al. 2023). This model produces a segmentation mask corresponding to a target object described in a natural language sentence using various reference expressions. Initially, a segmentation mask is created and enhanced in the subsequent phase. Our dual-phase segmentation approach covers areas more effectively than traditional single-phase models. Additionally, our method refines these areas using linguistic data, allowing us to anticipate the generation of masks that accurately represent both the instructions and the object area.

To validate our approach, we created the SHIMRIE dataset, which met all the necessary criteria. This dataset included 4341 images and 11,371 sentences, encompassing a vocabulary of 3558 words, amounting to a total of 196,541 words, and an average sentence length of 18.8 words. Within the SHIMRIE dataset, there were 10,153 training, 856 validation, and 362 test samples.

Table 5.2 presents the quantitative comparisons between the baseline method (i) and the variations in our proposed method. These results include the mean values and standard deviations of the five experiments. Table 5.2 indicates that the mean Intersection over Union (mIoU) for method (i) was 24.27%, whereas that for method (ii) was 30.19%. Thus, method (ii) achieved a 5.92-point improvement over method (i). Furthermore, method (ii) matches or surpasses method (i) in both the object Intersection over Union (oIoU) and precision at the k (P@k) metrics. The mIoU for method (iii) reached 34.40%, exceeding that of method (i) by 10.13 points, and method (iii) performed better than method (i) in terms of oIoU and P@k.

Figure 5.19 shows the qualitative results. This figure compares the prediction results of the baseline method (panels (a) and (c)) with those of the proposed method (panels (c) and (d)). The instruction for the lower images was "Go to the lounge and remove the small brown chair facing the counter," with the chair in panel (c) being the intended target. The baseline method incorrectly identified outside areas as relevant, whereas our proposed method produced a more accurate mask for the designated object than the baseline method.

5.4.2 Multimodal Language Understanding in "Carry" Tasks

In tackling multimodal language comprehension for "carry" tasks, we have developed various methods (Magassouba et al. 2021; Ishikawa and Sugiura 2022; Kaneda et al. 2023; Korekata et al. 2023). This task is called the dual-referring expression comprehension with fetch-and-carry (DREC-fc) task. In DREC-fc, upon receiving an instruction with referring expressions, the CA comprehends the intended object and its destination among various everyday items or furniture, and then transports the object to that location. Consequently, this task was divided into two segments: understanding the language and performing actions. The terms used in this subsection are defined as follows:

Table 5.2 Quantitative results on comparative experiments. "ME" refers to the multimodal encoder in the intermediate training step

[%]	Diff. step	ME	mIoU	oIoU	P@0.5	P@0.6	P@0.7	P@0.8	P@0.9
(i) LAVT [3]			24.27 ± 3.15	22.25 ± 2.85	21.27 ± 5.66	13.37 ± 3.74	5.97 ± 2.50	0.94 ± 0.38	0.00 ± 0.00
(ii) Ours		✓	30.19 ± 3.98	27.08 ± 2.89	31.66 ± 6.52	23.04 ± 4.66	10.33 ± 1.63	1.55 ± 1.36	0.00 ± 0.00
	✓	✓	$\mathbf{34.40 \pm 3.79}$	$\mathbf{31.59 \pm 3.03}$	$\mathbf{36.63 \pm 6.14}$	$\mathbf{27.79 \pm 5.28}$	$\mathbf{16.30 \pm 2.98}$	$\mathbf{6.41 \pm 1.19}$	$\mathbf{0.66 \pm 0.62}$

The best score is in bold

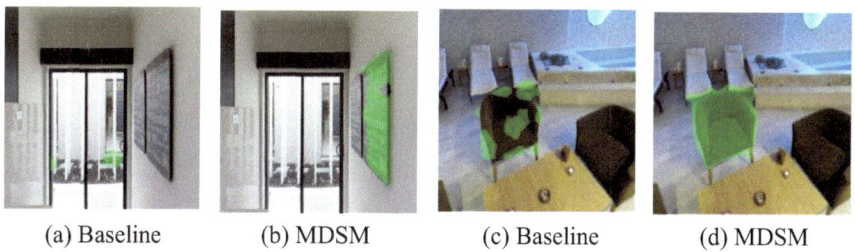

(a) Baseline   (b) MDSM   (c) Baseline   (d) MDSM

Fig. 5.19 Qualitative results on the SHIMRIE dataset. Panels **a** and **b** "Go to the laundry room and straighten the picture closest to the light switch." Panels **c** and **d** "Go to the lounge and remove the small brown chair facing the counter."

- Destination: furniture piece on which the target object is to be placed.
- Destination candidate: A furniture piece that the model predicts matches the destination.

To address the DREC-fc task, we developed the Switching Head–Tail Funnel UNITER (SHeFU) (Korekata et al. 2023), which is capable of identifying both target objects and destinations independently using a unified model. Figure 5.20 illustrates the structure of the SHeFU.

The operational complexity of the SHeFU is $O(M + N)$, as opposed to $O(M \times N)$, where M represents the number of potential target objects and N is the number of possible destinations. Unlike existing methods, our approach implements a switching head–tail mechanism, enabling the processing of both the target object and destination candidates through a single model. The switching head mechanism implicitly shares parameters within the model to predict both the target object and the destination. On the other hand, the Switching Tail mechanism facilitates the simultaneous learning of multiple tasks. These mechanisms consider both the visual and linguistic elements of the destination when predicting the target object, and vice versa. Moreover, separate models are not required because one model is sufficient for both tasks.

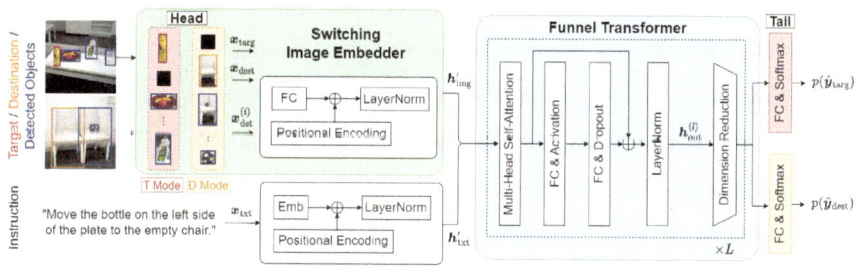

Fig. 5.20 The structure of SheFU, which consists of switching image embedder and funnel transformer

Table 5.3 Language comprehension accuracy on the ALFRED-fc dataset and the real-world environment

[%]	Method	ALFRED-fc	Real
(i)	Baseline (extended TDU [9])	79.4 ± 2.76	52.0
(ii)	Ours (W/o switching head)	78.4 ± 2.05	–
(iii)	Ours (W/o switching tail)	76.9 ± 2.91	–
(iv)	Ours (SHeFU)	**83.1 ± 2.00**	**55.9**

The best score is in bold

To validate the SHeFU, we performed both simulated and real-world experiments, as detailed in Korekata et al. (2023). Table 5.3 presents the quantitative outcomes. This table compares the precision of various methods on the ALFRED-fc dataset. According to Table 5.3, our proposed method (iv) attained a success rate of 83.1%, compared to 79.4% for the foundational method (i). Hence, our approach outperformed the baseline method by a margin of 3.7 points.

Figure 5.21 illustrates the qualitative outcomes of the physical experiments. In the upper panels, the identified target object and destination are the red chips and white table with the soccer ball, respectively. The CA picked the chips precisely and positioned them adeptly on the table. Similarly, in the lower panels, the identified target object and destination were the green cup and blue bin, respectively. The CA picks up the cup precisely and places it in the bin.

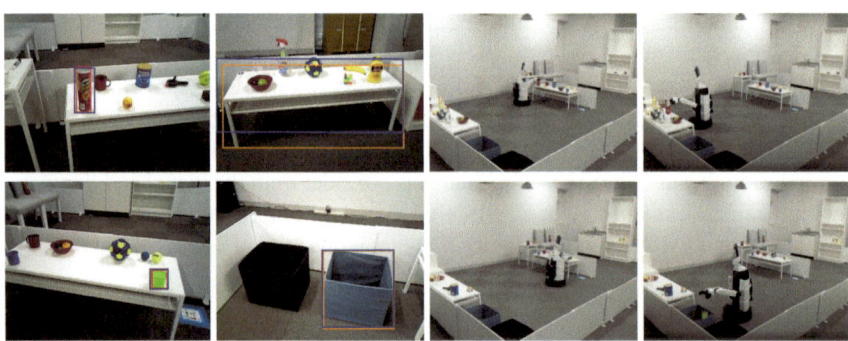

Fig. 5.21 Qualitative results of the physical experiments. From left to right: target object candidate, destination candidate, scene of object grasping, and scene of object placing. Red, orange, and blue bounding boxes represent the ground truth region of the target object, the ground truth region of the destination, and the target object or destination candidates, respectively

5.5 Research and Development of Daily-Physical-Support CAs

Cybernetic Avatars (CAs) are expected to provide daily physical support in various service environments such as offices, hospitals, and homes. These specific CAs are referred to as Daily-Physical-Support (DPS) CAs. To effectively perform DPS tasks, alleviate user fatigue, reduce workload, and enhance productivity, DPS-CAs must possess a degree of autonomy; they should be semi-autonomous. CAs require physical bodies for exploring the real world and interacting with objects and people. Consequently, DPS-CAs are essentially remotely teleoperated semi-autonomous robots. Furthermore, it is crucial for these robots to comprehend instructions to accomplish tasks on behalf of humans. This section introduces research and development in the field of DPS-CAs, focusing on acquiring and applying the necessary knowledge for task execution in daily environments and understanding context-based languages.

DPS-CAs should be semi-autonomous. Consider remote work scenarios in which DPS-CAs operate from home but are physically located in the workplace. Simple physical teleoperation is inadequate because it neither reduces workload nor enhances productivity. For instance, imagine a scenario in which a remote robot is tasked with retrieving a specific package to be delivered to an office. Directly manipulating a robot's arms, hands, wheels, and visual systems through remote controllers such as joysticks or motion capture systems is time consuming and requires an unreasonable additional workload for remote operators.

An alternative approach might be to instruct the DPS-CA as one would a colleague or secretary: "Please go to the reception on the first floor, ask Emily to give you the package that arrived yesterday, open the box, and pass the contents to Jochen." The aim of implementing DPS-CAs is to reduce the workloads of remote workers. Therefore, a direct-control teleoperation method does not satisfy this objective sufficiently or effectively improve productivity.

The ability of robots to understand and interpret semiotic (e.g., linguistic) information during interactions is vital. The DPS-CAs must comprehend user requests by leveraging their knowledge of specific environments and tasks. In a service environment, DPS-CAs must interpret instructions from remote users and possess local workplace knowledge. Remote robots must grasp the nuances of requests, which often involve complex physical and semantic operations. For instance, the robot must understand the locations of "the reception," the significance of "the baggage," the identities of "Emily" and "Jochen," and the action implied by "pass the content." Prior learning of this local semiotic knowledge, including the names of places and people, room locations, meanings of commonly used phrases, map layouts, and typical human behavioral patterns, is essential for effective service in such scenarios.

Taniguchi et al. (2021) highlighted the necessity of semiotically dynamic adaptations, both onsite and online, to local environments, enabling robots to perform DPS tasks when instructed via natural language from remote locations. They refer to this capability as semiotically adaptive cognition. Local semiotic knowledge pertains to the association between linguistic information and the experiences or activities in specific environments. This knowledge cannot be acquired before a robot is deployed in the workplace. Foundation models trained on extensive pre-existing datasets (Bommasani et al. 2021) have recently become popular and widely used (Ahn et al. 2022; Liang et al. 2023; Huang et al. 2022). However, the integration of foundation models alone is insufficient because it only allows the system to learn globally applicable knowledge. Local knowledge, such as the layout and appearance of specific rooms in an office, is not available in open datasets, and thus cannot be pre-learned.

Therefore, the development of practical machine learning methods that enable service robots to acquire local semiotic knowledge in situ and in real time is a significant area of research.

This section addresses the three key topics and related studies on the development of DPS-CAs. First, it discusses the integration of processes of local and global knowledge. Second, we explore active methods for acquiring local semantic knowledge. Third, it discusses the application of this knowledge to context-based language understanding, particularly in resolving exophoras. This section concludes the paper with a summary of the topics.

5.5.1 Integration of Common Sense and Local knowledge

The development of semiotically adaptive machine cognition that integrates local and global knowledge (common sense) presents a significant challenge. In the aforementioned example, the concept of "reception" depends on the specific environment but typically shares certain visual characteristics across different contexts. The former and the latter correspond to local and global knowledge, respectively. The effective merging of these two types of knowledge, ideally within a unified theoretical and software framework, is essential.

Recent advancements in deep and self-supervised learning have enabled the use of large language and foundation models (Bommasani et al. 2021). Knowledge independent of a specific environment is referred to as global knowledge or common sense. However, this alone is insufficient for effective functioning of DPS-CAs in specific environments. For instance, while many items fall under the category of "baggage," the specific piece of baggage referred to in an instruction is unique. The context is the key to identifying the intended object. This scenario exemplifies the need for local knowledge, in which the DPS-CA must be aware of the specific location of the target baggage.

The integration of global and local knowledge can be viewed as a form of transfer learning, which has been explored to enable highly accurate predictions using limited data. Transfer learning, or knowledge transfer, is pivotal for onsite and online learning using a service robot, allowing the model to adapt to new environments and acquire knowledge from minimal data. This approach facilitates swift acquisition of local knowledge. In service robotics, this integration can be modeled within the framework of a probabilistic generative model, considering the relationship between a latent variable and its prior. Hagiwara et al. (2021) developed a hierarchical Bayesian model for spatial concept formation that used global knowledge as a latent variable to expedite learning in new environments. Katsumata et al. (2020) applied generative adversarial networks (GANs) to model complex knowledge in semantic mapping across various home environments.

Hasegawa et al. (2023a) proposed a novel method that combines probabilistic logic with multimodal spatial concepts, enabling a robot to quickly learn the relationships between places and objects in a new environment. Figure 5.22 illustrates a cognitive system integrating global and local knowledge. They employed probabilistic logic to represent common sense knowledge of place-object relationships. In their experiments, the robot searched for daily objects, including those without predefined locations. The effectiveness of this method was demonstrated by the reduced number of place visits required for the robot to locate all objects. Furthermore, they enhanced their method by incorporating LLMs, such as GPT-4 (OpenAI 2023), which not only provides common sense but also aids DPS-CAs in planning sequential actions (Hasegawa et al. 2023b, c).

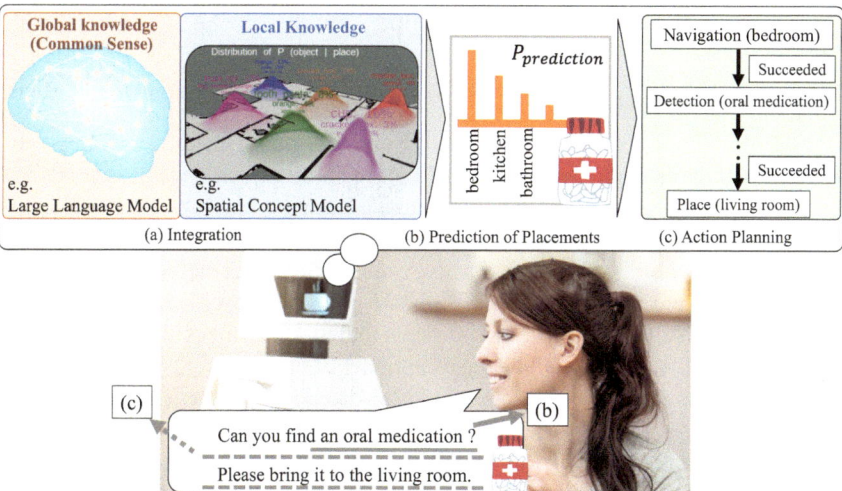

Fig. 5.22 Integration of global knowledge (common sense) and local knowledge, essential in specific task environments. By utilizing both types of knowledge, service robots can comprehend human linguistic instructions and execute specific tasks in varied environments

5.5.2 Active Semantic Mapping and Exploration in Local Environment

Acknowledging the importance of acquiring local knowledge through DPS-CAs, the effort required by users and developers to collect sufficient data from each local environment is significant. Therefore, methods that minimize the human effort in this learning process (ideally to zero) are highly desirable.

Active exploration is a key strategy for accelerating onsite and online adaptations while reducing user involvement. For a robot to quickly adapt to a new environment, it is crucial to acquire local knowledge, such as spatial concept formation (Taniguchi et al. 2016, 2017, 2020a, b), while actively exploring the environment. Many active learning methods utilize criteria that represent uncertainties, such as information gain, mutual information, and expected free energy (Friston et al. 2016).

For object identification and categorization, methods have been proposed for active perception and exploration, enabling robots to observe information selectively (Taniguchi et al. 2018; Yoshino et al. 2021). Active exploration in simultaneous localization and mapping (active SLAM) has been extensively studied (Stachniss 2005; Thrun et al. 2005; Chaplot et al. 2020). In active SLAM, the robot actively selects the next destination because it simultaneously generates a map and estimates its position within the environment.

In our project, we aimed to develop an active learning method for spatial concept formation (Taniguchi et al. 2023a). This approach allows robots to learn multimodal place categories and spatial lexicons through online learning, thereby enabling them to understand the semantic information of a place based on human linguistic instructions, even from remote locations. Such active decision-making can significantly alleviate the operational burden on humans.

Ishikawa et al. (2023) proposed an active semantic mapping method for household robots to facilitate rapid indoor adaptation and reduce the user burden (Fig. 5.23). They introduced Active-SpCoSLAM, a system that enables a robot to actively explore unexplored areas, using CLIP for image captioning to provide a flexible vocabulary as a substitute for human instructions. The robot's actions were determined by calculating the information gain by integrating both semantic and SLAM uncertainties. Their results demonstrated that this method allowed rapid coverage of the environment and efficient gathering of data for object discovery tasks, thus reducing user effort and enhancing the adaptability of the robot.

5.5.3 Understanding Situated Language

A key question is how the DPS-CA can leverage local knowledge and situate multimodal sensory information to comprehend the linguistic instructions of human users. A prime example is exophora resolution (Yu et al. 2019, 2021). In contrast to

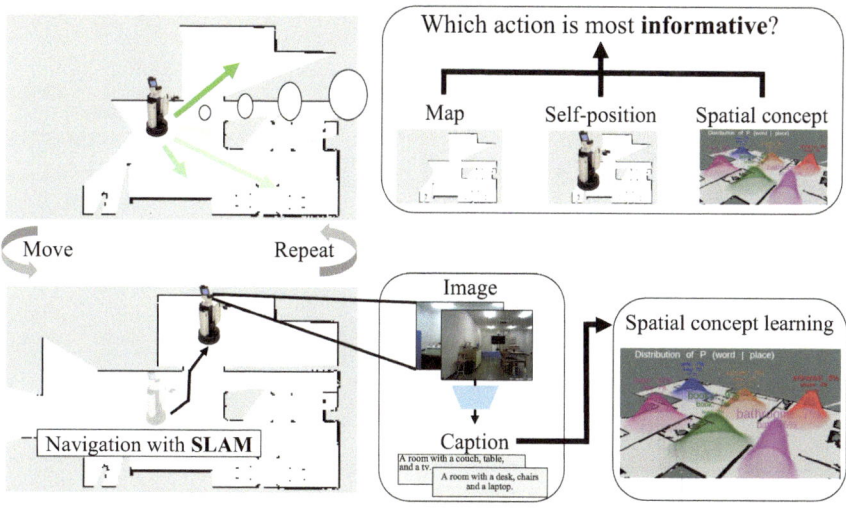

Fig. 5.23 Active semantic mapping service robots (Ishikawa et al. 2023). The robot autonomously explores the environment to build a map that assigns semantic meaning to each place and locates objects without human intervention

endophora resolution, which can be addressed using text data, exophora resolution requires external contextual information (Park and Kim 2023). For instance, if a user says, "Please take that bottle and bring it to her," DPS-CA needs to identify the specific bottle in the external world and the person to whom it should be brought. Several sources of information beyond text must be utilized in this process. The word "that" suggests that the bottle was located at a distance. If the user is pointing, the direction indicated may provide DPS-CA with additional clues for identifying the target object (Chen et al. 2021). Moreover, local knowledge about the placement of objects and their usage frequency can aid in decision-making. Naturally, the appearance of the object, or its "bottle-ness," its bottleneck, also plays a role.

Oyama et al. (2023) developed a method and system for DPS-CA that resolved ambiguity in language instructions containing demonstratives through exophora resolution using real-world multimodal information. The CA achieves this by incorporating three types of information: (1) likelihoods of object categories, (2) demonstratives, and (3) pointing gestures, along with knowledge of objects acquired from the robot's prior exploration of the environment (Fig. 5.24).

5.5.4 Summary

This section has focused on the challenges and strategies involved in acquiring semiotic knowledge in local environments for the development of Daily-Physical-Support

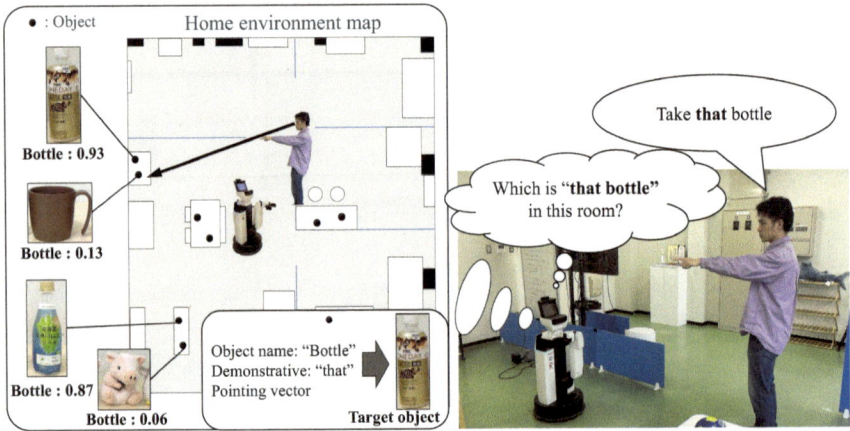

Fig. 5.24 Exophora resolution in linguistic instructions with a demonstrative in a service robotics scenario (Oyama et al. 2023). The robot identifies the object specified by a user through ambiguous verbal instructions and information gathered during prior exploration of the environment. Unlike endophora resolution, exophora resolution requires real-world situated multimodal information

Cybernetic Avatars (DPS-CAs). The discussion begins by identifying the motivations and challenges in this area of research. Subsequently, three critical topics were addressed, each pivotal to the advancement of DPS-CAs.

First, the integration of local and global knowledge was examined. This integration is essential for the effective functioning of DPS-CAs across diverse environments, allowing them to apply broad, universally applicable concepts, while adapting to specific local contexts.

Second, the significance of active exploration in acquiring local semantic knowledge was highlighted. Active semantic mapping and exploration enable DPS-CAs to learn about their environments autonomously, thereby reducing the need for intensive human involvement in data collection.

Third, the use of local knowledge in situated language understanding, particularly in exophora resolution, is explored. This functionality is crucial for DPS-CAs to interpret and respond accurately to human instructions, particularly when these instructions are context-dependent or contain ambiguous elements.

Although recent advancements in pretrained foundation models, especially LLMs (OpenAI 2023), are noteworthy, their isolated application falls short in actual service environments. A more comprehensive approach that involves the integration of cognitive modules and the amalgamation of global and local knowledge is required. This is not merely a technical requirement for performing tasks in a physical environment, but a vital aspect of effective situated language understanding and task execution by DPS-CAs.

In summary, the development of DPS-CAs, as presented in this section, represents a complex challenge that transcends the boundaries between traditional AI and robotics. The ongoing exploration and enhancement of these dimensions are key to the development of CAs, ushering in a new era of sophistication and functionality.

5.6 Tactile Sensing and Control for CA Manipulation

5.6.1 Overview of CA Manipulation

5.6.1.1 Expected Physical Support of CA

Physical work in the real world is an area in which CAs that operate remotely or autonomously are expected to play an active role. By enabling physical work at a distance, CAs will be able to not only communicate with people and monitor equipment, but also help with household chores and plant operations. The ability of many people to be physically involved in social activities through CAs will lead to both the maintenance of social infrastructure and a sense of fulfillment.

One of the most sophisticated examples of remote physical work is robotic medical surgery. In such cases, a person with specialized skills is required to move the remote body accurately with detailed knowledge of the worksite. On the other hand, there are simple but process-intensive physical tasks such as cleaning a messy room. What is required here is ease, speed, and the ability to accomplish the intended task with simple operations. If the operation is simple, multiple tasks can be performed in parallel.

It is important for CA to become popular in society that it be easy (and even fun) to handle without special training and without stress, and this research is developing technology to improve CA's hand skills.

The key to this research is the implementation of sensing capabilities in CA's hands and the generation of primitive (rational and reflexive) movements based on the acquired information.

For a remotely operated CA, the operator is equivalent to a higher-level planner of the system. If the most difficult operations involving physical contact are supported by the tactile-sensing function of the CA, (1) the operator will not have to concentrate on each task. (2) The operator does not have to stare at the monitor and become tired. (3) Multiple operations (such as handling two arms for different purposes) can be performed simultaneously. In addition, (4) even if there is a communication delay between the operator and CA, damage due to collision can be avoided.

5.6.1.2 Sensing Function for the Hand of Physical Support CA

Here, we consider the sensing functions necessary for physical support CA. The mechanical reproduction of human tactile-sensing functions has long been an active research area in robotics. Tactile sensors in robotics capture mechanical actions, such as the occurrence and loss of contact, force, and vibration, as changes in electrical signals. Various structures have been proposed, including resistive, capacitive, and optical types, which do not need to be based on the same principle as the human tactile sense. It is important to note that the structure and output information of the sensor must be compatible with the purpose of its use, operating environment, structure, and movement of the body on which it is mounted.

Although the body structures of physical support CAs are diverse, we consider cooperative robots equipped with multifingered hands and mobile manipulators. They are constructed of rigid materials and driven by electromagnetic motors through reduction gears. Although they provide the speed, accuracy, and payload required for work, they lack the softness of body structure. For these CAs to safely and physically interact with the world while demonstrating their motor skills, the adoption of proximity sensation as a spatially extended tactile sensation is appropriate.

Proximity sensing uses optical reflection and capacitance changes to provide tactile-like information about the presence of nearby objects in a non-contact manner (Navarro et al. 2022). The CA's hand equipped with proximity sensors begins to detect the object to be grasped once it is approached by the operator. Trajectory correction based on proximity sensing allows the hand pose of the CA to become appropriate for grasping an object with a simple operation.

This section outlines the techniques developed for CA hand manipulation. Section 5.6.2 describes the structure and characteristics of the proximity sensor. The sensor can find an object, follow the hand to the object, and provide useful information until the hand grasps the object. Section 5.6.3 describes the grasp stability prediction based on proximity sensing. It describes a method for the real-time selection of contact points to stably constrain an object, even if an accurate shape model of the object is not provided in advance. Section 5.6.4 describes the semi-autonomous teleoperation system that incorporates proximity sensing. This describes how the system realizes safe and secure object grasping while reflecting the intent of the operator.

5.6.2 Proximity Sensor for CA's Hand

5.6.2.1 Principle and Structure

A proximity sensor uses the reflection of infrared light to detect the surfaces of nearby objects. It comprises infrared light emitters (LEDs), receivers (phototransistors), and electronic circuits that perform computation and amplification.

To integrate the sensor into the hemispherical fingertip surface, its electronic components were mounted on a dedicated flexible substrate and positioned three-dimensionally using positioning components fabricated using a 3D printer. As shown in Fig. 5.25, it has a layered structure inside the fingertip with a radius of 20 mm: from the outside, a transparent coating layer, an LED layer, and a phototransistor layer. The layered structure allows spatially dense optical element placement and

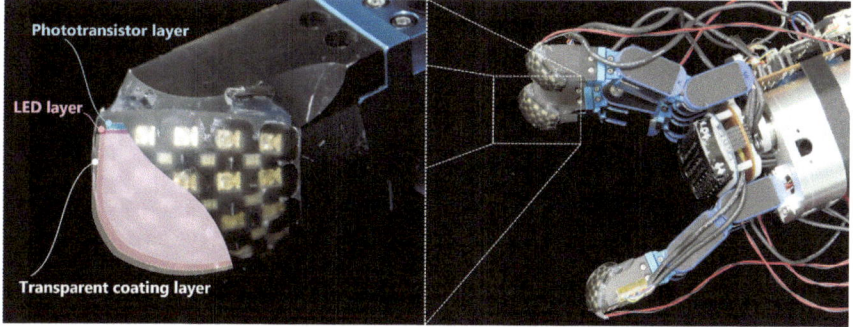

Fig. 5.25 Proximity sensor embedded on the fingertips of CA's hand

prevents direct incidence from the LEDs to the phototransistors. The transparent coating protected the sensor surface and increased the friction coefficient with little interference from light transmission.

5.6.2.2 Specifications

The specifications and functions of the fingertip proximity sensor are presented in Table 5.4 and Fig. 5.26.

The sensor can estimate the distance d_m and orientation (pitch angle θ_p and roll angle θ_r) of the nearest local surface from the fingertip at 1000 Hz. This information was extracted from the photocurrent distribution of the phototransistors by analog computation and was output as an analog voltage. The detection range is approximately $0 \leq d_m \leq 30$ [mm] and $-40 \leq \theta_p, \theta_r \leq 40$ [°].

The sensor can also estimate the approximate curvature ρ of the local surface at 500 Hz. This information was obtained from the raw value of the photocurrent distribution via serial SPI communication. The range that can be estimated is $-0.033 \leq \rho \leq 0.033$ [mm^{-1}] (i.e., convex and concave surfaces with radii greater than 30 mm).

Table 5.4 Specifications of the fingertip proximity sensor

Type of proximity sensing	Optical reflected light intensity (O-RLI)
Available information	Distance, orientation, local curvature, and point cloud
Range (distance)	0–30 mm
Range (orientation)	− 40–40°
Range (curvature)	− 0.033–0.033 mm^{-1}
Sampling rate	1000 Hz (for distance and orientation)
	500 Hz (for curvature)

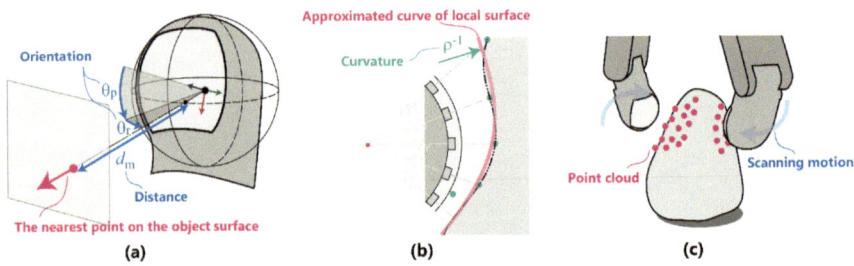

Fig. 5.26 Proximity sensing function: **a** distance and orientation estimation, **b** curvature estimation, and **c** point cloud generation

In addition, the sensor can generate a point cloud of an object by non-contact scanning. This is achieved by combining the sensing described above: distance estimation for fingertip feedback control and local shape estimation for 3D shape reconstruction.

The characteristics of each sensing function are described in Suzuki (2021, 2022, 2023).

5.6.2.3 Summary and Future Perspectives on Sensing

From the viewpoint of a CA that performs physical work through remote control, the features of the proximity-sensing function can be summarized as follows:

- Objects can be detected several tens of millimeters before the fingertip touches the object, thus avoiding unintended collisions owing to operator errors.
- For operators with blind spots, the system can provide information on the environment and location of the objects that complement them.
- When approaching an object and determining its grasping pose, the position of the future contact point and local shape can be predicted. This helps to determine whether it is safe to perform a grasping action (see Sect. 5.6.3).
- Because the detection range includes a distance $d_m = 0$, it also functions as a tactile sense to determine the occurrence and disappearance of contact. In other words, it can determine whether contact is correctly maintained during object grasping.
- A wider range of proximity sensors can also be mounted on the palm to support the maneuvers of the operator toward an object.

5.6.3 Operational Assistance Based on Proximity Sensing

5.6.3.1 Strategy for Operational Assistance

For operators who perform remote manipulations with the CA's hand, it is difficult to determine the optimal hand pose for grasping a target object. For example, when

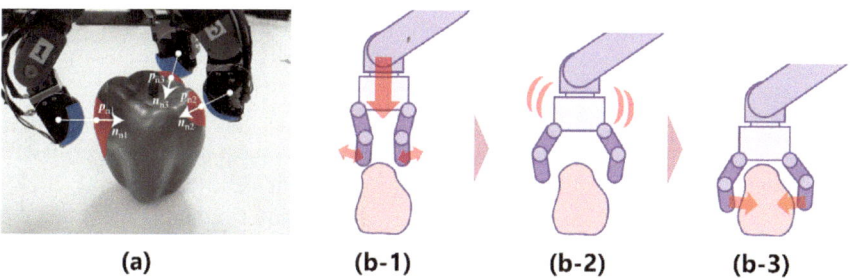

Fig. 5.27 Operational assistance based on grasp stability prediction: **a** prediction of future contact points and their normal vectors, and **b** strategy of the assisted grasp with **b-1** approaching with avoiding collision, **b-2** adjusting of the hand position, and **b-3** making contacts at the pose with high mechanical stability

looking down at an object from the head of the CA where a camera is installed, the distance to the object is not intuitively recognizable. However, this approach tends to be too short to avoid collisions. Consequently, the hand cannot apply the force in the direction required to lift the object.

To enable safe and easy maneuvering by the operator, it is desirable to provide operational support such that the hand pose that realizes stable grasping is selected naturally and rationally during the approach process. To achieve this based on the proximity-sensing function, this research proposes a non-contact, real-time grasp stability prediction method (Fig. 5.27).

Grasp stability is a metric of the degree to which the equilibrium of forces and moments acting on an object, or the constraint of object motion, is achieved mechanically or geometrically. The proposed method is based on the mechanical stability.

To evaluate the equilibrium of the forces and moments acting on an object, information regarding the positions of all contact points between the object and the fingertips and the orientations of their tangent planes is required.

In general, if the shape model of an object is not provided in advance, this information is only available after contact has been made. Therefore, re-grasping must be performed until a stable grasp is obtained. On the other hand, the developed proximity sensing enables the estimation of the shape of the object surface before contact. Therefore, at each time step of the approach, it is possible to predict "whether starting the grasp now would produce a good result." This allowed the selection of a combination of contact positions with high mechanical stability, even if the object was being grasped for the first time.

5.6.3.2   Primitive Motions Based on Grasp Stability Prediction

(1) **Grasp Stability Prediction**

While the fingertip is in motion in the vicinity of an object, the system always predicts the position of a point on the surface of the object that the fingertip will contact and the direction perpendicular to the tangent plane (Suzuki et al. 2022).

Next, the grasp wrench space (GWS) generated by combining the predicted contact points is calculated. A wrench is a set of forces and moments, and the GWS is the set of all wrenches that can be applied to an object via contact points. The GWS is calculated as the Minkowski sum of the wrench space that can be generated by the forces at each contact point.

Finally, the radius of the largest sphere (6D) contained in the GWS centered at the origin is determined as the evaluated value of grasp stability. The larger this metric is, the more it can withstand an external wrench in any direction.

(2) **Motion Control Primitives**

Keeping Distance and Posture: The fundamental and consistently used motion control in the proposed method is a feedback control that maintains the fingertip at a constant distance and posture from the surface of nearby objects. Here, the torques at the finger joints are determined such that the distance is a constant value, and the posture is such that the center of the fingertip is directly opposite the surface of the nearby object (Fig. 5.28a).

Approaching/Contacting: We also introduced a motion control for approaching the fingertip to the target contact point and applying a contact force, which should be generated when the grasp stability metric is high. The torque for the approaching motion is determined to provide the target velocity of the fingertip to the nearest point (Fig. 5.28b). The torque for the contact motion is determined to generate the target force at the contact point on the fingertip (Fig. 5.28c).

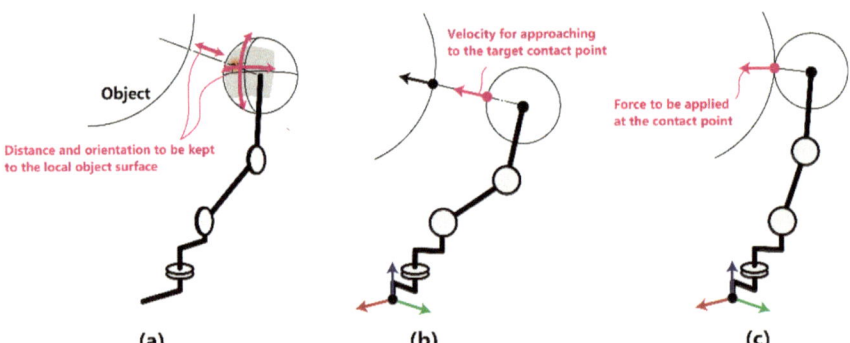

Fig. 5.28 Motion control primitives: **a** keeping distance and orientation, **b** approaching, and **c** contacting

5.6.3.3   Summary and Future Perspectives on Operational Assistance

Considering the difficulties in remote manipulation by the CA, the key points of operational assistance are as follows:

- Primary decision-making, such as when and what to approach and from what direction, is reflected by the operator with semi-autonomous assistance for difficult to maneuver parts that involve contact.
- The motion control for operational support is implemented as a primitive for purposeful local feedback.
- All primitives are in a torque-control-based form, which can be superimposed on the operator's operational input (Sect. 5.6.4.2 (4)).

5.6.4   Integration with Teleoperation System

5.6.4.1   Essentials of Integrating Teleoperation and Assistance

The target task is to grasp an object randomly placed in front of a teleoperated CA. The primary player is the operator, and the operational support should generate additive movements to improve the success rate of the task while adhering to the operator's intentions.

The key points in the integration of teleoperation and assistance are (Fig. 5.29).

- The trajectory of the hand and opening/closing of the fingers are proactively determined by the operator. These are inputted via the interface for control, with the former in the form of velocity of the wrist and the latter in the form of torque of the finger joints.
- The low-level controller for assistance interprets the operator's intent from the operational input while estimating the state between the hand and the object from the proximity sensory information. It then applies corrections to the velocity of the wrist and the torque of the finger joints.
- The operator can switch between the activation and deactivation of assistance at any time.

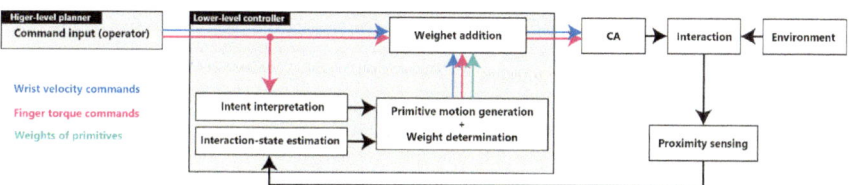

Fig. 5.29 Pipeline of the teleoperation of CA manipulation integrated with proximity-based operational assistance

5.6.4.2 Teleoperation System

(1) **Overall Configuration**

The overall configuration of the teleoperation system built for manipulation of the CA is shown in Fig. 5.30. The operator uses a control pad as an interface to operate the CA remotely. The operation input is sent to a remote computer via a VPN server. The body of the CA consists of a robotic arm with a three-fingered hand and proximity sensors mounted on the fingertips and palm. Additionally, a web camera is installed at a position corresponding to the head and transmits images in front of the CA to the operator. Computers and controllers at the remote site process the operational inputs and sensor information to control the motion of the robot arm and hand.

Note that time is required from the operator's operation until the corresponding motion of the CA can be confirmed in the image. The time lag depends on the experimental environment and is approximately 1 s in the worst cases.

(2) **Operational Input via Control Pad**

Only two analog sticks and two buttons (one digital and the other analog) are used as operational inputs. Two sticks (left and right) are used to generate the velocity command values for the CA wrist in three directions. Command values are provided in the wrist frame.

While pressing the digital button, the velocity command is converted to an angular velocity command around the tool center point (TCP). The position of the TCP is set on the central axis of the hand at the average position of the three fingertips.

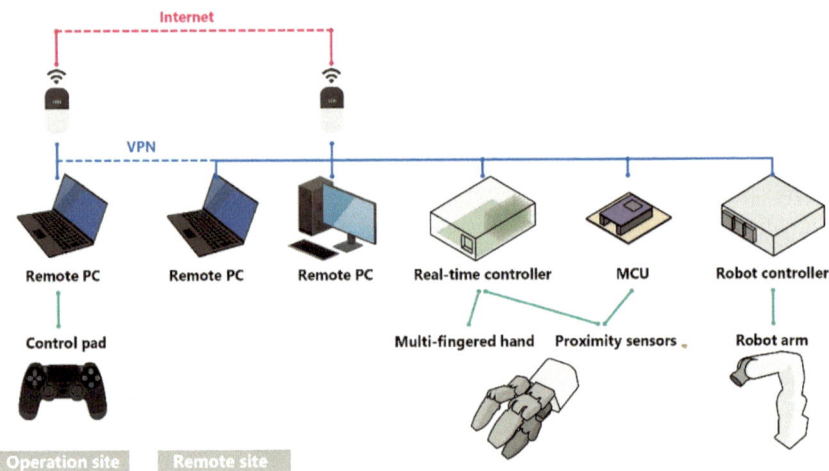

Fig. 5.30 Remote operation system for CA manipulation

The analog button generates a torque command value for the finger joint in the flexion direction. The magnitude of the torque is determined to be proportional to the amount of indentation.

This operation can be performed within a few minutes because of practice thanks to its simplicity, although individual differences can be observed.

(3) **Integrated Control of Wrist Velocity**

Two types of assistance were provided for the command values of the wrist velocity based on the operational input.

One is the correction of the hand position using the palm proximity sensor. The sensor is placed along the central axis of the hand. When an object deviates from this axis, velocity is generated to move the hand in the direction of the object. In addition, when the distance from the palm to the object is too small, velocity is generated to move the hand away from the object.

The second is the correction of hand position and posture using the fingertip proximity sensors. As the three fingertips approach the object, they independently adjust their position and posture with the local surface of the object using the control described in Sect. 5.6.3.2 (1). At this time, velocity is generated to move the wrist such that the flexion angles of the finger joints are balanced.

With this assistance, the operator can direct the hand position and posture toward the target object by easy operation.

(4) **Integrated Control of Finger Joint Torque**

For the command value of the finger joint torque, three types of assistance were applied: maintaining distance and posture, approaching, and making contact (Sect. 5.6.3.2 (2)).

Initially, only the motion for maintaining distance and posture is active. This assistance prevents unintended fingertip collisions during hand movements. In addition, if the operator increases the finger joint torque in an inappropriate hand pose (i.e., the evaluated value of the grasp stability is low), the assistance system resists this and does not lead to contact.

As the evaluated value of grasp stability increases, the repulsion is weakened, and the fingertip is brought into contact. Simultaneously, to generate the proper approach trajectory and contact force, the lower-level controller weakens the torque of the operation input and produces motion controls for approaching and contacting.

That is, when assistance is activated, the torque command value by the operator is interpreted as the strength of the grasping intention by the lower-level controller, which finally determines the torque by the weighted addition of the operational input and proximity-based assistance. An easy and practical operation that applies assistance involves providing the maximum torque command from the beginning and pushing the hand toward the object. The fingertip moves along the surface of the object, and when it reaches a mechanically stable pose, the grasp is automatically executed.

5.6.4.3 Experiment and Discussion

First, the teleoperation system was evaluated in a simple laboratory environment. There is a slight delay caused by communication, but the operator cannot see the CA directly. With this assistance, hand pose adjustment by the operator was rarely required, and grasping was successful in the first approach in many cases. The time required to lift an object was reduced by about half with the assistance.

The remote control system was also evaluated in a demonstration conducted at a remote site (over 200 km away). In this case, the delay caused by the communication cannot be ignored. With this assistance, the automatic adjustment of the hand position by the palm proximity sensor was effective. When a dangerous operation input that brought the hand too close to the object was provided owing to a delay, the hand was able to repel the input, avoid a collision, and contribute to the safe grasping of the object.

Although the above is a simple experiment, the findings and considerations from this study are as follows:

- Individual differences in proficiency were observed when operating with the control pad. Some people can perform grasping tasks smoothly once they become accustomed to them. However, the option of using a more intuitive interface is also necessary. An important advantage of semi-automated assistance is that it reduces the number of channels required for operational input. This makes it suitable for integration with motion capture systems and, in the future, with brain-machine interfaces.
- Operators should understand how semi-automated assistance works visually in advance. This finding was extracted from operators' impressions during a remote operation demonstration. In other words, knowing the performance of collision avoidance and adaptation by assistance in advance enables operators to maneuver the CA with greater security and sense of ownership.

5.7 Harmonized Control of Autonomous CAs Based on Voluntary BMIs

5.7.1 Brain–Machine Interfaces

BMIs are technologies that connect the brain to external devices. Implantable BMIs involve neurosurgical craniotomy to place electrodes in the skull, wirelessly transmitting precise brainwaves measured from within the skull to an external computer via a wireless implantable device, and using AI to decode these brainwaves to estimate "what the person is trying to do" and operate external devices such as robotic arms. If CAs can be operated using BMI, people with physical disabilities can operate CAs simply by thinking, which can restore impaired functions. This is expected to be an innovative technology for the social reintegration of people with physical disabilities.

5.7.2 Significance of Harmonized Control Between CAs and BMIs

We have been studying robot arm control using BMI for many years. As BMIs are controlled based on the decoding results of brain signals, they excel in terms of voluntariness. However, fine control is not always excellent and tends to fail in grasping objects, such as a claw machine. However, robots excel in accurate autonomous control, and if equipped with visual sensors, they can recognize objects and autonomously grasp them accurately and reliably. Therefore, accurate control of the CA can be achieved by performing a control that exploits the advantages of BMI, which excels in voluntariness, and CA, which excels in autonomy. Based on this idea, we have been developing a technology to harmoniously control the BMI and CA. In this section, we provide an overview of the harmonious control of BMI and CA, while explaining the surrounding technologies.

5.7.3 Implantable Brain–Machine Interfaces and Their Target Diseases

In diseases such as amyotrophic lateral sclerosis (ALS), spinal cord injury, amputation, and post-stroke paralysis, the quality of life of patients is significantly impaired owing to severe physical disabilities such as quadriplegia and speech disorders. Implantable BMI is expected to be a medical technology for reconstructing physical functions severely impaired due to these diseases. Among them, ALS is considered to be the disease that should be first applied to functional reconstruction by an implantable BMI because of its extremely severe physical disability.

5.7.3.1 ALS

ALS is a disease in which motor neurons selectively and gradually disappear, making movement and speaking difficult. It is a rare disease with an annual incidence of 1–2 per 100,000 people, and the number of patients worldwide is 100,000–300,000. Famous patients include baseball player Lou Gehrig (1903–1941) and physicist Dr. Stephen Hawking (1942–2018). As ALS becomes more severe, patients communicate with those around them by pressing switches with their fingers or by forehead muscle contractions that move only slightly. Usually, the last body part to move is the muscles around the eyes; however, when they cannot move, they enter a completely locked-in state, a state where they cannot communicate with others. Most patients require an artificial respiratory system because they cannot breathe within approximately 3–5 years of disease onset. ALS is a disease that gradually progresses, but functions other than motor function are relatively preserved. Therefore, patients with ALS can understand what people are talking about, and the sensation of pain remains. However, it is difficult to convey these intentions to others; therefore, it is a very stressful disease for both patients and their families who care for them.

5.7.3.2 The Dilemma Faced by Patients with ALS

The quality of life of patients with ALS is significantly impaired due to severe physical disabilities. To understand this situation, we conducted a nationwide questionnaire survey in collaboration with the Japan ALS Association. We sent questionnaires to 1640 patients and received responses from 468 (Kageyama et al. 2020). The following results were obtained:

(1) Desire to live

Most patients with ALS lose hope of living at least once, and there is an incidence of assisted suicide. However, the questionnaire revealed that most patients had a very high desire to live.

(2) Anxiety about being locked in

Most patients with ALS are very worried about becoming locked in.

(3) Anxiety about burdening caregivers

ALS is designated an intractable disease, and patients with ALS in Japan can receive various healthcare services, including home-visit care and medical services, with small monthly payments. Nevertheless, many patients worried about imposing a great burden of care on their families. This is a dilemma between the desire to live and desperation due to anxiety about being locked in and the caregivers' burdens. As a result, when the disease progresses and breathing becomes difficult, only approximately 20% of patients choose to undergo tracheostomy or artificial respiration (invasive artificial respiration), which is the best way to prolong life. Whenever patients want to live alone, it is difficult for them to choose to live alone. This is a social pain.

5.7.3.3 Many Patients with ALS Hope to Use Implantable BMI

In the questionnaire survey, we also asked the question, "If you can communicate with others and manipulate robots to do daily various things, would you undergo a surgical procedure to implant a BMI device?" (Fig. 5.31). Consequently, half the patients hoped to use their BMI. Among patients who have already undergone (or plan to undergo) invasive artificial respiration, about 70% of patients hoped to use the BMI. Even in patients who were hesitant about undergoing invasive artificial respiration and those who had decided not to undergo artificial respiration, approximately 20% of patients hoped to use BMI. Considering that this judgment changes decisions related to life and death, 20% is a large number, and implantable BMIs can be considered a significant technology that can provide hope for patients with ALS.

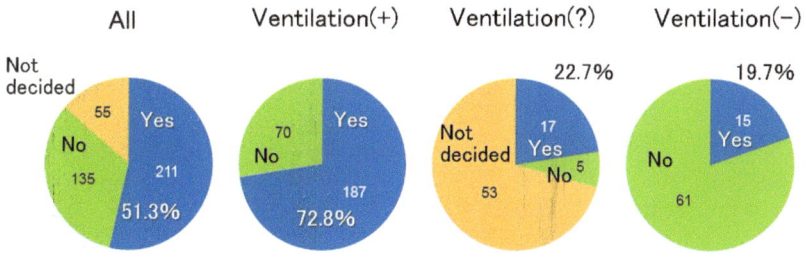

Fig. 5.31 Desire to use implantable BMI in patients with ALS

5.7.4 Core Technology of Implantable BMI

Two core technologies are available for implantable BMI. One is the technology to implant electrodes within the skull and measure accurate brain signals, and the other is the technology to decode brain signals using AI.

5.7.4.1 Core Technology 1: Implantable Device

The most important technology in implantable BMIs is the device technology, which can accurately measure brain signals. If brain signals can be measured accurately, AI can decode their meaning. The CA can be manipulated accurately based on the decoding results. However, because the device is implanted within the body, it is a difficult technology. Neural decoding technology can be developed on an annual basis, whereas the development of implantable devices requires a long period of time, on the order of decades. AI is evolving rapidly and used worldwide. Many of the latest AIs are open to use, and even the latest AI technology is easy to use. However, the development of implantable devices is time consuming. Because it is implanted in the body, it requires high reliability and safety, development costs a huge amount of money, the hurdle for approval as a medical device is the highest, it damages the image of the company, and liability for compensation in the event of trouble is large; therefore, many companies hesitate to enter the implantable device market. This tendency is particularly strong in Japanese companies. Cardiac pacemakers are good examples of the special characteristics of medical technology. Cardiac pacemakers have hardly changed in function or size for as long as 50 years since they were put into practical use in the 1970s. In other words, once an implantable device such as a cardiac pacemaker is approved, it can be used for a long period. Electrical stimulation technology used in cardiac pacemakers is basically a technology used to treat the body by stimulating it with electricity, but this electrical stimulation technology is now also applied not only to heart diseases but also to neurological disorders such as

Parkinson's disease and intractable epilepsy. Therefore, it is expected that once the technology of implantable BMI is established, it will become a technology that can be used for a long period of time, including expanded clinical applications for other diseases.

Depending on their location, various types of electrodes are used in implantable BMIs, such as intracranial needle, cortical, and intravascular electrodes. Each electrode is briefly described as follows.

Intracranial Needle Electrodes

Intracranial needle electrodes involve the insertion of several small needle-like electrodes into the brain. It was originally used in the field of animal neuroscience. Research using monkeys has shown that neurons in the motor cortex respond differently depending on the direction in which the monkey tries to move its hand (Georgopoulos et al. 1986). This property is known as directional tuning. This property implies that it is possible to estimate the direction in which the hand is moved and by which neurons are activated. Using this property, the direction of the hand movement can be reliably estimated. This method is now widely used for decoding movements from brain signals, particularly in the United States. "Braingate" is the largest BMI research group (Rubin et al. 2023).

Intracranial needle electrodes can accurately measure brain signals; however, they can cause brain damage. This damage causes chronic inflammatory reactions and problems with gradual performance degradation over time.

In 2017, Elon Musk, the founder of Tesla Motors, established the startup company Neuralink (Musk 2019). He invested over 100 million dollars of his own money to develop a high-performance BMI. More than 500 million dollars has been invested in this field, and research and development are rapidly advancing. In animal experiments using monkeys, it has been reported that monkeys can play a ping pong game just by thinking about movement, and they have recently obtained approval from the FDA to start a clinical trial. The BMI that Musk is developing is a type of intracranial needle electrode that uses a special robot to insert as many as 1000 ultrathin thread-like electrodes into the cortex, similar to a sewing machine. However, because it is a needle type, it has the disadvantage of damaging the brain. In addition, surgery is complicated and requires a dedicated precision robot, raising concerns about cost. While Musk is working on invasive BMI, Facebook was developing a non-invasive wearable BMI using an entire research building, but was already withdrawn. They judged that non-invasive BMI for functional restoration is not realistic for the time being because it is currently impossible to accurately measure brain signals without noise contamination.

Intravascular Electrodes

In 2021, a technology was developed that placed a mesh-like electrode in a large vein called the superior sagittal sinus in the parietal region of the skull by passing a thin tube, called a microcatheter, into blood vessels (Oxley et al. 2016). The device, called Stentrode, does not require craniotomy; therefore, FDA approval for clinical trials was obtained relatively quickly, and trials have begun. It seems less invasive

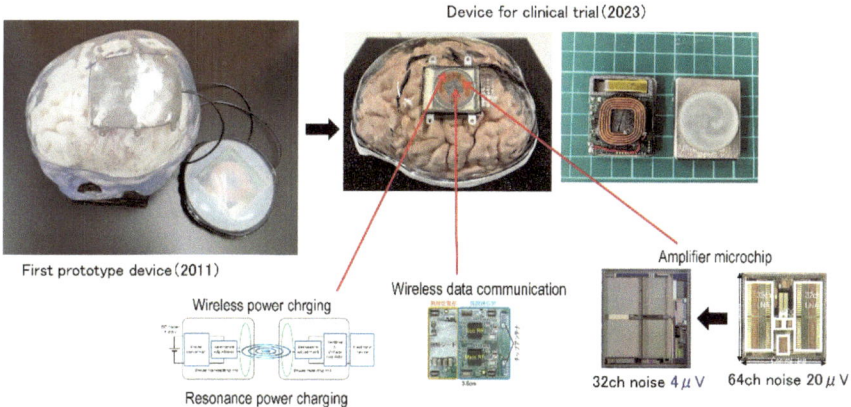

Fig. 5.32 Implantable BMI device

because it does not require a craniotomy. However, there is a risk of thrombus formation; therefore, the overall degree of invasiveness remains controversial. In addition, because the electrode can only be placed inside a blood vessel and is difficult to place in the optimal cortical area, there are disadvantages such as difficulty in achieving high performance compared with other invasive methods, despite its invasiveness.

Intracranial Cortical Electrodes
We have been developing implantable BMI devices using an intracranial electrode sheet placed on the brain surface for approximately 15 years (Hirata et al. 2011; Matsushita et al. 2018; Yan et al. 2020, 2022, 2023a, b). These electrode sheets are used in neurosurgery to identify epileptic foci based on their accurate recording performance. We modified the electrodes to increase the inter-electrode spacing to a few millimeters. The idea was to achieve a high-performance BMI by accurately measuring brain signals using this intracranial cortical electrode. The recording device must be miniaturized and implanted within the body. Therefore, we developed sufficiently small electronic modules including a wireless power supply module using resonant power supply technology, a wireless digital communication module, and an integrated amplifier microchip that can amplify brain waves at the 100-channel level with a high signal-to-noise ratio. We repeatedly evaluated these through animal experiments and developed an implantable BMI device of approximately 3 cm × 1 cm × 1 cm (Fig. 5.32). This sheet-like intracranial cortical electrode requires craniotomy but does not damage the brain itself, and can be used stably for a long time, which is an advantage over needle electrodes.

5.7.4.2 Core Technology 2: Neural Decoding
Decoding of Neuronal Activity
In 2008, the University of Pittsburgh reported BMI enabled 3-dimensional robot arm control using nonhuman primate experiments (Velliste et al. 2008). After a

microneedle electrode array was inserted into the motor cortex of a monkey, the monkey was trained to control the robot arm using only the decoded brain signals. After several months of training, the monkey could operate the robot arm freely, grab food, bring it to its mouth, and eat it. They had already discovered about 20 years ago that when a monkey tries to move its hand in various directions, specific neurons in the motor cortex react differently depending on the direction (Georgopoulos et al. 1986). Using this property, it is possible to estimate the direction in which the hand is moved and the neurons that are activated. Therefore, they inserted several microneedle electrodes into the monkey's motor cortex, measured the activity of neurons, estimated the direction in which the hand was moving, and moved the robot arm in that direction. Four years later, in 2012, they reported the results of a clinical research on 3D robot control based on BMI in humans (Collinger et al. 2013). A 52-year-old female patient with quadriplegia due to spinocerebellar degeneration participated in this study. Many microneedle electrodes were inserted into the motor cortex, the direction of the upper limb movement was estimated from the brain signals, and the patient was able to control the robot arm by thinking, bringing chocolate to her mouth, and eating it. In the same year, Harvard University reported that a patient with quadriplegia due to a brainstem stroke was able to drink juice by controlling a robotic arm with only brain signals (Hochberg et al. 2012).

Decoding of Intracranial Brain Waves
In neurosurgery, it is important to ensure that surgeries can be performed safely without impairing the function of the language and motor areas. In addition, it is important to accurately identify epilepsy foci before surgery. For this purpose, a craniotomy is performed, and a 4×5 cm electrode sheet is placed on the surface of the brain to accurately investigate epilepsy foci and functional localization as a preoperative examination. Using this intracranial electrode, we examined brain activity during various movements and found that subtle neural activities in the high-frequency band (60–200 Hz) could be measured, and that they accurately indicated the cortical areas corresponding to their functions (Yanagisawa et al. 2011, 2012). These high-frequency activities are difficult to measure using non-invasive scalp electrodes. It was also found that, to instantly and accurately infer the movements that a person is going to perform, it is essential to measure these high-frequency brain activities. In other words, it has become clear that in order to achieve a practical BMI for the purpose of functional reconstruction, it is necessary to implant electrodes within the skull.

Neural Decoding Using AI
Here, we describe how an AI automatically distinguishes brain signals using a two-dimensional model for ease of understanding (Fig. 5.33). The brain performs different functions depending on its location, and specific functions are localized to specific brain regions. For example, when a person grasps and opens his/her hand, there may be differences in the areas where the brain is activated. If the brain signals are measured using one electrode, there may be differences in the measured brain signals between the two movements. If we describe brain signal patterns using a

two-dimensional plane by corresponding brain signals from an electrode and an evaluation function to the *x*- and *y*-axes, respectively, we can plot each movement data on this two-dimensional plane. The AI can learn a sufficiently large amount of data for both movements and find a straight line ($y = a \times x + b$) that becomes the boundary between the two movements. For example, a support vector machine: a standard machine learning algorithm sets this boundary line to maximize the distance between two movements. As a result, the AI can automatically distinguish whether to grasp or open a hand based on this straight line. Similarly, more axes could be defined for more electrodes. Therefore, a straight line ($y = ax + b$) is represented by matrices **a** and **b**. This AI concept is referred of AI called machine learning. In fact, recent AI classifies on a scale of "tens of thousands of dimensions layered approximately ten times", which is called deep learning. Therefore, the AI must learn a large amount of data, but it can achieve high performance.

Decoding of Natural Speech

In 2019, the University of California, San Francisco, reported successfully inferring a person's natural speech content from intracranial brainwaves only (Anumanchipalli et al. 2019). When a person tries to speak, brain commands related to the movement of the mouth, tongue, jaw, and pharynx during speech are issued from the corresponding motor cortex. If we decode these brain signals, we can infer mouth movement. Here, technology to estimate speech content from mouth movement has already been established. Therefore, by combining the two methods, we can infer speech content only from brain signals. This research is significant in that it shows that not only the movement function of the limbs, but also speech function can be practically reconstructed using BMI technology.

Decoding of Handwriting

In 2021, Stanford University reported a technology that allows intended characters to be displayed on a screen when human subjects imagine the action of writing the characters they want to write (Willett et al. 2021). The AI inferred the alphabet that the subject intended to write from the patient's neuronal signals. The subjects were able to spell more than 110 characters per minute with an accuracy of over 90%.

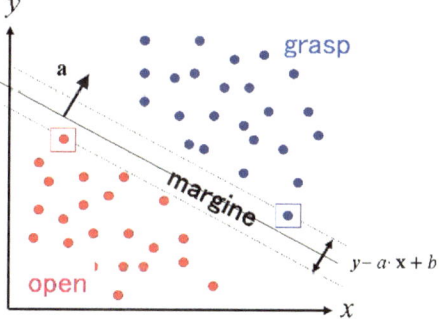

Fig. 5.33 How AI automatically distinguishes brain signals?

This research is also significant because it shows that writing can be performed by BMIs at a practical level.

5.7.5 Harmonized Control of Autonomous CA and Voluntary BMI

As mentioned in Sect. 5.4.7.2, by harmonizing the autonomous control of the CA and the voluntary control of BMI, it is possible to enhance the overall performance of CA control by BMI. We define the following formula to represent this harmonized control:

$$v_{cmd} = w_p \times v_p + (1 - w_p) \times v_a$$

v_{cmd} velocity of harmonized control
v_p velocity of voluntary BMI control
v_a velocity of the autonomous avatar control
w_p weight.

By compensating for the voluntary BMI control with autonomous avatar control, we aim to optimize the overall performance through harmonized control.

5.7.5.1 Harmonized Control of CA's Upper Limb Function

Based on the aforementioned concept, we developed a harmonized control method for upper limb function of the CA. We designed a BMI manipulation task using the CA arm to grab and move PET bottles. In this task, arm control was performed based on BMI decoding until it approached the PET bottle. When the arm approaches the PET bottle and the operator's intention to grasp the bottle is detected by neural decoding, fine controls, such as the fine approaching movement of the arm to the bottle, the fine grasping movement of the hand, the fine opening movement of the hand, and the grip, are autonomously controlled by the arm of the CA based on the results of AI-based environmental recognition using vision sensors (Fig. 5.34). This study showed that the harmonized control of CA and BMI enables the reconstruction of practical upper limb function.

5.7.5.2 Harmonized Control of CA-Assisted Conversation

In addition to upper limb function, harmonized control of the CA and BMI may enhance the overall conversation performance. For the implantable BMIs that we developed, a flick operation is currently being developed (Fig. 5.34). However, conversation using the flick operation is still slower than that using natural speech.

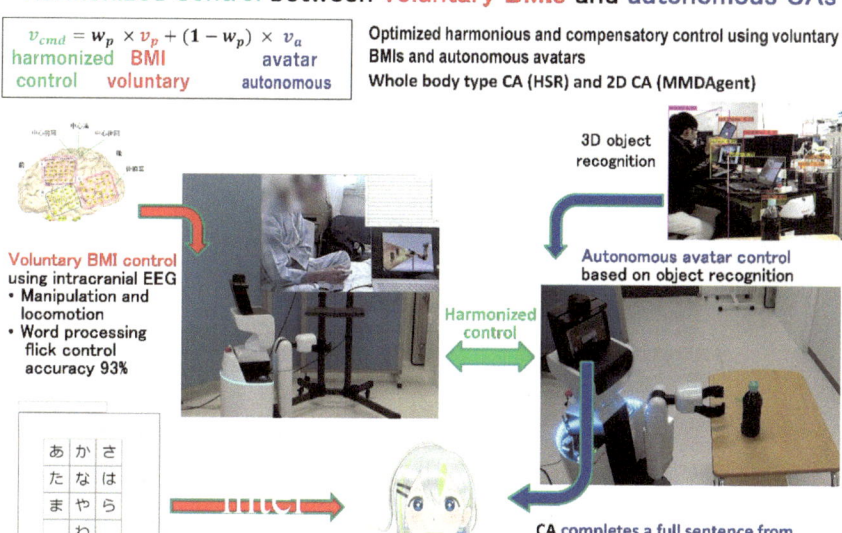

Harmonized control between **voluntary BMIs** and **autonomous CAs**

$$v_{cmd} = w_p \times v_p + (1 - w_p) \times v_a$$

harmonized   BMI   avatar
control   voluntary   autonomous

Optimized harmonious and compensatory control using voluntary BMIs and autonomous avatars
Whole body type CA (HSR) and 2D CA (MMDAgent)

3D object recognition

Voluntary BMI control
using intracranial EEG
• Manipulation and locomotion
• Word processing flick control accuracy 93%

Harmonized control

Autonomous avatar control based on object recognition

あ	か	さ
た	な	は
ま	や	ら
	わ	、

CA completes a full sentence from keywords and autonomously express emotion from the sentence

Fig. 5.34 Harmonized control of autonomous CA and voluntary BMI

Therefore, we are developing technology that allows conversation at a speed close to that of natural conversation. The CA autonomously creates a natural conversation sentence from a keyword generated by the BMI-based flick operation, in addition to considering the context of the conversation. We realized this by using the prompting function of a generative AI to create an appropriate response sentence that includes, for example, the keyword 'interesting' against the previous sentence "How do you feel fishing?". Furthermore, we developed a technology that estimates the emotional factors inherent in the generated response sentence using generative AI and allows the CA to respond with an emotional expression that appropriately reflects the content of the response sentence. Thus, even in conversation, it is possible to enhance the overall performance by harmonizing voluntary BMI control and CA autonomous control.

5.7.6   Extending Diversity of Devices Connected to BMI

A BMI can be connected to various external devices (Fig. 5.35). At the current technology level, it is possible to connect smart devices and use them with flick operations. Even bedridden people with physical disabilities can be active in the metaverse or Internet world, similar to healthy people. It is now also possible to freely control home appliances, nursing beds, electric wheelchairs, etc. Ultimately, the body's movement

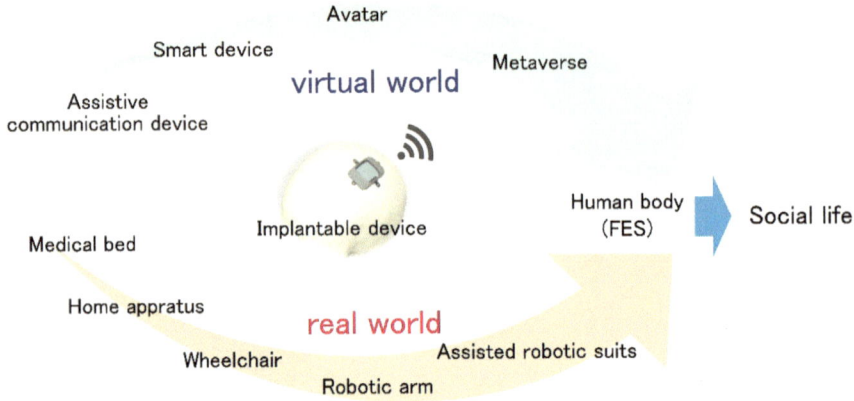

Fig. 5.35 Extending diversity of devices connected to BMI

can be fully restored to its original level using implantable functional electrical stimulation (a technology to move one's muscles by electrical stimulation). Currently, the development of implantable devices for BMI is progressing, clinical trials are being conducted, and prospects for practical applications have been established. If this becomes practical, the progress of functional restoration is expected to accelerate. In the future, if the measurement performance of BMI is further improved, the accumulation of high-quality brain big data will progress, and the neural decoding performance of AI will also improve, making it possible to further enhance the performance with the autonomous control of the CA. Furthermore, various devices can be connected to the patient's brain, and the CA can be moved more freely in daily life as well as in the virtual world, and it is expected that the era will come when physical disabilities can be overcome.

References

Ahn M, Brohan A, Brown N et al (2022). Do as I can, not as I say: grounding language in robotic affordances. arXiv preprint arXiv:2204.01691

Ahn M, Brohan A, Brown N, Chebotar Y, Cortes O, David B, Finn C, Fu C, Gopalakrishnan K, Hausman K, Herzog A, Ho D, Hsu J, Ibarz J, Ichter B, Irpan A, Jang E, Ruano RJ, Jeffrey K, Jesmonth S, Joshi NJ, Julian R, Kalashnikov D, Kuang Y, Lee KH, Levine S, Lu Y, Luu L, Parada C, Pastor P, Quiambao J, Rao K, Rettinghouse J, Reyes D, Sermanet P, Sievers N, Tan C, Toshev A, Vanhoucke V, Xia F, Xiao T, Xu P, Xu S, Yan M, Zeng A (2023) Do as I can, not as I say: grounding language in robotic affordances. In: Proceedings of machine learning research

Anderson P, Chang A, Chaplot DS, Dosovitskiy A, Gupta S, Koltun V, Kosecka J, Malik J, Mottaghi R, Savva M, Zamir AR (2018a) On evaluation of embodied navigation agents

Anderson P, Wu Q, Teney D, Bruce J, Johnson M, Sunderhauf N, Reid I, Gould S, Van Den Hengel A (2018b) Vision-and-language navigation: interpreting visually-grounded navigation instructions in real environments. In: Proceedings of the IEEE computer society conference on computer vision and pattern recognition, pp 3674–3683

Anumanchipalli GK, Chartier J, Chang EF (2019) Speech synthesis from neural decoding of spoken sentences. Nature 568:493–498. https://doi.org/10.1038/s41586-019-1119-1

Bommasani R, Hudson DA, Adeli E, Altman R, Arora S, von Arx S, Bernstein MS, Bohg J, Bosselut A, Brunskill E, Brynjolfsson E (2021) On the opportunities and risks of foundation models. arXiv preprint arXiv:2108.07258

Chaplot DS, Gandhi D, Gupta S et al (2020) Learning to explore using active neural SLAM. In: Proceedings of the international conference on learning representations (ICLR), virtual conference, formerly Addis Ababa, Ethiopia

Chen Y, Li Q, Kong D, Kei YL, Zhu S-C, Gao T, Zhu Y, Huang S (2021) YouRefIt: embodied reference understanding with language and gesture. In: IEEE/CVF international conference on computer vision (ICCV), pp 1385–1395

Collinger JL, Wodlinger B, Downey JE, Wang W, Tyler-Kabara EC, Weber DJ, McMorland AJ, Velliste M, Boninger ML, Schwartz AB (2013) High-performance neuroprosthetic control by an individual with tetraplegia. Lancet 381:557–564. https://doi.org/10.1016/S0140-6736(12)618 16-9

Friston K, Fitzgerald T, Rigoli F et al (2016) Active inference and learning. Neurosci Biobehav Rev 68:862–879. Available from: https://doi.org/10.1016/j.neubiorev.2016.06.022

Georgopoulos AP, Schwartz AB, Kettner RE (1986) Neuronal population coding of movement direction. Science (1979) 233:1416–1419. https://doi.org/10.1126/science.3749885

Hagiwara Y, Taguchi K, Ishibushi S et al (2021) Hierarchical Bayesian model for the transfer of knowledge on spatial concepts based on multimodal information. Adv Robot 36(1–2):33–53

Hagiwara Y, Furukawa K, Taniguchi A, Taniguchi T (2022) Multiagent multimodal categorization for symbol emergence: emergent communication via interpersonal cross-modal inference. Adv Robot 36. https://doi.org/10.1080/01691864.2022.2029721

Hasegawa S, Taniguchi A, Hagiwara Y, El Hafi L, Taniguchi T (2023a) Inferring place-object relationships by integrating probabilistic logic and multimodal spatial concepts. In: Proceedings of the IEEE/SICE international symposium on system integration (SII), pp 1–8

Hasegawa S, Yamaki R, Taniguchi A, Hagiwara Y, Taniguchi T (2023b) Understanding language instructions that include the vocabulary of unobserved objects by integrating a large language model and a spatial concept model. In: Proceedings of the 37th annual conference of the Japanese Society for Artificial Intelligence (JSAI)

Hasegawa S, Ito M, Yamaki R, Sakaguchi T, Hagiwara Y, Taniguchi A, El Hafi L, Taniguchi T (2023c) Leveraging a large language model and a spatial concept model for action planning of a daily life support robot. In: Proceedings of the 41st annual conference of the Robotics Society of Japan (RSJ)

Hirata M, Matsushita K, Suzuki T, Yoshida T, Sato F, Morris S, Yanagisawa T, Goto T, Kawato M, Yoshimine T (2011) A fully-implantable wireless system for human brain-machine interfaces using brain surface electrodes: W-HERBS. IEICE Trans Commun E94-B:2448–2453. https://doi.org/10.1587/transcom.E94.B.2448

Hochberg LR, Bacher D, Jarosiewicz B, Masse NY, Simeral JD, Vogel J, Haddadin S, Liu J, Cash SS, van der Smagt P, Donoghue JP (2012) Reach and grasp by people with tetraplegia using a neurally controlled robotic arm. Nature 485:372–375. https://doi.org/10.1038/nature11076

Huang W, Xia F, Xiao T et al (2022) Inner monologue: embodied reasoning through planning with language models. arXiv preprint arXiv:2207.05608

IAADP (2023) Tasks performed by guide, hearing & service dogs. https://iaadp.org/membership/iaadp-minimum-training-standards-for-public-access/tasks-performed-by-guide-hearing-and-service-dogs/. Accessed 29 Nov 2023

Iioka Y, Yoshida Y, Wada Y, Hatanaka S, Sugiura K (2023) Multimodal diffusion segmentation model for object segmentation from manipulation instructions

Iocchi L, Holz D, Ruiz-Del-Solar J, Sugiura K, Van Der Zant T (2015) RoboCup@Home: analysis and results of evolving competitions for domestic and service robots. Artif Intell 229:258–281. https://doi.org/10.1016/j.artint.2015.08.002

Ishikawa S, Sugiura K (2021) Target-dependent UNITER: a transformer-based multimodal language comprehension model for domestic service robots. IEEE Robot Autom Lett 6:8401–8408. https://doi.org/10.1109/LRA.2021.3108500

Ishikawa S, Sugiura K (2022) Moment-based adversarial training for embodied language comprehension. In: Proceedings—international conference on pattern recognition

Ishikawa T, Taniguchi A, Hagiwara Y, Taniguchi T (2023). Active semantic mapping for household robots: rapid indoor adaptation and reduced user burden. In: Proceedings of the IEEE conference on systems, man, and cybernetics (SMC)

Kageyama Y, He X, Shimokawa T, Sawada J, Yanagisawa T, Shayne M, Sakura O, Kishima H, Mochizuki H, Yoshimine T, Hirata M (2020) Nationwide survey of 780 Japanese patients with amyotrophic lateral sclerosis: their status and expectations from brain–machine interfaces. J Neurol 267:2932–2940. https://doi.org/10.1007/s00415-020-09903-3

Kaneda K, Korekata R, Wada Y, Nagashima S, Kambara M, Iioka Y, Matsuo H, Imai Y, Nishimura T, Sugiura K (2023) DialMAT: dialogue-enabled transformer with moment-based adversarial training. CVPR 2023 embodied AI workshop

Katsumata Y, Taniguchi A, El Hafi L et al (2020). Spcomapgan: spatial concept formation-based semantic mapping with generative adversarial networks. In: Proceedings of the IEEE/RSJ international conference on intelligent robots and systems (IROS), pp 7927–7934

Khandelwal A, Weihs L, Mottaghi R, Kembhavi A (2022) Simple but effective: CLIP embeddings for embodied AI. In: Proceedings of the IEEE/CVF computer vision and pattern recognition conference (CVPR), pp 14829–14838

Kim T, Ahn S, Bengio Y (2019) Variational temporal abstraction. In: Advances in neural information processing systems

Korekata R, Kambara M, Yoshida Y, Ishikawa S, Kawasaki Y, Takahashi M, Sugiura K (2023) Switching head-tail funnel UNITER for dual referring expression comprehension with fetch-and-carry tasks. In: IEEE/RSJ international conference on intelligent robots and systems (IROS)

Liang J, Huang W, Xia F et al (2023) Code as policies: language model programs for embodied control. In: Proceedings of the IEEE international conference on robotics and automation (ICRA), pp 9493–9500

Magassouba A, Sugiura K, Kawai H (2018) A multimodal classifier generative adversarial network for carry and place tasks from ambiguous language instructions. IEEE Robot Autom Lett 3:3113–3120. https://doi.org/10.1109/LRA.2018.2849607

Magassouba A, Sugiura K, Quoc AT, Kawai H (2019) Understanding natural language instructions for fetching daily objects using GAN-based multimodal target-source classification. IEEE Robot Autom Lett 4:3884–3891. https://doi.org/10.1109/lra.2019.2926223

Magassouba A, Sugiura K, Kawai H (2020) A multimodal target-source classifier with attention branches to understand ambiguous instructions for fetching daily objects. IEEE Robot Autom Lett 5:532–539. https://doi.org/10.1109/LRA.2019.2963649

Magassouba A, Sugiura K, Kawai H (2021) CrossMap transformer: a crossmodal masked path transformer using double back-translation for vision-and-language navigation. IEEE Robot Autom Lett 6:6258–6265

Matsushita K, Hirata M, Suzuki T, Ando H, Yoshida T, Ota Y, Sato F, Morris S, Sugata H, Goto T, Yanagisawa T, Yoshimine T (2018) A fully implantable wireless ECoG 128-channel recording device for human brain-machine interfaces: W-HERBS. Front Neurosci 12:511. https://doi.org/10.3389/fnins.2018.00511

Mordatch I, Abbeel P (2018) Emergence of grounded compositional language in multi-agent populations. In: 32nd AAAI conference on artificial intelligence, AAAI 2018

Murray M, Cakmak M (2022) Following natural language instructions for household tasks with landmark guided search and reinforced pose adjustment. IEEE Robot Autom Lett 7:6870–6877. https://doi.org/10.1109/LRA.2022.3178804

Musk E (2019) An integrated brain-machine interface platform with thousands of channels. J Med Internet Res 21:e16194. https://doi.org/10.2196/16194

Nagano M, Nakamura T, Nagai T, Mochihashi D, Kobayashi I, Kaneko M (2018) Sequence pattern extraction by segmenting time series data using GP-HSMM with hierarchical Dirichlet process. In: IEEE international conference on intelligent robots and systems

Nagano M, Nakamura T, Nagai T, Mochihashi D, Kobayashi I (2022) Spatio-temporal categorization for first-person-view videos using a convolutional variational autoencoder and Gaussian processes. Front Robot AI 9. https://doi.org/10.3389/frobt.2022.903450

Nakamura T, Nagai T, Mochihashi D, Kobayashi I, Asoh H, Kaneko M (2017) Segmenting continuous motions with hidden semi-Markov models and Gaussian processes. Front Neurorobot 11. https://doi.org/10.3389/fnbot.2017.00067

Nakamura T, Nagai T, Taniguchi T (2018) SERKET: an architecture for connecting stochastic models to realize a large-scale cognitive model. Front Neurorobot 12. https://doi.org/10.3389/fnbot.2018.00025

Navarro SE, Muhlbacher-Karrer S, Alagi H, Zangl H, Koyama K, Hein B, Duriez C, Smith JR (2022) Proximity perception in human-centered robotics: a survey on sensing systems and applications. IEEE Trans Robot 38:1599–1620. https://doi.org/10.1109/TRO.2021.3111786

Obata S, Aoki T, Nagai T (2023a) Task execution by multiple robots using large-scale language models. In: Annual conference of the Robotics Society of Japan, p 2K302

Obata S, Aoki T, Nagai T (2023b) User goal intention estimation for shared control of mobile robots with BMI. In: Annual conference of the Robotics Society of Japan, p 1J102

OpenAI (2023) GPT-4 technical report. arXiv preprint arXiv:2303.08774

Otsuki S, Ishikawa S, Sugiura K (2023) Prototypical contrastive transfer learning for multimodal language understanding

Oxley TJ, Opie NL, John SE, Rind GS, Ronayne SM, Wheeler TL, Judy JW, McDonald AJ, Dornom A, Lovell TJH, Steward C, Garrett DJ, Moffat BA, Lui EH, Yassi N, Campbell BCV, Wong YT, Fox KE, Nurse ES, Bennett IE, Bauquier SH, Liyanage KA, van der Nagel NR, Perucca P, Ahnood A, Gill KP, Yan B, Churilov L, French CR, Desmond PM, Horne MK, Kiers L, Prawer S, Davis SM, Burkitt AN, Mitchell PJ, Grayden DB, May CN, O'Brien TJ (2016) Minimally invasive endovascular stent-electrode array for high-fidelity, chronic recordings of cortical neural activity. Nat Biotechnol 34:320–327. https://doi.org/10.1038/nbt.3428

Oyama A, Hasegawa S, Nakagawa H, Taniguchi A, Hagiwara Y, Taniguchi T (2023). Exophora resolution of linguistic instructions with a demonstrative based on real-world multimodal information. In: Proceedings of the 32nd IEEE international conference on robot and human interactive communication (RO-MAN), pp 2617–2623. https://doi.org/10.1109/RO-MAN57019.2023.10309487

Park SM, Kim YG (2023) Visual language integration: a survey and open challenges. Comput Sci Rev 48:100548

Parvizi J, Kastner S (2018) Promises and limitations of human intracranial electroencephalography. Nat Neurosci 21

Radford A, Kim JW, Hallacy C, Ramesh A, Goh G, Agarwal S, Sastry G, Askell A, Mishkin P, Clark J, Krueger G, Sutskever I (2021) Learning transferable visual models from natural language supervision. In: Proceedings of machine learning research

Rubin DB, Ajiboye AB, Barefoot L, Bowker M, Cash SS, Chen D, Donoghue JP, Eskandar EN, Friehs G, Grant C, Henderson JM, Kirsch RF, Marujo R, Masood M, Mernoff ST, Miller JP, Mukand JA, Penn RD, Shefner J, Shenoy KV, Simeral JD, Sweet JA, Walter BL, Williams ZM, Hochberg LR (2023) Interim safety profile from the feasibility study of the BrainGate neural interface system. Neurol 100:e1177–e1192. https://doi.org/10.1212/WNL.0000000000201707

Sethuraman J (1994) A constructive definition of Dirichlet priors. Stat Sin 4

Shah D, Osiński B, Ichter B, Levine S (2023) LM-Nav: robotic navigation with large pre-trained models of language, vision, and action. In: Proceedings of machine learning research

Shridhar M, Thomason J, Gordon D, Bisk Y, Han W, Mottaghi R, Zettlemoyer L, Fox D (2020) ALFRED. a benchmark for interpreting grounded instructions for everyday tasks. In: Proceedings of the IEEE Computer Society conference on computer vision and pattern recognition, pp 10740–10749

Shridhar M, Manuelli L, Fox D (2021) CLIPort: what and where pathways for robotic manipulation. In: Proceedings of the 5th conference on robot learning (CoRL), pp 894–906

Stachniss C (2005) Information gain-based exploration using Rao-Blackwellized particle filters. In: Robotics: science and systems, Cambridge, MA

Suzuki Y (2021) Proximity-based non-contact perception and omnidirectional point-cloud generation based on hierarchical information on fingertip proximity sensors. Adv Robot 35:1181–1197. https://doi.org/10.1080/01691864.2021.1969268

Suzuki Y, Yoshida R, Tsuji T, Nishimura T, Watanabe T (2022) Grasping strategy for unknown objects based on real-time grasp-stability evaluation using proximity sensing. IEEE Robot Autom Lett 7:8643–8650. https://doi.org/10.1109/LRA.2022.3188885

Suzuki Y, Yoshida R, Tsuji T, Nishimura T, Watanabe T (2023) Local curvature estimation and grasp stability prediction based on proximity sensors on a multi-fingered robot hand. J Robot Mechatron 35:1340–1353. https://doi.org/10.20965/jrm.2023.p1340

Takeuchi K, Yamazaki Y, Yoshifuji K (2020) Avatar work: telework for disabled people unable to go outside by using avatar robots "orihime-d" and its verification. In: ACM/IEEE international conference on human-robot interaction

Taniguchi A, Taniguchi T, Inamura T (2016) Spatial concept acquisition for a mobile robot that integrates self-localization and unsupervised word discovery from spoken sentences. IEEE Trans Cogn Dev Syst 8(4):285–297

Taniguchi A, Hagiwara Y, Taniguchi T et al (2017) Online spatial concept and lexical acquisition with simultaneous localization and mapping. In: Proceedings of the IEEE/RSJ international conference on intelligent robots and systems, pp 811–818

Taniguchi T, Yoshino R, Takano T (2018) Multimodal hierarchical Dirichlet process-based active perception by a robot. Front Neurorobot 12:22

Taniguchi A, Hagiwara Y, Taniguchi T et al (2020a) Improved and scalable online learning of spatial concepts and language models with mapping. Auton Robot 44(6):927–946. https://doi.org/10.1007/s10514-020-09905-0

Taniguchi T, Nakamura T, Suzuki M, Kuniyasu R, Hayashi K, Taniguchi A, Horii T, Nagai T (2020b) Neuro-SERKET: development of integrative cognitive system through the composition of deep probabilistic generative models. New Gener Comput 38. https://doi.org/10.1007/s00354-019-00084-w

Taniguchi T, El Hafi L, Hagiwara Y, Taniguchi A, Shimada N, Nishiura T (2021) Semiotically adaptive cognition: toward the realization of remotely-operated service robots for the new normal symbiotic society. Adv Robot 35(11):664–674

Taniguchi A, Tabuchi Y, Ishikawa T, El Hafi L, Hagiwara Y, Taniguchi T (2023a) Active exploration based on information gain by particle filter for efficient spatial concept formation. Adv Robot 20:1–31

Taniguchi T, Yoshida Y, Matsui Y, Le Hoang N, Taniguchi A, Hagiwara Y (2023b) Emergent communication through Metropolis-Hastings naming game with deep generative models. Adv Robot 37:1266–1282. https://doi.org/10.1080/01691864.2023.2260856

Thrun S, Burgard W, Fox D (2005) Probabilistic robotics. Intelligent robotics and autonomous agents series. The MIT Press, Cambridge

Velliste M, Perel S, Spalding MC, Whitford AS, Schwartz AB (2008) Cortical control of a prosthetic arm for self-feeding. Nature 453:1098–1101. https://doi.org/10.1038/nature06996

Wei J, Wang X, Schuurmans D, Bosma M, Ichter B, Xia F, Chi EH, Le QV, Zhou D (2022) Chain-of-thought prompting elicits reasoning in large language models. In: Advances in neural information processing systems

Willett FR, Avansino DT, Hochberg LR, Henderson JM, Shenoy KV (2021) High-performance brain-to-text communication via handwriting. Nature 593:249–254. https://doi.org/10.1038/s41586-021-03506-2

Yamamoto T, Terada K, Ochiai A, Saito F, Asahara Y, Murase K (2019) Development of human support robot as the research platform of a domestic mobile manipulator. Robomech J 6:1–15. https://doi.org/10.1186/s40648-019-0132-3

Yan T, Kameda S, Suzuki K, Kaiju T, Inoue M, Suzuki T, Hirata M (2020) Minimal tissue reaction after chronic subdural electrode implantation for fully implantable brain-machine interfaces. Sensors 21:178. https://doi.org/10.3390/s21010178

Yan T, Suzuki K, Kameda S, Maeda M, Mihara T, Hirata M (2022) Electrocorticographic effects of acute ketamine on non-human primate brains. J Neural Eng 19. https://doi.org/10.1088/1741-2552/ac6293

Yan T, Suzuki K, Kameda S, Kuratomi T, Mihara M, Maeda M, Hirata M (2023a) Intracranial EEG recordings of high-frequency activity from a wireless implantable BMI device in awake nonhuman primates. IEEE Trans Biomed Eng 70:1107–1113. https://doi.org/10.1109/TBME.2022.3210286

Yan T, Suzuki K, Kameda S, Maeda M, Mihara T, Hirata M (2023b) Chronic subdural electrocorticography in nonhuman primates by an implantable wireless device for brain-machine interfaces. Front Neurosci 17:1260675. https://doi.org/10.3389/fnins.2023.1260675

Yanagisawa T, Hirata M, Saitoh Y, Goto T, Kishima H, Fukuma R, Yokoi H, Kamitani Y, Yoshimine T (2011) Real-time control of a prosthetic hand using human electrocorticography signals. J Neurosurg 114:1715–1722. https://doi.org/10.3171/2011.1.JNS101421

Yanagisawa T, Hirata M, Saitoh Y, Kishima H, Matsushita K, Goto T, Fukuma R, Yokoi H, Kamitani Y, Yoshimine T (2012) Electrocorticographic control of a prosthetic arm in paralyzed patients. Ann Neurol 71:353–361. https://doi.org/10.1002/ana.22613

Yoshino R, Takano T, Tanaka H et al (2021) Active exploration for unsupervised object categorization based on multimodal hierarchical Dirichlet process. In: Proceedings of the IEEE/SICE international symposium on system integrations (SII), Fukushima, Japan

Yu X, Zhang H, Song Y, Song Y, Zhang C (2019) What you see is what you get: visual pronoun coreference resolution in dialogues. In: Conference on empirical methods in natural language processing and International joint conference on natural language processing (EMNLP-IJCNLP), pp 5123–5132

Yu X, Zhang H, Song Y, Zhang C, Xu K, Yu D (2021) Exophoric pronoun resolution in dialogues with topic regularization. In: Conference on empirical methods in natural language processing (EMNLP), pp 3832–3845

Zeng H, Shen Y, Hu X, Song A, Xu B, Li H, Wang Y, Wen P (2020) Semi-autonomous robotic arm reaching with hybrid gaze-brain machine interface. Front Neurorobot 13. https://doi.org/10.3389/fnbot.2019.00111

Chapter 6
Development of the CA Platform

Takahiro Miyashita, Akira Utsumi, and Takashi Yoshimi

Abstract The Cybernetic Avatar (CA) platform serves as a software foundation that facilitates connections between CAs and teleoperators and empowers both of them to deliver services to people. This chapter introduces the CA platform and emphasizes its crucial attributes: scalability, customizability, and interoperability. We describe the CA platform's conceptual framework and social field experiments that leverage its capabilities and discuss its international standardization efforts to maintain the interoperability of the platform's specifications. This comprehensive overview will provide readers with a deeper understanding of the CA platform's pivotal role in advancing CAs and their applications.

6.1 Introduction

In this chapter, we describe a Cybernetic Avatar (CA) platform that enables active participation in society for everyone. The CA platform is a software platform that connects CAs and teleoperators and enables both to provide services to people. Any person can actively engage and participate in social activities with CAs. Those who previously struggled to engage in such daily activities as shopping or working in physical spaces will be able to use CAs through the CA platform to effortlessly participate in social life regardless of age, physical or cognitive state, appearance, or gender hindrances.

Some companies (Zachiotis et al. 2018) have already begun to provide services that support social participation by teleoperated robots and CG agents, including

T. Miyashita (✉) · A. Utsumi
Advanced Telecommunications Research Institute International, Seika-cho, Soraku-gun, Kyoto, Japan
e-mail: miyasita@atr.jp

A. Utsumi
e-mail: utsumi@atr.jp

T. Yoshimi
Shibaura Institute of Technology, Koto, Tokyo, Japan
e-mail: yoshimit@sic.shibaura-it.ac.jp

© The Author(s) 2025
H. Ishiguro et al. (eds.), *Cybernetic Avatar*,
https://doi.org/10.1007/978-981-97-3752-9_6

Fig. 6.1 Example scene of avatar symbiotic society

teleconference and teleguidance systems. Most such companies have independently developed a CA that provides services. To enable social participation, various interfaces must be tailored to the situations of teleoperators, and different types of CAs must be reconfigured to services and customers as well as many kinds of service applications. A ubiquitous network robot platform (UNR-PF) has been proposed in the field of networked robots (Kamei et al. 2012, 2017) for developing robots using the same concept. However, robot components and service applications on UNR-PF were created mainly for autonomous networked robots, not for remote operation robots such as CAs. In this chapter, we introduce the development and prospects of the CA platform that provides CA services. We outline the CA platform and its system structure and also discuss its social implementation and various types of interconnectivity for future prospects (Fig. 6.1).

6.2   What is a Cybernetic Avatar Platform (CAPF)?

6.2.1   Cybernetic Avatar Platform

The CA platform is a software foundation that connects multiple CAs and multiple teleoperators and enables CAs to provide services. Figure 6.2 outlines the CA platform structure and CA services. Our research group is currently developing a CA platform that possesses these four basic functions:

Fig. 6.2 Outline of CA platform structure and CA services

1. *CA monitoring*: Recording the data necessary for improving CA control and functions such as the activity records of CAs, teleoperators, and those around the CA.
2. *CA experience management*: Managing the CA experiences acquired by CA monitoring and reusing them for functional improvements.
3. *Multiple CAs control*: Managing various CA services by linking CAs based on the level of semi-autonomous technologies.
4. *Teleoperator assignment*: Combining many teleoperators and CAs depending on the CA services provided.

A CA platform must have the following three characteristics to serve as a software platform that connects CAs and teleoperators:

1. *Scalability*: To function as a software platform for providing various CA services, it must handle situations where multiple CAs are used by multiple teleoperators at various service locations.
2. *Customizability*: For allowing teleoperators in various situations to use the system, it must be designed so that teleoperators can customize the interface for CA operations. Additional programs required for specific CAs and CA services should be made available in combination with the CA platform's basic functionality.
3. *Interoperability*: To make the CA platform a software platform that connects CAs and teleoperators, different types of CAs (developed by various companies) must be connected.

There is one more important concept for controlling multiple CAs for the CA platform: semi-autonomous technology. Semi-autonomous technology simplifies operations that require a teleoperator to perform multiple steps. For example, one of the goals of this project is to enable a single teleoperator to manipulate a large number of CAs (Fig. 6.3). This capability has become possible by implementing semi-autonomous technologies.

The nature of service provisions with CA depends on the type of CAs, the situations of the teleoperators and the CA service provision locations, and the kinds of CA services. We are developing a CA platform that meets the requirements of these diverse service provision conditions in various combinations. Functional social demonstrations are progressing.

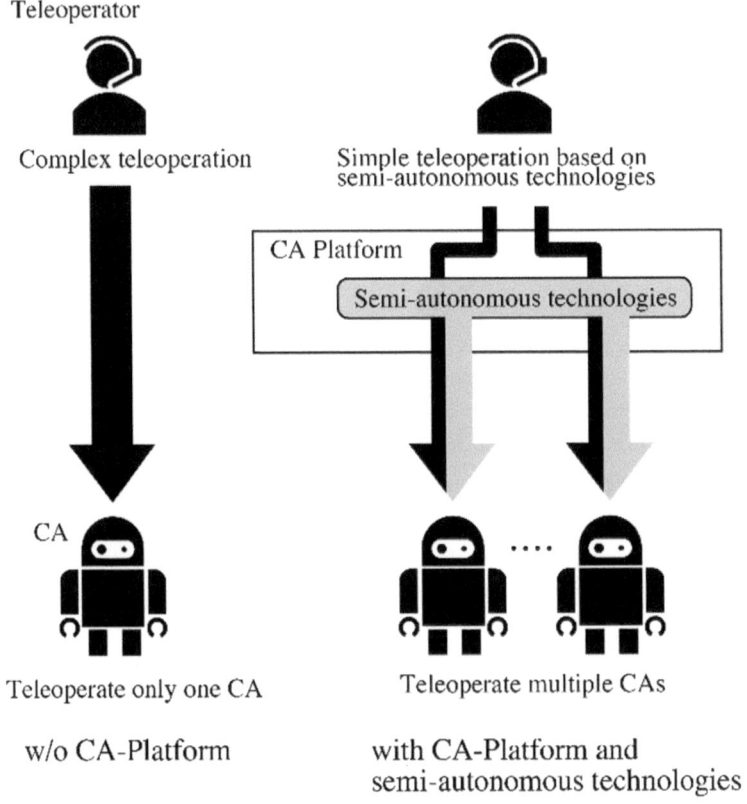

Fig. 6.3 Concept of controlling multiple CAs

6.2.2 *Implementation of Cybernetic Avatar Platform*

This section describes a social implementation example for the CA platform and its connection with CAs. Figure 6.4 shows the implementation's overall structure. To create a communication framework that is robust to user numbers and locations, we used Amazon Web Services (AWSs). We used the WebRTC of Kinesis Video Streams (KVSs) in AWS as a framework for connecting video and audio between teleoperators and CAs and WebSocket for sending control commands to CAs.

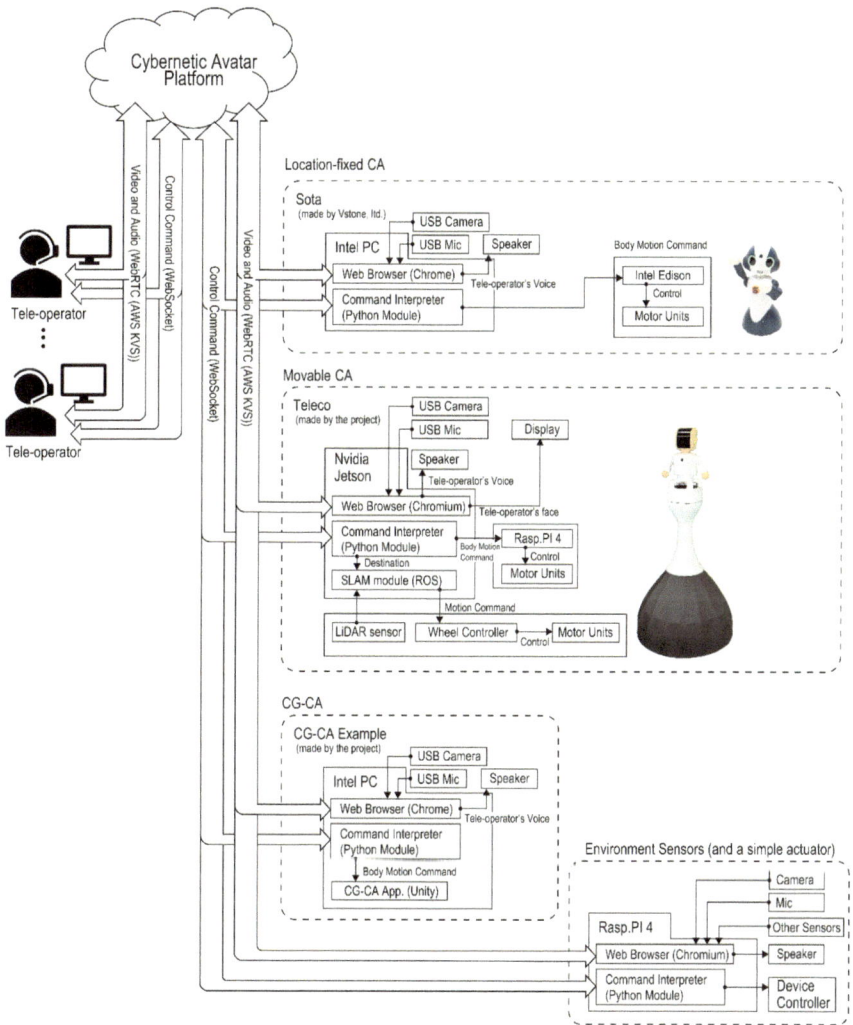

Fig. 6.4 Implementation of CA platform

Here, we assume three types of CAs for connecting them to the CA platform. They are capable of actively providing services in both cyber and physical societies. We explain these types of CAs and describe implementation examples as well as the employed environment sensors as follows:

(1) Location-Fixed CA

A location-fixed CA is designed for tasks that do not require any ambulation, such as reception, counter services, and product explanations. An example of this type is Sota, made by Vstone, Ltd. (Fig. 6.4). A teleoperator communicates through a camera and a microphone with people around Sota via WebRTC of KVS; Sota's operation commands are sent via WebSockets.

(2) Movable CA

A movable CA is intended for operations that are performed while going from place to place, such as road guidance, security guards at facilities, and explanations of multiple products in stores. An example of this type is Teleco, made through our project. Video and audio taken from cameras and microphones on Teleco are sent by WebRTC of KVS, and operation commands are sent via WebSocket in the same way as with Sota in a location-fixed CA. Teleco is given a destination on a map to which it moves autonomously with the simultaneous localization and mapping (SLAM) method. While the system is moving, the teleoperator can communicate with those around the Teleco via KVS.

(3) CG-CA

A CG-CA is a kind of system that can approach the physical world, even from the cyber world. Examples include CAs intended for tasks such as facility guidance and product introduction to people in the physical world via digital signage. Standard digital signage is not incorporated in a CA. A CG-CA can provide information to customers along with a teleoperator's hospitality from within its digital signage. CG-CA implementation is identical as a location-fixed CA and a movable CA in the physical world (Fig. 6.4). The physical world's video and audio are acquired by the camera and microphone on such CG-CA display devices as digital signage and sent via a WebRTC of KVS. The operation commands of CG-CA are sent to Unity via WebSocket.

(4) Environment Sensors

In addition to the above three types of CAs, environment sensors for obtaining environmental information must also be connected to the CA platform. Information about the environment surrounding the CA's activities is essential to facilitate teleoperator operations. On the CA platform, environment sensors can be linked using the same connection method as that for CA to transmit the sensors' acquired environmental information to the teleoperator. Figure 6.4 shows how the sensors are connected.

In the previous section, we described the three characteristics required by a software platform. Each is developed in this CA platform:

1. *Scalability*: The CA platform service is provided as a cloud service that utilizes AWS on a web browser (Chrome). The number of CAs and teleoperators can be handled by adjusting the number of virtual servers for the CA platform.

2. *Customizability*: Video, audio, and robot commands between the CA and the teleoperator on the CA platform can be used by external programs through virtual devices on each PC for the CA and the teleoperator. In addition, we are currently developing a function that allows the windows and buttons of the CA platform to be rearranged like a dashboard to simplify usage.
3. *Interoperability*: We connect various types of CAs to the CA platform using WebRTC and WebSocket, which are already de facto and de jure communication protocols, and the Robotic Interaction Service (RoIS) Framework, a description method of software modules for robots that is standardized in the Object Management Group (OMG) (Object Management Group 2018a). We are also working on the international standardization of the description method mentioned above.

In our project, we are conducting a field experiment on receptionist performance at our company's entrance using a location-fixed CA, environment sensors, and the CA platform. Specifically, we are conducting experimentation on the functional and social verifications of the "CA reception work" executed by a receptionist who is teleoperating the CA. As shown in Fig. 6.5, the Android ERICA, which combines teleoperation and semi-autonomous technologies, is used as an actual location-fixed CA in a workplace environment and is prepared for the remote control of reception services in an entrance area (approximately 200 m^2). We have been regularly conducting reception demonstrations for about a year and have confirmed that reception work can be accomplished through a CA.

Fig. 6.5 CA reception work in field experiment

6.2.3   Discussion

We already described the concept of the CA platform, which is a software platform that connects CAs and teleoperators to simplify participation in society through CAs for people in various locations. We introduced our social implementation trial using a CA platform that connects three types of CAs and environment sensors that assist teleoperators. Our research group aims to create a society in which various people can sustainably benefit from using CAs. In this section, we introduce a CA platform that technically connects teleoperators and CAs. Further development is required to use them in society. For example, a network is essential to connect teleoperators and CAs, although sometimes it may be disconnected. Even in such cases, a mechanism is needed to safely provide services via CAs. A mechanism is also required that involves companies as CA developers and as service providers that implement CAs. We will develop a CA platform, CAs, and a CA marketplace in the same manner, for example, as Android OS, Android devices, and Google Play. We will introduce these elements to the market and create a symbiotic society where CAs will actually be active.

6.3   User/CA Activity Monitoring and Management

6.3.1   What is Activity Monitoring and Management?

Activity monitoring and management are one of the major components of the CA platform. Since all human activities using a CA are performed via the CA platform, they can be monitored directly or indirectly[1] by it. Data retrieved from the monitoring include CA operation commands issued by human operators, the behaviors of CAs and interaction partners, and sensory data detected by CA systems (e.g., positions, human positions, and geometrical information of CA activity areas). In addition, higher levels of information can be stored, such as the content of utterances, facial expressions, and the emotions of humans obtained by analyzing video and audio streams. Such information can improve the present and future CA services.

As mentioned above, the CA platform's role is to deliver a variety of services by dynamically connecting humans and various CAs. As shown in Fig. 6.2, in the CA society, a person becomes an operator of a CA(s) to provide (as a servicer) CA-based services to another person(s). At other times, the same person may be a service user who receives a service provided by a CA(s) controlled by others. Nor is the connection between humans and CAs always 1 to 1; it can be 1 to N or N to M. One operator may simultaneously use a large number of CAs to provide a service to a single person or multiple people. Alternatively, multiple operators at different

---

[1] A separate monitoring program, such as WebRTC, must be prepared for the content of communications conducted through P2P.

locations may collaboratively control multiple CAs to achieve a single service. Therefore, the time, location, number, and type of CAs to be assigned to a service might differ for each use and can change dynamically. To handle such various service conditions, the "CA manager" in the CA platform dynamically assigns appropriate service participants (operator(s) and CA(s)) to every new service session based on a service's particular demands. To accomplish the assignments, the "CA manager" refers to the several databases built into the platform that store information on operators (servicers, "operator database"), service users ("user database"), CAs ("CA database"), and the properties of the CA activity space ("space database"). These databases are updated and maintained through service sessions based on the information provided by service applications. Furthermore, the CA activities controlled by operator(s) in the service sessions are stored in the "CA activity database." This stored information contains CA behavior selected by a human operator in every specific situation appearing in the service. An analysis of the dataset extracts the service's essential knowledge. In our project, the extracted information is used for improving service. In other words, the knowledge and expertise of a service operator is relayed to future operators of the service, a step that is expected to enhance the collection and sharing of human experience and expertise about user activities.

6.3.2   Roles of Activity Monitoring and Management

The CA activity data stored in the activity database on the CA platform can be used in many tasks for building a CA society, including supporting the operators and service recipients and the development of CA-based services and the CA itself. We next briefly explain these roles below.

(1)   Support of Operators
    Reducing the burden on CA operators is one critical purpose for monitoring CA activities. Since monitoring places a heavy burden on the operators who remotely manipulate a CA with high degrees of motion and appearance freedom (Rea Daniel and Seo Stela 2022), reducing the degree of operation freedom and the frequency of operation demands is crucial.
    A promising approach to solving this problem is predicting an operator's behavior based on past recorded activities. The system can thus predict the behavior that the person should or is likely to do next and assist the operator by presenting predicted behaviors as candidate actions.
    Sharing activity history among multiple operators is also important for supporting operators. By using the activity data, the operation talents of skillful operators can be transferred to novice operators.
(2)   Support for Recipients of CA Services
    Services based on interactions with CAs, which are often very different from conventional face-to-face services, may cause difficulties for service recipients. On the other hand, much detailed activity data on CA services have become

available on the CA platform. We can use these data to identify the source of problems and overcome the difficulties faced by those who rely on CA-based services.

(3) Service Development

Activity data provide an essential element in developing services as well. For example, the preferences of service users can be reflected to improve services. Activity data can also lead to the discovery and development of new services. Generally, CA services have a hierarchical structure, and data gathered in lower-level services can also be used for higher-level services. Sharing activity data among different services based on this service hierarchy enhances the quality of every related service.

(4) CA Development

Activity monitoring provides CA developers with rapid feedback on CA usage in real-world situations.. Moreover, CA activity data that accumulate in the CA platform can be used for efficient design, implementation, and verification in a CA development cycle. The requirements received from CA servicers can uncover real needs, which contribute to the development of CAs with new functions. Furthermore, the CA platform defines a set of common functions and protocols that CAs should hold, and such structure guides CA developers to follow an open standard. This fuels the ability to guarantee the interoperability of CAs.

In addition to the above roles, activity data can be used for more general purposes. CA services should cover a wide range of human activities. Along with the increasing number of domains using CA-based services, activity monitoring can store a wide range of human activities through avatars as common format data. These data should contain various human activities using CAs worldwide. This critical set of data will undoubtedly lead to the preservation of human knowledge, skills, and culture for future generations.

6.3.3 Mechanism that Supports Activity Monitoring

Figure 6.6 shows the configuration of a typical CA-based service. Here, one operator and one avatar are connected to a CA platform (as mentioned, this connection can be $N \times M$), and the communication paths of WebRTC (video and audio communication) and WebSocket (commands and data communication) are established between them via the CA platform. The operator pilots the remote CA by these communication channels and interacts with the interaction partner(s) at the site where the CA operates. During the interactions, the observed data can be analyzed in real time to support the interactions. For instance, showing speech recognition results increases the operator's understanding. Detection of the interaction partner's attention to the CA is also helpful so that the operator appropriately reacts to the partner. Thus, various types of

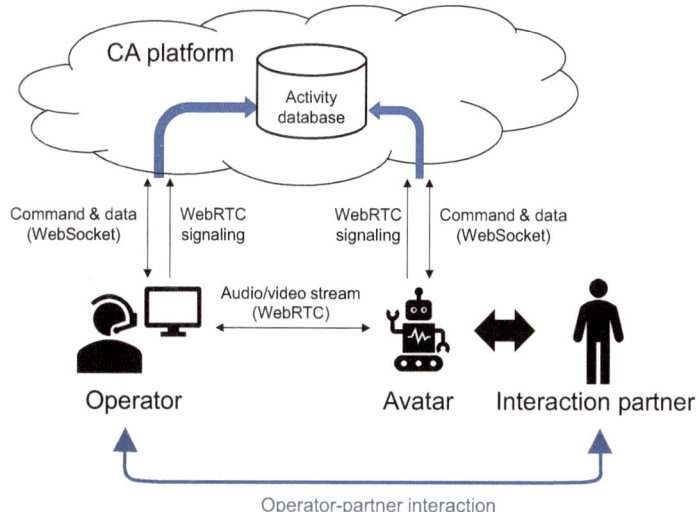

Fig. 6.6 Activity monitoring

information supporting the use of activity data are expected to facilitate interactions toward improving CA-based services.

To enhance the process, the CA platform manages all the available services in a hierarchical structure that involves inheritance relationships among services, i.e., a complex service that can be developed by combining multiple basic services. The CA activity database also reflects the hierarchical structure and stores information based on it. The collection of human experience based on this mechanism will contribute to productive advances in society.

There are two types of CA services: One is where the servicer(s) controls the CA(s), and the other type is where a service user controls the CA(s). In the former type, CA services are provided by CA(s) controlled by a human operator (servicer). In other words, these services extend human work styles beyond physical and spatial restrictions using CAs. For instance, in a reception service, the assigned operator engages in necessary conversation with visitors using a CA placed at the reception desk. In the latter type, service users themselves control the CA(s) to receive services. In a remote shopping service, a service user at home can remotely window shop, talk with shop staff, and select goods via a CA(s) in an actual shopping mall. A user of a sightseeing service can enjoy local activities through a CA without any travel time. In both types of CA services, CA monitoring is a promising way to enhance present and future service quality by using stored avatar and human behaviors.

Figure 6.7 shows an example sequence of monitored activities in a CA platform. The recorded activity data contain commands for controlling the CA as well as the motions of an interaction partner. In addition, the data contain speech recognition results for both the operator's utterances and those of the interaction partner.

```
20211125153404.538329,4,away
20211125153404.597913,4,stopfar
20211125153404.662679,4,away
20211125153404.783529,4,stopfar
20211125153408.118676,4,look
20211125153409.042907,2,bow
20211125153411.376651,4,nolook
20211125153412.689909,4,look
20211125153413.572671,4,nolook
20211125153409.332361,1,お待たせいたしました。中山が参りました。,4.089715
20211125153417.311798,4,away
20211125153417.440729,4,stopfar
20211125153417.626587,4,away
20211125153417.686703,4,stopfar
20211125153417.815578,4,away
20211125153417.877026,4,stopfar
20211125153432.417870,4,approach
20211125153432.418868,4,look
20211125153432.858955,2,bow
```

Fig. 6.7 Observed CA activities (example)

6.3.4 Application of Activity Monitoring

Next, we introduce two example applications where activity data are used. The first one is an application for evaluating the operator's actions based on activity data. The second is a behavior-based recommendation to assist the operator during such actions.

First, we introduce operator evaluations based on activity monitoring. As mentioned in Sect. 6.3.1, an operator's actions can always be monitored in the CA platform, and we can evaluate the operator's behavior in his/her CA operations. Here we describe demonstration experiments held at a commercial facility (Asia & Pacific Trade Center, Osaka, Japan) in July 2023. In the experiment, visitors to the facility freely participated in the experiments as operators of a CA as well as interaction partners (Fig. 6.8, left). We feed backed the evaluation results of the operation to participants to influence their motivation (Fig. 6.8, right). This evaluation was performed based on time duration, number and type of operation commands used in the sessions, operator's utterances, and the number of CAs they operated.

The second example is a behavior-based recommendation of an operator's speech and task in an avatar-based reception scenario. The following four types of information can be observed in avatar-based interactions:

- Speech of avatar operator.
- Speech of interaction partner.
- Avatar behavior (= a series of actions invoked by the operator).
- Behavior of interaction partner.

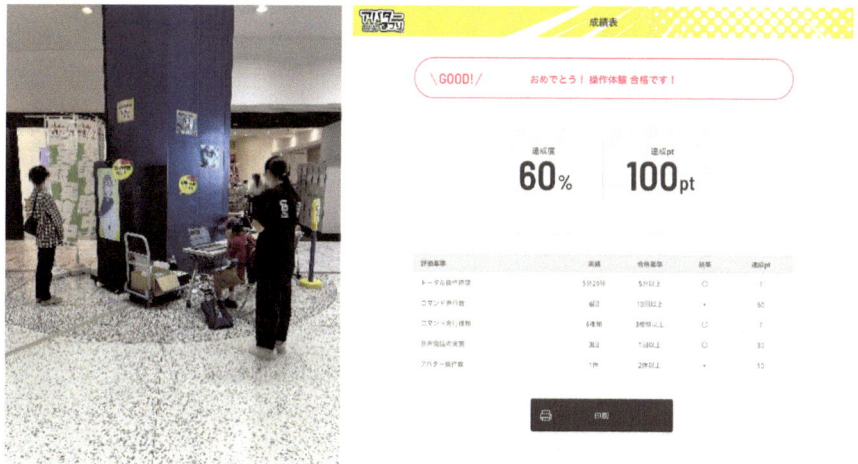

Fig. 6.8 Operator evaluation using activity data: left: experimental site, right: evaluation results (example)

All four types of information mentioned above consist of time-series information. Such information observed in a time window is expressed as a bag-of-words model, i.e., a vector representation of the frequency of word occurrences.

Here, "words" are not only verbal bits of spoken language in an interaction context but also "behavioral" words corresponding to avatar control commands and the detected behavior of the interaction partner. Every time the human operator invokes a speech or control command for the CA, and a combination of the observed bag-of-words and the invoked speech or control command is stored in the database.

In the estimation phase, we employ a Naive Bayes approach to calculate the scores for each behavior (speech and operation). After calculating the scores for all the behaviors stored in the database, those with high scores are selected as candidate behaviors that the operator should select in the current situation. These candidates form a recommendation that is presented as GUI buttons to the operator who can easily invoke the behaviors.

The recommendation accuracy for the speech is shown in Fig. 6.9. The score rankings of the recommended candidates correctly matched the operator's actual speech. Almost 80% of the operator's speech was correctly recommended within the top-five ranking.

Figure 6.10 summarizes the differences in the operation styles between the with/without recommendation conditions. The ratio of speeches registered as GUI buttons

Fig. 6.9 Recommendation accuracy

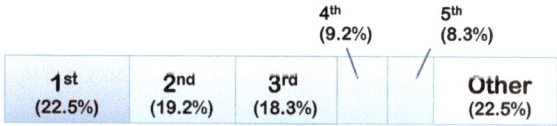

| 1st (22.5%) | 2nd (19.2%) | 3rd (18.3%) | 4th (9.2%) | 5th (8.3%) | Other (22.5%) |

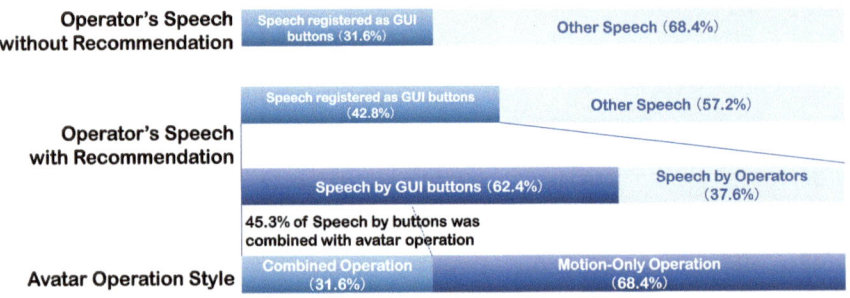

Fig. 6.10 Changes in operation style due to recommendations

increased from 31.6 to 42.8% in the "with recommendation" condition, an improvement that suggests a behavioral change in the operators toward using more registered speech.

As for the registered speech, operators frequently selected speech using the playback buttons (62.4%), which is obviously a much larger portion than the speech made by themselves (37.5%). About 31.6% of the avatar operations were done using the combination button (i.e., speech playback and avatar operation invoked through a single button). These results show clear changes in operation style due to the recommendations.

6.3.5 Discussion

In this section, we describe activity monitoring and management in the CA platform. The activity data observed in various CA-based services, which are stored in the platform's activity database, can be used to enhance such services. This mechanism, which must be efficacious, is essential to establish a society of CAs.

On the other hand, privacy considerations are also important in accumulating activity data. Although access to video and audio information is restricted on the CA platform, it is always necessary to consider how to properly anonymize such data during the accumulation process. Another critical problem in CA services is CA "spoofing," which greatly changes the appearance of the people who operate them, even though strict user authentication is required to prevent unauthorized use. These topics are discussed in Chap. 9.

To accelerate the use of activity monitoring as well as CA-based services, APIs and data formats common to all CAs are required. As described above, activity data have a hierarchical structure that reflects a service hierarchy. Therefore, the adopted data format requires the capability to effectively represent such properties. Standardization of CA platforms is discussed in the next section.

6.4   Standardization for Service Robots and Cybernetic Avatars

In the systemization of robots and services using them, the ability to widely reuse the developed elemental technologies is crucial. Therefore, modularization is progressing from both hardware and software approaches, and robots and robot services are being achieved their combination. Since 2004, the Object Management Group (OMG; https://www.omg.org/index.htm), a private standardization organization targeting the design and implementation of distributed systems, has been standardizing robotics technology. In 2006, the Robotics Domain Task Force (Robotics DTF; https://www.omg.org/robotics/), a technical task force for robot technology, was established, and specifications have been issued, including the Robotic Technology Component (RTC) (Object Management Group 2012a) and the Robotic Interaction Service (RoIS) (Object Management Group 2018a). Also, in ISO, TC299 (Robotics) (ISO 2015)/WG6 (Modularity for Service Robots) is proceeding with the formulation of specifications for modularization in terms of both hardware and software. A Robot Operating System (ROS) (https://www.ros.org/), which has been mainly used in research fields, has emerged as a de facto standard.

These specifications are solving the problem of interface specification design for interconnecting and reusing components/modules that modularize robot functions toward the achievement of robot services. However, the method, which describes what functions these components provide, what environments they can be used in, and how they affect the outside world, remains unclearly defined. Service or component designers are limited to writing in natural language. To provide assorted robot services in varied environments, various robot functions (components) must be dynamically linked. For this purpose, a means is needed to verify the effectiveness and safety of the components involved in the dynamic configuration of services. This problem is common in many fields, and the definition of ontology is being promoted in multiple technical fields in OMG. Here an ontology is a set of frameworks and specifically defined vocabularies for formally describing things in a certain field and their relationships as knowledge. Ontology specifically refers to knowledge that is defined to be verified by computation. Furthermore, the specifications of the ontology platform that supports them have also been formulated. In the field of robotics, the IEEE Robotics and Automation Society (RAS) published the Core Ontology for Robotics and Automation (CORA) (IEEE Robotics and Automation Society 2015), which is formulating multiple field-specific ontologies. In addition, the OMG Robotic Service Ontology (RoSO) (Object Management Group 2018b) was launched to formulate an ontology specifically for robot services; its standardization is currently underway.

On the other hand, the goal of the R&D program of the Moonshot Research and Development Project (https://avatar-ss.org/en/index.html), launched in Japan in FY2020, is achieving a society that frees people from the constraints of their bodies, brains, space, and time by 2050.

This program promotes the research and development of CA technology that exploits the advanced use of a set of technologies known as cyborgs and avatars to augment human physical, cognitive, and perceptual abilities. Its goal is to create a symbiotic society within which avatars play an active role. To implement a wide variety of CAs in parallel and operate and simultaneously manipulate many of them, their functions must be defined and implemented in a standardized manner. Therefore, at OMG, this project is conducting international standardization activities to create CA functions and establish a CA-based protocol as a global standard and promote modularization and reusability. This section introduces the standardization activities of robot-related technology being promoted at OMG and describes the strategy of CA's proposed international standardization efforts, based on a policy of expanding the existing information communication and robot standardization.

6.4.1   Standardization of Robot Technology at OMG

6.4.1.1   OMG Robotics DTF

Based on their targets, OMG's technical committees are roughly divided into two types. One is the Platform Task Force (PTF), which belongs to the Platform Technical Committee (PTC) that targets technologies that can be commonly used in various fields. The other is the Domain Task Force (DTF), which belongs to the Domain Technical Committee (DTC) that targets specific technical fields. The standardization of robot technology is currently being discussed at the Robotics DTF, which was established in 2006. So far, the Robotics DTF has standardized the following: the Robotic Technology Component (RTC) (Object Management Group 2012a), the Robotic Localization Service (RLS) (Object Management Group 2012b), the Robotic Interaction Service (RoIS) (Object Management Group 2018a), and the Finite State Machine Components for RTC (FSM4RTC) (Object Management Group 2016).

6.4.1.2   Standardization of Component Technology

The standardization of component functions has been promoted by the Infrastructure Working Group within the Robotics DTF. RTC is a middleware specification that defines functions commonly required for components in the field of robot technology based on distributed component technology. It was initially discussed in the OMG Middleware and Related Services (MARS) PTF and published as RTC1.0 in 2008. Distributed component technology enables software components running on multiple computers to operate cooperatively through communication. Currently, RTC1.1, which was revised in 2012, is the latest version (Object Management Group 2012a).

Proposals from Japan are mainly promoted by the National Institute of Advanced Industrial Science and Technology (AIST). The specifications of RT middleware

(OpenRTM-aist), developed and distributed by AIST, are based on RTC. Implementations based on RTC specifications are provided in addition to OpenRTM-aist, and their interoperability has been confirmed.

6.4.1.3 Standardization of Robot Function Services

RLS and RoIS are specifications designed to define the functions required for robots. High-level service-related functions have been standardized by the Robotic Functional Service Working Group in the same DTF. RLS defines the representation format and the interface of location information in a generic format that is independent of specific devices and algorithms. RLS 1.0 was published in 2010. Revised specifications were published as RLS 1.1 in 2012, with modifications that added a posture information description method and the generalization of a coordinate system description (Object Management Group 2012b). RoIS defines a framework for standardizing interfaces for using HRI functions (functions of various robots, such as human detection, individual identification, and speech recognition) from service applications. By using this framework, the same service application can work on different robots. RoIS 1.0 was published in 2013, and the most current specification is RoIS 1.2, revised in 2018 (Object Management Group 2018a). The RLS and RoIS specifications were submitted to OMG by the Japan Robot Association (JARA) and the Electronics and Telecommunications Research Institute (ETRI). Eleven organizations, including AIST and Advanced Telecommunications Research Institute International (ATR), collaborated to develop these specifications. Implementation examples are proposed by ETRI and Advanced Telecommunications Research Institute International (ATR).

6.4.1.4 Ontology Standardization for Robot Services

A service robot provides resources to humans in the environments where they are active. The services, which include both those that involve physical interaction and others that interactively provide information to people, are mainly designed to support human activities based on human interaction. Service robots are usually designed for consumers, not for industrial purposes. They make decisions and act semi-autonomously or fully autonomously to provide a given service in an uncontrolled or unpredictable living environment. To do so, various interactions will be targeted between such components as sensors and actuators that comprise the robot and the service environment and various other elements in the environment. A formal framework is needed to describe the conditions necessary for each element to operate as well as for how the results of that action affect other elements. Robotic Service Ontology (RoSO), which was launched at OMG in 2018, defines an ontology for robot services as a framework for this purpose.

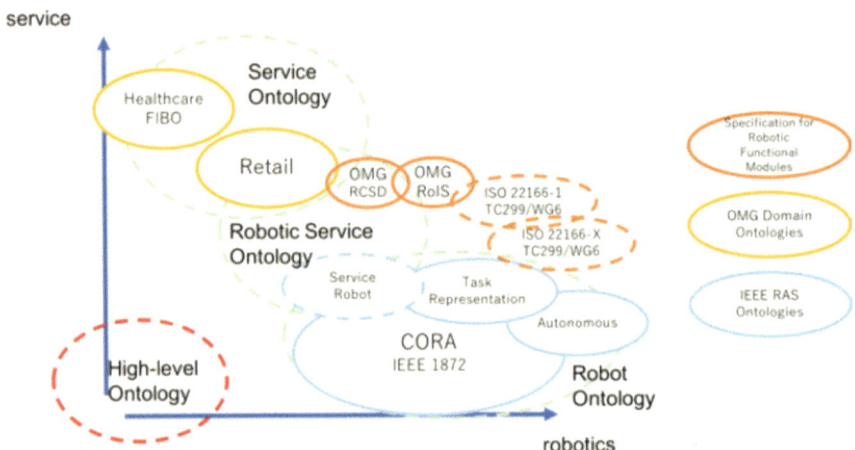

Fig. 6.11 Relationship between RoSO's target areas and other standards. *Source* Robotic Service Ontology RFP (Object Management Group 2018b)

RoSO provides a formal framework, which describes the component functions of the preceding RoIS specifications, and offers a basis for future extensions. RoSO standardization activities are advancing based on the RoSO Request for Proposal (RFP) document issued by OMG in December 2018 (Object Management Group 2018b). Figure 6.11 shows the relationship between RoSO's target areas and other standards. Regarding ontologies in the field of robotics, IEEE 1872 2015 (CORA; Core Ontology for Robotics and Automation) published by IEEE-RAS (IEEE Robotics and Automation Society 2015) defines the vocabulary that forms the basis of robotics technology.

Based on CORA, IEEE-RAS is formulating a definition of high-level ontology for multiple domains such as autonomous and collaborative robots. On the other hand, OMG is defining ontologies in various service areas. The definition of ontology in robot interaction/retail services is particularly relevant to robot technology. The regions of interest for RoSO are located at these nodes. Therefore, task forces will continue to exchange opinions with other fields within OMG and such other standardization organizations as IEEE-RAS and ISO TC299/WG6 and proceed with specification formulation.

6.4.2 Standardization for Cybernetic Avatars

For achieving social participation support services that utilize CAs by implementing a wide variety of real and virtual CAs in parallel and simultaneously operating and manipulating many of them, their functions must be defined and implemented by standardized methods. CA's basic technologies are being standardized as measures

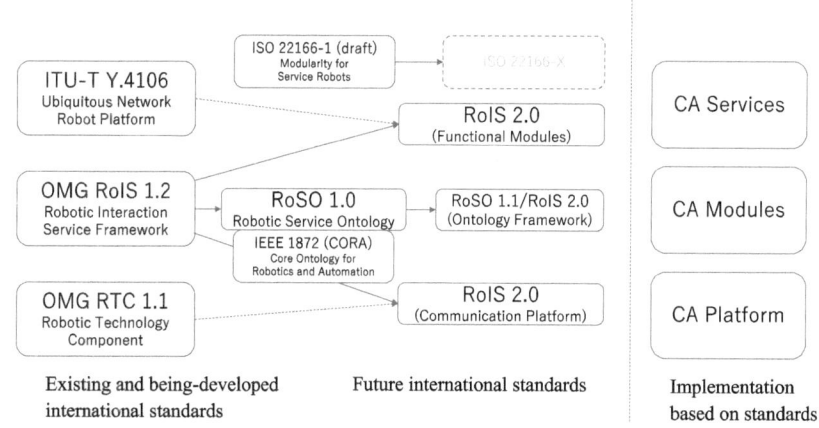

Fig. 6.12 International standards related to our project and our implementation

for widely disseminating throughout society the CA platform that was proposed in the R&D program of the Moonshot Research and Development Project. These measures include developed CA functions and CA platform protocols as international standard specifications. To standardize CA's basic technologies, CA platform's international standardization activities are promoted through OMG and the International Organization for Standardization (ISO), both of which have a proven track record for standardizing robot software. CA's platform also referred to the discussions at IEEE-RAS (Robotics and Automation Society). The strategy of this activity is to expand the existing standardization of information communication and robots to CAs.

Figure 6.12 shows the relationship among international standards: those which exist, those which have been developed, and those which will be developed in the future as well as the implementation planned for this project. By promoting international standardization activities based on the expansion of current and being-developed international standards with track records of activities, the discussions are expected to proceed promptly and efficiently to promote international standardization. The proposed strategy for the international standardization for CA's basic technologies is introduced in the following parts.

6.4.2.1 International Standardization of Description Method of CA Service Function Specifications

Robotics DTF, a technology committee subgroup of OMG, is promoting the international standardization of the modularization of robot technologies. Robotic Technology Component (RTC) specification 1.1 stipulates specifications for implementing the functional modules of robots based on distributed components (Object Management Group 2012a). Robotic Interaction Service (RoIS) specification 1.2

stipulates a specification-description method for robot functional components for interactive services as well as the specifications for specific common components (Object Management Group 2018a). OMG Robotics DTF is currently developing a Robotic Service Ontology (RoSO) (Object Management Group 2018b) specification that stipulates the ontology for describing the functional requirements of robot services for the extension of RTC and RoIS specifications. The specifications, which describe the proposed CA service functions, are expected to be incorporated into RoSO to achieve the international standardization of a description method of CA service function specifications.

Related to robot technology, IEEE-RAS is currently promoting the cooperation of various ontologies that are being studied in parallel by various organizations. Related meetings are being held at IROS and ICRA, both of which are influential international conferences in the field of robots. The progress of the specification development of OMG RoSO is reported at these meetings, and the international standardization of such specifications is promoted based on these discussions. The international standardization of a description method of the CA service function specifications is achieved by incorporating them into the final proposal of the OMG RoSO 1.0 specifications.

6.4.2.2   International Standardization of CA Platform Specifications

The lower layer parts of the above RTC and RoIS specifications are defined as distributed component technologies, and the robot function platform specifications are defined in the upper layer of these technologies. As an example, ITU-T Recommendation Y.4106 (requirements and functional model for a ubiquitous network robot platform that supports ubiquitous sensor network applications and services) defines a platform for network robots with reference to RoIS (International Telecommunication Union 2013).

In a roadmap for extending the RTC and RoIS specifications, OMG is defining the middle layer of the RoIS specification as RoSO and reconsidering the platform technology of the lower layer and the definition of the specific components of the upper layer. Therefore, as a proposed strategy, the international standardization of the CA platform specifications is incorporating them into the revised specifications of the lower layer of OMG RoIS 2.0 (RoIS 2.0 is the tentative name of the next RoIS version).

6.4.2.3   International Standardization of CA and Its Operation Interface Functions

As with the common component functions of the robot dialogue service defined as the upper layer of the OMG RoIS, the functions required by the CA and the operation interface are achieved by modularizing and defining them. This part is standardized as CA's international specifications and its operation interface functions.

Similar to the previous part, OMG is considering the revision of RoIS by redefining its upper layer using RoSO. As a proposed strategy, the international standardization of CA and its operation interface functions are carried out by proposing definitions of the modularized CA functions and incorporating them into the revised specifications of the upper layer of OMG RoIS 2.0.

6.4.3   Standardization for Cybernetic Avatars and Their Social Implementation

This section introduces the standardization activities of robot-related technology that are being promoted at OMG. It also describes the proposed strategy of CA's international standardization based on a policy of expanding the existing information communication and robot standardization in the Moonshot R&D Project. From the flow of various de facto and de jure standards related to robot services so far, the flow of defining the ontology of robot services is inevitable. In the future, modularization is expected to be related to robot service development and the improvement of its reusability simplification. In addition, by incorporating CA standardization into RoSO, more general-purpose robot-related technology will be internationally standardized.

The Moonshot project aims to devise advanced uses for CA technology to improve people's physical, cognitive, and perceptual abilities for achieving a society where people and CAs coexist. To that end, the project is promoting CA's social implementation. Four types of people advance its social implementation: those who build and provide CA systems, those who directly operate CAs, those who provide services using them, and those who receive services provided by them. The goal is that everyone can easily use CAs in the same way, and each function must be modularized and prepared as tools. Since standardizing, defining, and implementing these tools are effective, international standardization is obviously one efficacious means of achieving this goal.

These activities are expected to contribute to the creation of robot services and CA markets that will inevitably flourish in the post-coronavirus pandemic period.

References

IEEE Robotics and Automation Society (2015) Core ontology for robotics and automation (CORA). https://standards.ieee.org/ieee/1872/5354/. Accessed 5 Jan 2024

International Telecommunication Union (2013) Recommendation Y.4106: requirements and functional model for a ubiquitous network robot platform that supports ubiquitous sensor network applications and services. https://www.itu.int/rec/T-REC-Y.4106/. Accessed 5 Jan 2024

ISO (2015) TC299 (Robotics). https://www.iso.org/committee/5915511.html. Accessed 5 Jan 2024

Kamei K, Nishio S, Hagita N, Sato M (2012) Cloud network robotics. IEEE Network Mag 26(3):28–34. https://doi.org/10.1109/MNET.2012.6201213

Kamei K, Zanlungo F, Kanda T, Horikawa Y, Miyashita T, Hagita N (2017) Cloud networked robotics for social robotic services extending robotic functional service standards to support autonomous mobility system in social environments. In: Proceedings of 2017 14th international conference on ubiquitous robots and ambient intelligence (URAI), pp 897–902. https://doi.org/10.1109/URAI.2017.7992862

Object Management Group (2012a) Robotic technology component (RTC) version 1.1 (formal/2012-09-01). http://www.omg.org/spec/RTC/. Accessed 5 Jan 2024

Object Management Group (2012b) Robotic localization service (RLS), version 1.1 (formal/2012-08-01). http://www.omg.org/spec/RLS/. Accessed 5 Jan 2024

Object Management Group (2016) Finite state machine component for RTC (FSM4RTC), version 1.0 (formal/2016-04-01)

Object Management Group (2018a) Robotic interaction service (RoIS) frame-work, version 1.2 (formal/2018-05-04). http://www.omg.org/spec/RoIS/. Accessed 5 Jan 2024

Object Management Group (2018b) Robotic service ontology (RoSO) request for proposal (robotics/2018-05-04). https://www.omg.org/cgi-bin/doc.cgi?robotics/2018-12-3. Accessed 5 Jan 2024

Rea Daniel J, Seo Stela H (2022) Still not solved: a call for renewed focus on user-centered teleoperation interfaces. Front Rob AI 9. https://doi.org/10.3389/frobt.2022.704225

Zachiotis G, Andrikopoulos G, Gornez R, Nakamura K, Nikolakopoulos G (2018) A survey on the application trends of home service robotics. In: Proceedings of 2018 IEEE international conference on robotics and biomimetics (ROBIO). IEEE, pp 1999–2006. https://doi.org/10.1109/ROBIO.2018.8665127

Chapter 7
Multidisciplinary Investigation on How Avatars and Devices Affect Human Physiology

Shinpei Kawaoka, Yoshihiro Izumi, Keisuke Nakata, Masahiko Haruno, Toshiko Tanaka, Hidenobu Sumioka, David Achanccaray, and Aya Nakae

Abstract The development of Cybernetic Avatars (CAs) will change our lives dramatically. Such a rapid evolution of new technologies is advantageous but also raises concerns, such as addiction. Herein, we present a new scientific issue regarding how to investigate the effects that using CAs throughout our lives has on our physiology. In this chapter, we discuss how avatars and devices affect human physiology at multiple levels, from gene expression to brain activity. We insist that our multidisciplinary investigation of how CAs affect us will be critical for the further development of our CA society.

S. Kawaoka (✉)
Tohoku University, Sendai, Miyagi, Japan
e-mail: kawaokashinpei@gmail.com

Kyoto University, Kyoto, Kyoto, Japan

Y. Izumi · K. Nakata
Kyushu University, Fukuoka, Fukuoka, Japan
e-mail: izumi@bioreg.kyushu-u.ac.jp

K. Nakata
e-mail: knakata@bioreg.kyushu-u.ac.jp

M. Haruno · T. Tanaka
National Institute of Information and Communications Technology, Suita, Osaka, Japan
e-mail: mharuno@nict.go.jp

T. Tanaka
e-mail: toshiko_t@nict.go.jp

H. Sumioka · D. Achanccaray · A. Nakae
Advanced Telecommunications Research Institute International, Seika-cho, Soraku-gun, Kyoto, Japan
e-mail: sumioka@atr.jp

D. Achanccaray
e-mail: dachanccaray@atr.jp

A. Nakae
e-mail: ayanakae@atr.jp

© The Author(s) 2025
H. Ishiguro et al. (eds.), *Cybernetic Avatar*,
https://doi.org/10.1007/978-981-97-3752-9_7

7.1   Introduction

The development of Cybernetic Avatars (CAs) will markedly change our lives. By using CAs, we can communicate with people wherever we are. We can talk to more than two people simultaneously with the help of CAs. CAs enable us to work even after we become older and less active, thereby increasing productivity. CAs will be everywhere around us, improving our quality of life and impacting our economy.

Such a rapid evolution of new technologies is advantageous but also raises concerns, such as addiction. Smartphones are an example. Smartphones are useful everywhere and have changed the style of communication. Many human activities rely on smartphones; however, smartphones have led to new issues, including smartphone addiction. For example, it is a matter of debate how parents should control their children's use of smartphones. As discussed previously, it is common for new technologies to bring about new aspects and concerns. The same can be said of CA technology.

Herein, we present a new scientific issue regarding how to investigate the effects that using CAs throughout our lives has on our physiology. Subjectively evaluating the positive and negative aspects of the use of new technologies in humans is challenging. Do CAs affect the metabolism, gene expression, and brain activity? If yes, how? This chapter introduces the challenges in addressing this critical question.

In Sect. 7.2, we discuss how avatars and devices affect human physiology at multiple levels. In Sect. 7.3, we introduce a metabolomic analysis that measures various metabolites. In Sect. 7.4, we discuss how avatars affect brain activity. In Sect. 7.5, we present how avatars affect psychological and physiological aspects. In Sect. 7.6, we discuss how to establish health standards for the use of CAs. We insist that our multidisciplinary investigation of how CAs affect us will be critical for further developing our CA society.

7.2   Integrative Bioanalytics on Human–Avatar Interactions

7.2.1   Homeostasis in Biology

Homeostasis maintains an organism's condition within a certain range, enabling survival in an ever-changing environment. The crucial function of homeostasis is indispensable in biology and medicine.

Homeostasis is dynamic. Consider the blood glucose level as an example (Fig. 7.1) (Röder et al. 2016). Fasting blood glucose levels range from 70–99 mg/dL in Japan. This range is narrow. When we eat meals, blood glucose levels rapidly elevate, and the pancreas eventually senses the elevated blood glucose levels. In the pancreas, β cells in islets secrete insulin in response to the increase in blood glucose levels. In turn, insulin stimulates peripheral cells such as hepatocytes to take glucose into the cells, consequently reducing the blood glucose levels. Blood glucose maintenance

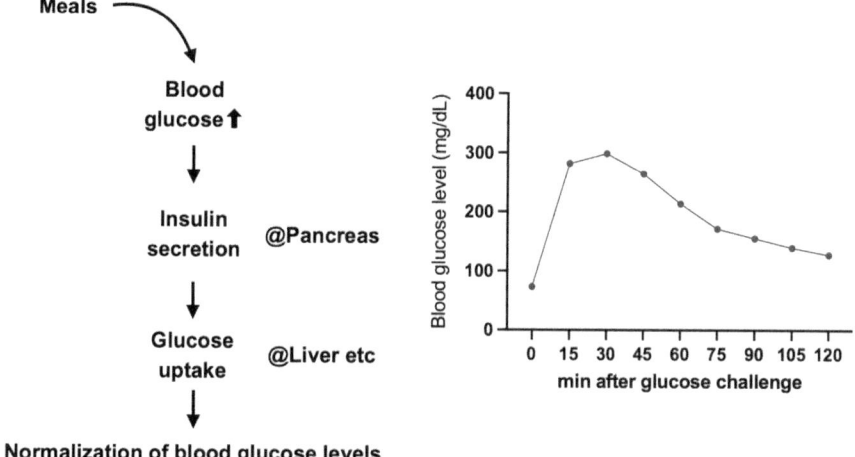

Fig. 7.1 Dynamic glucose homeostasis

is an example of dynamic homeostasis. This dynamic mechanism maintains blood glucose levels within a certain range.

Homeostasis can be disrupted by diseases. Consider the example of glucose homeostasis. Patients with diabetes are often unable to control their blood glucose levels (Donath et al. 2019). In type I diabetes, self-reactive T cells eliminate pancreatic β cells, resulting in an inability to produce insulin. Under these conditions, blood glucose levels cannot be reduced after meals. In type II diabetes, insulin is present; however, peripheral cells cannot respond to insulin. This anomaly makes it impossible for organisms to normalize their blood glucose levels. The inability to normalize blood glucose levels involves different molecular mechanisms.

Glucose homeostasis is one of the best-studied mechanisms of homeostasis. It is known which cell types are critical for glucose homeostasis, and we are aware of biological molecules that regulate glucose homeostasis. The timescale of the biochemical reactions involved in glucose homeostasis is also known. Hence, we can confidently distinguish between normal and abnormal glucose homeostasis during regular health check-ups. Various medications are used to normalize abnormal glucose homeostasis. Understanding the details of glucose homeostasis is crucial for detecting abnormalities and developing therapeutics for this essential biological pathway.

Let us generalize our knowledge of glucose homeostasis (Fig. 7.1). Glucose homeostasis is our body's mechanism for maintaining glucose levels (i.e., a key biological parameter) in response to stimuli (i.e., foods), and it involves various types of cells and molecules. We use similar formulas to understand our responses to drugs, stress, infections, etc. This view is one of the critical bases of biology and medicine.

As expected, biological homeostasis has been extensively studied in developmental biology and disease research. However, our knowledge of the homeostatic regulation of other common daily activities remains limited. For example, what happens to homeostasis while listening to music? What types of molecules fluctuate during a verbal conversation? How does the use of CAs affect our homeostasis? Can homeostatic regulation during CA use be understood in a manner similar to that of glucose homeostasis? Does everyone respond similarly to such stimuli? Does each person respond differently to stimuli? This section discusses these topics by introducing a multi-omics analysis, which is a powerful tool for answering these questions.

7.2.2 Introduction into Multi-omics Analyses

Glucose and insulin cannot be ignored when explaining glucose homeostasis because we know that these molecules are essential for understanding this phenomenon. On the other hand, we do not know the entire effects of CAs on homeostasis. We are unsure of the cell types, molecules, and other factors that play a role in the response to CAs. One of the main reasons for the difficulty in understanding the homeostatic response of the body to CAs is the large number of genes, metabolites, and cells involved in this process. Although we can establish the hypothesis that a particular gene is involved in this process, we have more than 20,000 genes. We have also more than 1000 metabolites. Further, there are more than 30 trillion cells and 200 hundred cell types. Given the large number of factors that must be considered, it is not easy to formulate an appropriate hypothesis (i.e., choose a factor) when we do not know the whole picture regarding our responses to the use of CAs.

Multi-omics analysis plays a prominent role in this situation (Fig. 7.2) (Hasin et al. 2017). Omics is a derivative of the Greek word "ome," which means mass, many, and whole. The word "genome" is the best example. This word comes from "gene" and "ome," representing whole genes. Similarly, the word "transcriptome" comes from "transcript" and "ome." Likewise, "proteome" comes from "protein" and "ome," and "metabolome" comes from "metabolite" and "ome." These different layers form a multi-omics network in cells. Omics analysis refers to the measurement and data analysis of these molecules. Omics analyses are useful for thoroughly characterizing biological phenomena, and these techniques enabled us to observe forests and trees simultaneously.

Omics layers are intrinsically connected (Fig. 7.2) (Ille et al. 2022). Let us now explain how the typical protein-coding gene X is regulated. Core information is written into the genome as DNA sequences.

We have four different letters, namely, A, T, G, and C, in our genome, and we can "read" genome sequences in laboratories. When appropriate, information on gene X is "transcribed" into messenger RNAs (mRNAs), which are also called transcripts. These transcripts are further "translated" into protein X by ribosomal translational machinery. Translated proteins are often modified (e.g., phosphorylated) in various ways. Protein X, if it is an enzyme, catalyzes biochemical reactions to consume

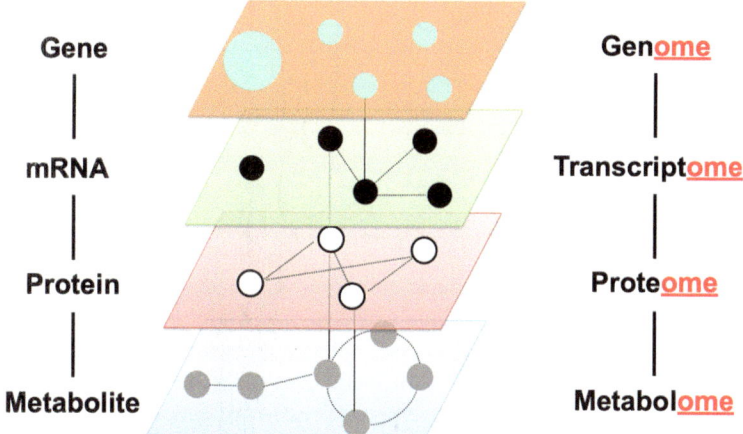

Fig. 7.2 Multi-omics network in biology

metabolite X and produce metabolite Y. The process of producing proteins from information encoded by genes is called gene expression. Our genome contains more than 20,000 genes that are differentially expressed in different cell types under different conditions. Surprisingly, these cells can regulate gene expression in response to complex external and internal cues to maintain homeostasis.

7.2.3 Transcriptomics

This subsection introduces the omics technique known as transcriptomics (Longo et al., 2021). Transcriptome analysis measures transcripts (mRNAs). The number of transcripts indicates the number of mRNA copies of the genes transcribed from the genome. However, caution must be exercised when using transcripts to measure gene activity (e.g., enzymatic activity). It is known that the abundance of transcripts sometimes does not always correlate linearly with enzymatic activity. Thus, it is safe to consider the transcriptome as an indirect measure of gene activity. The transcriptome is a useful and powerful method for inferring cellular status from various biological aspects.

For the transcriptome analysis, we collect mRNAs (transcripts) from the cells of interest. Using complex procedures, we generate complementary DNA (cDNA) from the transcripts and sequence them. Sequencing experiments give rise to numerous sequence reads, for example, 20 million reads from the prepared samples, depending on the purpose of the analysis. We then count the number of reads per gene to calculate the number of transcripts transcribed from the genes in the genome and constructed a gene expression matrix (Fig. 7.3).

Fig. 7.3 Examples of transcriptome datasets

Person A	Pre	Post	Fold change
Gene X	2.2	11	5
Gene Y	8	9.1	1.1375
Gene Z	20	8	0.4

Fig. 7.3 Examples of transcriptome datasets

The transcriptome analysis of many cell types is referred to as "bulk" transcriptome analysis. In a bulk transcriptome analysis, we obtain gene expression scores averaged from a series of cell types. For example, peripheral mononuclear blood cells (PBMCs) contain T cells, B cells, neutrophils, and so on. Gene expression scores obtained from PBMCs indicate the average expression of genes in the different cell types. Recent advancements in bioinformatics have given us the capability to deduce the abundances of these different cell types in PBMCs, thereby allowing for the deconvolution of the bulk transcriptome.

We now present the key to this subsection. When we perform transcriptome analyses before and after stimuli, such as the use of CAs, we are able to calculate gene expression changes in response to the stimuli (Fig. 7.3). Such measurements allow us to analyze how cells, organs, and people respond to stimuli. Such analyses cannot be performed with single-timepoint measurements. Moreover, omics measurements do not require firmly defined hypotheses. We measure everything possible, according to the methods used. The goal of our Moonshot group is to perform multi-omics analyses to measure how people respond to CAs, thereby opening a new avenue for research in this field.

We also have different types of transcriptome experiments. The single-cell transcriptome provides gene expression matrices for each cell type (Longo et al. 2021). This is an experimental deconvolution of the bulk transcriptome, which is powerful for identifying cell types within samples. The procedure for single-cell transcriptome analysis is much more complicated than that for bulk transcriptome analysis. Thus, it is important to obtain a bulk transcriptome and a single-cell transcriptome to increase the validity of experiments. Spatial transcriptome analysis is an emerging method that simultaneously performs gene expression and histological analyses. Both bulk transcriptome and single-cell transcriptome analyses destroy tissue structures. Therefore, it is impossible to know what types of cells neighbor the cells of interest. The spatial transcriptome resolves this technical issue, further strengthening the power of transcriptome analyses in biology and medicine (Longo et al. 2021; Vandenbon et al. 2023).

Despite having different methods, other omics techniques, including proteomics and metabolomics, have similar concepts. These techniques measure their targets in an unbiased manner. By comparing measurements taken before and after an interaction with CAs, we can describe responses to CAs without an a priori hypothesis.

7.2.4 Confounding Factors

When performing omics experiments, we must be aware of the various parameters that affect the results. When comparing cancerous mice with healthy mice, the effects of cancer are expected to be strong. In this circumstance, we may not be overly worried about confounding factors. In contrast, the effects of CAs on humans may be weaker than those of cancer. In this case, it is important to identify possible confounding factors before conducting experiments.

For example, the circadian rhythm is the mechanism that confers an approximately 24-h cycle to biological phenomena. If we do not consider circadian rhythms and do not control the time of sampling, our data will be affected by circadian differences that can mislead our interpretation. For example, differences caused by circadian rhythms can be larger than those caused by CAs, which can mislead us into thinking that CAs can cause a certain response in humans when the real cause is the circadian rhythm. This may not always be the case, but it is important to consider confounding factors in experiments. In most cases, we do not know all the confounding factors, but understanding confounding factors is crucial for successful measurements.

7.2.5 The Need for Omics Analyses in the Field of Cybernetic Avatars

In this section, we discuss the need for multi-omics analyses to investigate the effects of CAs on physiology. Historically, such investigations have relied on text-based surveys and measurements of a particular set of biological molecules. Text-based surveys are useful because they are feasible and easy to implement. However, these surveys tend to be subjective. Furthermore, data interpretation is challenging. For example, using POMS2, we calculate "emotions" from the results of questionnaires. One of the fundamental problems in this process is that there is no objective method for examining the validity of our interpretations. Despite their utility, text-based surveys have limitations in terms of data analysis.

Researchers measure hormones, including cortisol, to evaluate the effects of these devices and CAs in humans (Russell and Lightman 2019; Sumioka et al. 2013). Cortisol is considered a "stress hormone." The hypothalamus-anterior pituitary-adrenal cortex axis increases blood cortisol levels. Cortisol plays a role in the cellular stress response, and cortisol levels become elevated under stressful conditions.

However, does an elevation in cortisol always suggest "stress"? In the first place, what is "stress"? How can stress be defined firmly in this context? Is this a detrimental reaction or an adaptive response? Ignoring these basic questions is sometimes dangerous because it can lead researchers in the wrong direction. Cortisol cannot be oversimplified as a stress hormone, at least in the avatar context that we are discussing.

Omics experiments are potent for capturing network-wide changes in a dataset. Thus, we are able to investigate whether cortisol is the most prominently altered hormone in a certain situation. We determined whether other stress-response pathways change under the same conditions. The role of cortisol in a particular dataset can be evaluated to help interpret results. Oxytocin is another frequently measured hormone that is sometimes called the "happiness hormone." We agree that oxytocin is elevated under good conditions, but it may also be elevated under other conditions. How should we interpret the data when cortisol and oxytocin are simultaneously upregulated? Omics experiments can resolve these issues and help interpret our datasets to characterize the systemic status of the human body.

Studies on the effects of CAs on glucose homeostasis differ from those on glucose homeostasis. Our knowledge of this topic remains limited. Care must be taken to avoid the over-discussion and over-interpretation of results and to understand the entire picture of the effects before attempting a detailed discussion or interpretation.

7.2.6   Learning from Gaming Issues

The application of omics experiments to the field of CAs is a new endeavor. Hence, we do not have many examples, other than our ongoing experiments, to investigate the effects of CAs on human physiology. However, there are topics to which we can apply our principles.

In gaming, an avatar is used. Gaming is a large market that includes professionals (e.g., e-sports). However, excessive gaming is associated with health hazards. Excessive gaming exhausts people, deprives them of time, and may ultimately lead to addiction (Rosendo-Rios et al. 2022). Gaming addiction is prevalent worldwide and has become an issue in many countries. As a result, parents often establish rules for gaming, for example, prohibiting their children from gaming for more than two hours per day. We observed the same trend for newly emerging devices, such as smartphones. We expect to encounter similar issues regarding the use of CAs in the near future.

Let us establish questions based on the knowledge described previously. What is the nature of fatigue caused by gaming? Is it similar to fatigue caused by other activities, such as exercise? Does gaming fatigue differ from exercise fatigue? If yes, how? What is gaming addiction? How can gaming addiction be objectively described? How should rules for children's gaming be established scientifically?

The omics experiments can answer these questions. We can measure the gene expression in immune cells before and after gaming. We also can measure the levels of metabolites and hormones in blood samples. Again, we do not need to have a formal hypothesis. Instead, we thoroughly characterize what happens in gene expression and metabolism in the body during gaming. The same analyses can be performed for exercise to compare gaming and exercise. We could change the duration of gaming in our experiments and observe time-dependent changes. People with or without addiction to gaming are likely to have different multi-omics networks, even before gaming (i.e., different basal statuses).

Fig. 7.4 Individuality in omics datasets

Gene X	Pre	Post	Fold change
Person A	5	25	5
Person B	20	80	4
Person C	5	6	1.2

Furthermore, comparing pre-post datasets (i.e., before and after gaming) from individuals is important for considering individuality (Fig. 7.4). It is likely that healthy individuals respond differently to gaming. This is strikingly different from glucose homeostasis, in which healthy (non-diabetic) individuals respond similarly to glucose. Individual-level datasets can be mined to stratify different people into sub-categories, providing them with a "precision guide" for gaming.

Such omics experiments will eventually lead to the identification of a specific set of molecules that represent our responses to gaming. In the best case scenario, we will be able to identify a molecule that plays a central role in the human body's response to gaming, analogous to insulin, which plays a central role in glucose metabolism. The genes and metabolites that represent our response to gaming will be useful for evaluating the effects of gaming on our bodies. Such molecules can also be used to design the structures of gaming software, which has beneficial effects on us by design. Positive feedback between omics-based evaluation and device design will enhance the evolution of the avatar society in a better manner.

Omics experiments are common in biology and medicine but are still rare in the research area of human-device interaction. We are working intensively on this topic through the Ishiguro Moonshot Project to revolutionize our methods for evaluating and controlling the effects of CAs on humans, thereby improving our new avatar society from a human health perspective.

7.3 A Survey of Biomarkers to Assess the Impact of Cybernetic Avatars and Devices on Users

7.3.1 Metabolomics

Metabolomics is a term coined from the combination of "metabolite," the Greek word "ome," meaning "all and complete," and the suffix "ics," meaning "academics," and it refers to the comprehensive analysis of low molecular weight compounds with a molecular weight of less than 2000 (Fiehn 2002; Fukusaki and Kobayashi 2005). A metabolomic analysis comprehensively captures chemical changes in small enzyme-based molecules generated from genomic information through transcription and translation. Since Fiehn et al. identified silent mutations in *Arabidopsis thaliana* as metabolite phenotypes in 2000 (Fiehn et al. 2000), metabolomics has become one of the closest high-resolution phenotyping tools for genomic information. Metabolomes are potentially useful indicators of the phenotype of biological activity, which is complex and constantly changing, depending on genes or external factors such as drugs or the environment. Recent studies of cancer metabolism have shown that in addition to the enhanced "Warburg effect," in which glucose, a carbon source, is metabolized anaerobically in cancer cells (carbon shift), the metabolic system to utilize glutamine-derived nitrogen for DNA synthesis is enhanced in malignant cancers (nitrogen shift) (Kodama et al. 2020). This indicates that there is a strong relationship between disease and metabolism. In recent years, the importance of metabolomic analyses in medical research has been recognized, and metabolomic analyses have been applied worldwide for biomarker discovery and disease metabolism research (Yoshida et al. 2012). The advantages of metabolomics in medical research include (i) quantitative observation using unbiased methods, (ii) high-resolution phenotyping of disease phenotypes, and (iii) the ability to find links between etiology and metabolism from a bird's-eye perspective.

The U.S. National Institutes of Health (NIH) defines a biomarker as "a characteristic that is objectively measured and evaluated as an indicator of a normal biological process, a pathogenic process, or a pharmacological response to a therapeutic intervention." (Strimbu and Tavel 2010) Blood glucose is used as a metabolite biomarker for diabetes mellitus, and cholesterol and neutral lipids are used as metabolite biomarkers for dyslipidemia during routine health screenings (American Diabetes Association 2006; Xia et al. 2013). However, the currently used metabolites are limited to those with high in vivo concentrations (e.g., mg/dL). The use of metabolites at low concentrations as biomarkers is expected to enable the prevention and diagnosis of more diseases and lead to the development of precision medicine (Beebe and Kennedy 2016; Trivedi et al. 2017; Lange and Fedorova 2020).

7.3.2   New Technologies for Metabolomic Analysis

Each step of metabolomics (Fig. 7.5) involves elements that introduce errors, making it extremely difficult to establish standard techniques. Metabolomics, a multidisciplinary field in the life sciences, organic chemistry, analytical chemistry, and computer science, is still in its infancy in terms of both technology and operational method development. Metabolites have a wide range of polarities, from hydrophilic compounds, such as amino acids and organic acids, to hydrophobic compounds such as lipids, as well as charge characteristics, such as cationic, amphiphilic, anionic, and uncharged compounds, which result in a variety of physicochemical properties. In addition, because metabolites contain many structural and geometric isomers, separation analysis methods combining various types of chromatography with mass spectrometry (MS) are commonly used to comprehensively and accurately measure metabolites. However, because it is virtually impossible to measure all metabolites in

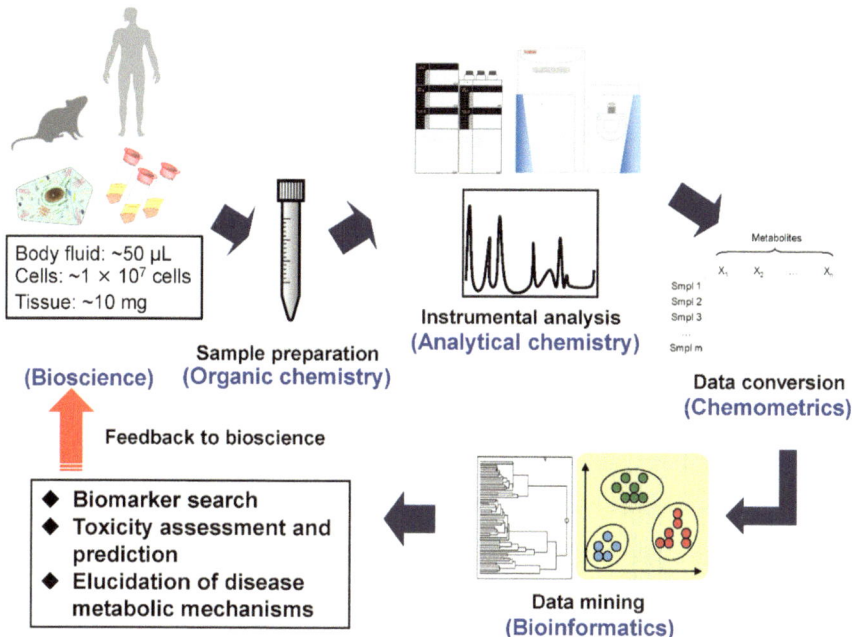

Fig. 7.5 Scheme of metabolomics

a single analysis, a combination of several analytical systems is a common practice for improving metabolite coverage (Izumi et al. 2019) (Fig. 7.6).

Key metabolic pathways (glycolysis, pentose phosphate pathway, citric acid cycle, nucleic acid metabolism, amino acid metabolism, etc.) are involved in the generation of vital energy, cell maintenance, and repair processes, and they are also the most important pathways for understanding metabolism in cancer and other diseases. Many metabolic intermediates are ionic or highly polar. Hydrophilic metabolites are mainly measured using capillary electrophoresis mass spectrometry (CE-MS) (Harada et al. 2008) and ion-pairing reversed-phase liquid chromatography mass spectrometry (IP-RP-LC/MS) (Kato et al. 2012). However, because of the wide range of physical properties of hydrophilic metabolites such as polarity, charge, and molecular weight, comprehensive and practical measurements have not yet been achieved, and the development of analytical methods as a first choice is still in its infancy. Recently, we developed a simple method that enables the comprehensive and simultaneous analysis of hydrophilic metabolites using unified-hydrophilic-interaction/anion-exchange liquid chromatography tandem mass spectrometry (unified-HILIC/AEX/MS/MS) with a polymer-based mixed amine column composed of methacrylate-based polymer particles and primary, secondary, tertiary, and quaternary amines as functional groups. Unified HILIC/AEX/MS/MS is the only method capable of the simultaneous chromatographic separation and MS detection

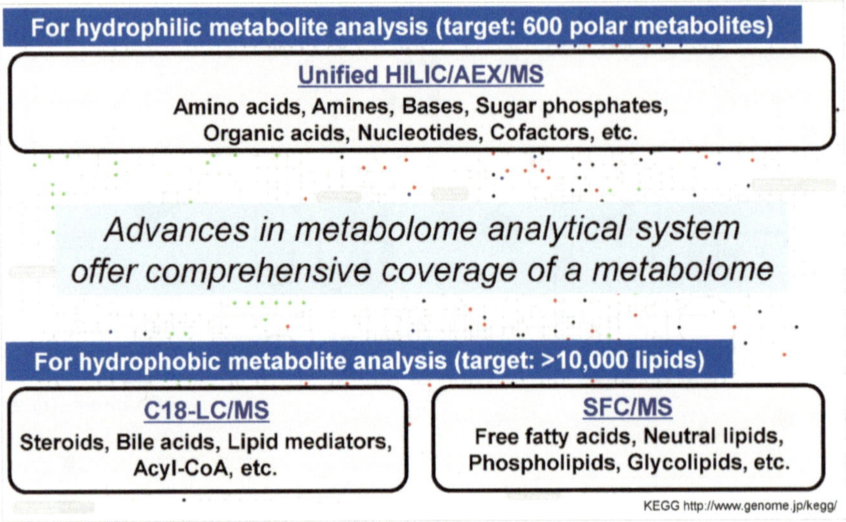

For hydrophilic metabolite analysis (target: 600 polar metabolites)

Unified HILIC/AEX/MS
Amino acids, Amines, Bases, Sugar phosphates,
Organic acids, Nucleotides, Cofactors, etc.

*Advances in metabolome analytical system
offer comprehensive coverage of a metabolome*

For hydrophobic metabolite analysis (target: >10,000 lipids)

C18-LC/MS	**SFC/MS**
Steroids, Bile acids, Lipid mediators, Acyl-CoA, etc.	Free fatty acids, Neutral lipids, Phospholipids, Glycolipids, etc.

KEGG http://www.genome.jp/kegg/

Fig. 7.6 Development of a metabolomics analysis platform

of cationic, zwitterionic, and anionic polar metabolites, making it the first choice for polar metabolomic analyses (Nakatani et al. 2022).

For comprehensive lipidome analyses, we developed a unique and widely targeted quantitative lipidomic analysis method using supercritical fluid chromatography tandem mass spectrometry (SFC/MS/MS) and an in silico multiple reaction monitoring (MRM) library (Takeda et al. 2018). To overcome the problem of quantification, which is a major weakness of MS, we simultaneously standardized the matrix effects of individual lipid molecules within a lipid class by establishing an internal standard for each lipid class and using SFC separation conditions under which the internal standard and individual lipid molecules co-elute. Specifically, the SFC separation mode was selected using a normal-phase column to achieve the chromatographic separation of each lipid class by polar head groups, and a triple quadrupole mass spectrometer (QqQMS) was used for MS and coupled to the SFC to allow for a comprehensive and highly sensitive analysis of the simultaneously eluted individual lipid molecules in each lipid class. MRM transitions derived from fatty acid side chains have been used to discriminate structural isomers (e.g., phosphatidylcholine [PC] 16:0 20:4 and PC 18:2 18:2) within each lipid class. However, the QqQMS method is a targeted analysis and therefore requires prior information about the lipid molecules present in the sample. Because the number of fatty acids constituting lipids in living organisms is limited and fragmentation by MS/MS is regular for each lipid class, hypothetical MRM transitions were created. Using the developed SFC-MS/MS method, human plasma was analyzed, and more than 500 lipid molecular species were successfully detected and quantified.

7.3.3 Examples of Biomarker Discovery Studies

Currently, most applications of metabolomics in biomedical research involve the discovery of biomarkers. Under various pathological conditions, metabolic changes caused by enzymes and proteins occur in disease-related cells and tissues, resulting in disease-specific patterns of metabolites that are reflected in the blood and urine. Furthermore, the metabolome is considered to be the culmination of upstream omics cascades, and metabolomic analyses may be able to detect the state of the organism before the phenotype appears as well as minute changes in metabolic pathways. Therefore, there is a growing movement to apply metabolomics to the diagnosis of diseases using blood, urine, and saliva, which can be collected relatively noninvasively.

Nishiumi et al. performed a large-scale metabolomic analysis of sera from more than 100 colorectal cancer patients using gas chromatography–mass spectrometry (GC/MS) (Nishiumi et al. 2012). A colorectal cancer prediction model was constructed using a multiple logistic regression analysis with a stepwise variable selection method for metabolites showing significant changes. The predictive model consisted of 2-hydroxybutyrate, aspartic acid, kynurenine, and cystamine with an area under the curve (AUC) of 0.9097, a sensitivity of 85.0%, a specificity of 85.0%, and an accuracy of 85.0%. Notably, the sensitivity for detecting stage 0–2 colorectal cancer was as high as 82.8%. The predictive model constructed via serum metabolomic analysis using GC–MS is useful for the early detection of colorectal cancer and may become a new screening test for colorectal cancer.

Sreekumar et al. analyzed 262 urine, blood, and tissue samples from patients with malignant and benign prostate cancers using metabolomics with liquid chromatography (LC/MS) and GC/MS (Sreekumar et al. 2009). The results confirmed the correlation between sarcosine molecules and cancer progression by comparing metastatic and non-metastatic cancer tissues. Cultured cells from invasive prostate cancer also had elevated levels of sarcosine compared with those from benign cancers. The knockout of the enzyme that converts glycine to sarcosine reduces cancer cell invasion. Based on these results, Sreekumar et al. concluded that increased sarcosine levels were involved in prostate cancer progression.

Although a relationship between depression and personality has long been suggested, biomarker studies of depression have largely been overlooked. Setoyama et al. obtained 63 plasma metabolite profiles from 100 drug-naïve patients with major depressive disorder (MDD) and 100 healthy controls through a metabolomic analysis using LC/MS (Setoyama et al. 2021). They found that the levels of tryptophan pathway-related plasma metabolites, including tryptophan, serotonin, and kynurenine, were significantly lower in patients with MDD. Further translational studies are needed to clarify the biological relationships among personality traits, stress, and depression.

Internet gaming disorder (IGD) is a psychiatric disorder induced by excessive and prolonged Internet gaming. It shares many pathological symptoms with attention-deficit hyperactivity disorder (ADHD). Cho et al. correlated plasma metabolites with the severity of Internet addiction in patients with IGD and compared potential biomarkers in combination with clinical parameters (Cho et al. 2017). The GC/MS metabolite profiling of 54 samples (healthy: 28; IGD: 24) identified 104 metabolites. The covariance regression of plasma metabolite sets (arabitol, *myo*-inositol, methionine, pyrrole-2-carboxylic acid, and aspartic acid) using the Internet Poisoning Severity Rating Scale revealed specific relationships. The identified metabolic traits, their associations with clinical parameters, and their biochemical associations are expected to support future studies on the etiology of IGD.

Sleep restriction and the disruption of the circadian clock have been associated with metabolic abnormalities such as obesity, insulin resistance, and diabetes. Davies et al. used LC/MS metabolomics to investigate the effects of acute sleep deprivation on plasma metabolite rhythms (Davies et al. 2014). Twelve healthy young male subjects were maintained under controlled laboratory conditions with respect to ambient light, sleep, diet, and posture during a 24-h wake/sleep cycle that was followed by a 24-h awake period. Plasma samples were collected every 2 h for 48 h for analysis via LC/MS. Of the 171 metabolites quantified, diurnal rhythms were observed in the majority, with 78 maintaining their rhythms during the 24 h of wakefulness but with reduced amplitude. Twenty-seven metabolites (tryptophan, serotonin, taurine, 8 acylcarnitines, 13 glycerophospholipids, and 3 sphingolipids) were significantly increased during sleep deprivation when compared with sleep. Increases in serotonin, tryptophan, and taurine levels may explain the antidepressant effects of acute sleep deprivation and require further study.

The mechanisms through which exercise benefits human health remain unknown. Morville et al. performed LC/MS metabolomic profiling of plasma from randomized within-subject crossover trials of endurance or resistance exercise, which are two types of skeletal muscle activities that have different effects on human physiology (Morville et al. 2020). The analysis of 836 metabolites showed that succinic acid, acylcarnitine, and ketone body 3-hydroxybutyrate levels increased during endurance exercise. The increase in succinic acid levels during endurance exercise suggests that it acts as a fuel for thermogenic adipocytes and increases energy expenditure. The increase in acylcarnitine and 3-hydroxybutyrate levels suggests a switch from sugar-fueled metabolism to metabolism fueled by fatty acids and ketone bodies. In contrast, nucleotide metabolism (inosine, hypoxanthine, and xanthine) and branched-chain amino acids (BCAA, leucine, isoleucine, and valine) increased during resistance exercise. These differences in the plasma metabolome of the different exercise modes clearly indicate distinct metabolic adaptations.

7.3.4 Metabolomics to Assess the Effects of Cybernetic Avatar Use

We have established an ultra-comprehensive metabolomic method combining several analytical systems, mainly unified-HILIC/AEX/MS/MS and SFC/MS/MS, that allows us to obtain quantitative information on 800–1000 metabolites from ~ 50 μL human plasma samples (Nishiumi et al. 2022) (Fig. 7.6). We are currently searching for new biomarkers that can detect the effects of using remote interaction systems, such as Zoom, conventional avatars, such as games, and CAs on an organism (happiness, satisfaction, stress, fatigue, etc.) with high sensitivity, speed, and accuracy through ultra-exhaustive measurements, including measurements of unknown metabolites (Fig. 7.7). Currently, we are discovering a group of metabolites that commonly fluctuate in subjects before and after each task. However, changes in blood metabolite patterns vary widely among individuals, depending on the type of task and the way people perceive the task. Based on our novel metabolomic analysis technology, we aim to contribute to the social implementation of CA by developing a next-generation personalized health monitoring system that uses multiple biomarkers as indicators and quantitatively evaluates the biological effects of CA use.

7.4 Brain Response Analysis During the Use of Avatars and Devices

The use of remote communication systems, robots, and avatars for various social activities is becoming increasingly common. Devices that facilitate these interactions include virtual reality (VR) headsets, augmented reality (AR) glasses, body suits, and teleoperated robots. However, the exploration of the impact of these devices on our brains and behavior during the physical embodiment and immersion of avatars is still in its infancy. Human brain responses can be measured using neuroimaging techniques such as functional magnetic resonance imaging (fMRI), electroencephalography (EEG), and near-infrared spectroscopy (fNIRS). However, little is known about brain changes during avatar use. Regarding behavioral changes, the Proteus effect, in which an individual's response changes in line with the impression that the avatar gives, is well-known (Banakou et al. 2013; Kilteni et al. 2012; Ries et al. 2008; Yee and Bailenson 2007). The Proteus effect is an interesting and crucial phenomenon when discussing the effects of avatar usage. We first introduce the Proteus effect, which is followed by a discussion on the gradual progress in the field of brain research.

7.4.1 The Proteus Effect

According to the first study on the Proteus effect, the more attractive the avatar, the more actively the participants participated in bargaining-related games (Yee and Bailenson 2007). Subsequent studies have shown the effects on our abilities: avatars with casually dressed skin can play percussion instruments rhythmically (Kilteni et al. 2013), Einstein's avatars improve scores on cognitive function tests, (Banakou

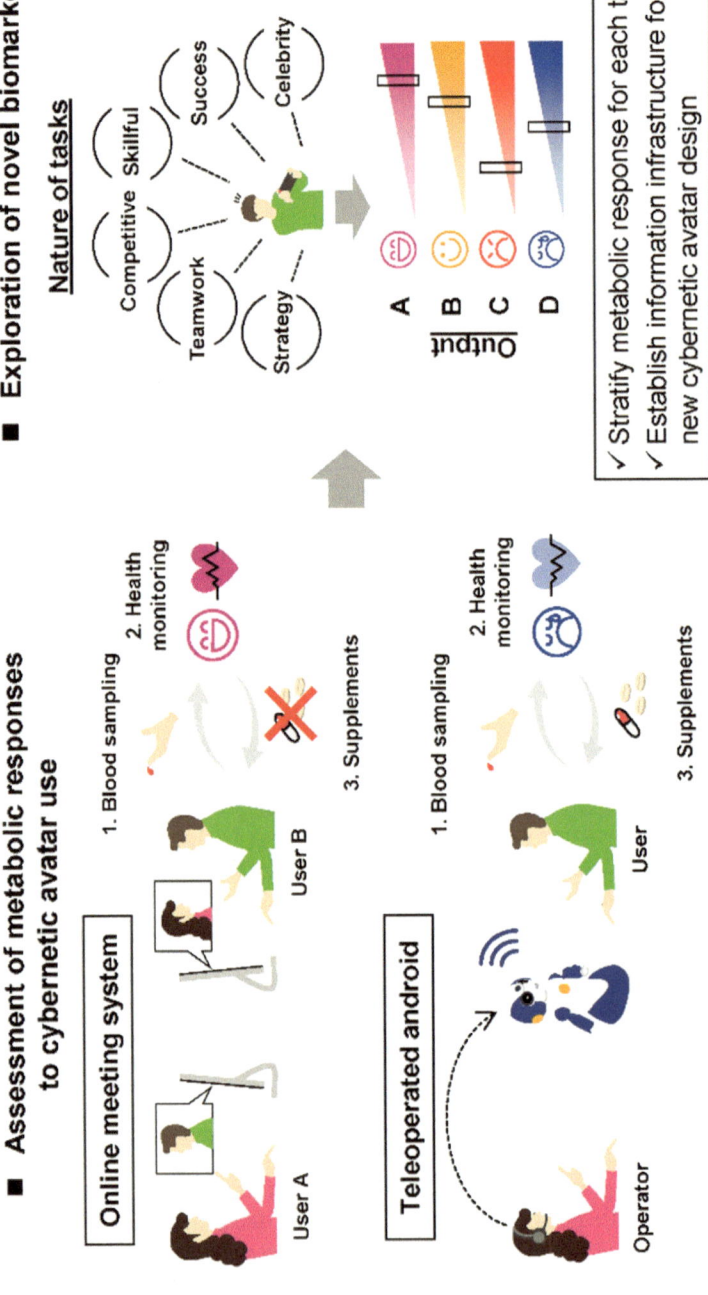

Fig. 7.7 Metabolomic analysis to study the effects of CA use on human body

et al. 2018), and avatars of older adults take longer to perform motor tasks (Beaudoin et al. 2020). Psychological effects have also been reported. Avatars of different races have been reported to reduce potential prejudice against race (Peck et al. 2013), and superhero avatar experiences have been reported to promote positive support for others (Rosenberg et al. 2013). Animal avatars can create a sense of physical possession, and the Proteus effect has also been observed in non-human avatar conditions. For example, the use of monster avatars changes gait (Ahn et al. 2016; Charbonneau et al. 2017; Krekhov et al. 2018; Oyanagi et al. 2021; Oyanagi and Ohmura 2017).

7.4.2 Brain in Virtual Environments

The question of what happens in the brain when immersed in and acting in a VR environment is a topic of this book chapter. However, our knowledge of the effects of VR on brain function is limited. Here, we present reports on the effects of VR environments on brain activity.

Safaryan and Mehta (2021) reported brain activity specific to a VR environment in animal studies. They compared the LFPs of rats running in VR and the real world (RW) and showed that the hippocampal theta rhythm differed between VR and RW. Theta rhythms were amplified twice as much in VR, and new eta rhythms were observed mainly in the hippocampal interneurons, suggesting that VR amplifies or regulates theta rhythms, reflecting changes in a wide range of neural processes in the brain. In a study in humans, Álvarez-Pérez et al. (2021) compared the effects of exposure to the causes of small animal phobia between VR and RW. They reported changes in the location of precuneus activity, which is increasingly implicated in the self-referential process, between VR and RW exposures; this may indicate that VR use can influence self-reflection and alleviate psychiatric symptoms.

It has also been suggested that the operation of avatars in VR involves specific brain networks. Adamovich et al. (2009) used fMRI to identify brain activity during the control of a virtual representation of one's hand in a first-person real-time view. They developed their own system capable of measuring complex hand and finger movements to directly investigate how controlling a virtual representation of the hand in real-time affects neural activation. Specifically, they compared brain activity when performing pre-observed and learned finger movements in a VR environment "while observing the movement of a virtual hand controlled by oneself (real-time feedback)" and "while observing the rotation of an ellipse." They found that brain activity was specific to the feedback from the VR hands. They also noted that observations within the VR space and in the real world caused similar brain activity (bilateral frontoparietal networks). Their findings suggest that the observation of virtual but realistic effectors may involve similar neural substrates as well as the existence of networks specific to controlling avatars in VR.

Furthermore, differences in brain activity have been reported according to the type of avatar used in VR environments (Iwasaki et al. 2023). They performed a gambling task in VR using two different avatars and compared the EEG results. The amplitude, which indicates attentional processing, increased when the avatar was appropriately dressed for a casino. The Proteus effect of changing avatars occurred before the participants became conscious, suggesting that avatar change affected the allocation of attentional resources.

7.4.3 Social Interaction in VR

Thus far, we explored the impact of immersing oneself in a VR environment and manipulating one's avatars. However, VR technology is not only a tool for individual use but also a revolutionary medium for social communication. In VR, we and our interaction partners can easily alter our identities and simultaneously assume multiple personalities. Our perception of the social world is multifaceted because of our ability to easily change perspectives. Despite the increasing prevalence of these diverse social interactions within VR, our understanding of their influence on behavior, decision-making processes, and brain function remains limited. Finally, we introduce our recent work on the changes that occur when confronted with avatars that are manipulated by others. To investigate whether behavior and brain activity change when the partner is an avatar or a real human, we examined the changes in gambling rates because of the partner's appearance during the gambling task. As shown in Fig. 7.8, we used a task in which participants performed a gambling task while being observed by a human. During the task, the partner's appearance was randomly changed to that of an avatar (the observer was always the same person; only the appearance differed). In this task, the partner's face was first presented in a video.

One second after the facial presentation, the options for gambling and safety were presented. The safe option was linked to a 100% small reward and the gambling option to a low-probability large reward. The gambling option's success reward was kept constant, and the probability varied, whereas the safe option was always 100%, and the reward amount varied. However, because of the negative image associated with the word "gamble," it was not used in the instructions, and the participants were asked to choose whether they would challenge the large reward. If they chose the gambling option, they were rewarded with the probability presented and received feedback from the observer in the form of praise. When they failed to receive the reward, they were given a look of contempt. An analysis of variance revealed a three-way interaction between the probability of success, safe reward amount, and observer condition for the gambling choice (p-value $= 5.7 \times 10^{-3}$ and $F = 1.88$). The differences in gambling rates between the two conditions are shown in Fig. 7.9. On the left, the gambling rate is plotted against the ratio of the expected value between the two options, and on the right, the gambling rate is plotted against the number of safe options. In all conditions, gambling rates were predominantly higher in the avatar condition than in the human condition ($p = 0.025$), with particularly significant differences shown in yellow shades (the p-values are 3.7×10^{-3} and 7.1×10^{-4} for the upper left and right, respectively). We found that people were more likely to gamble during the avatar observer condition, especially in the middle condition; that

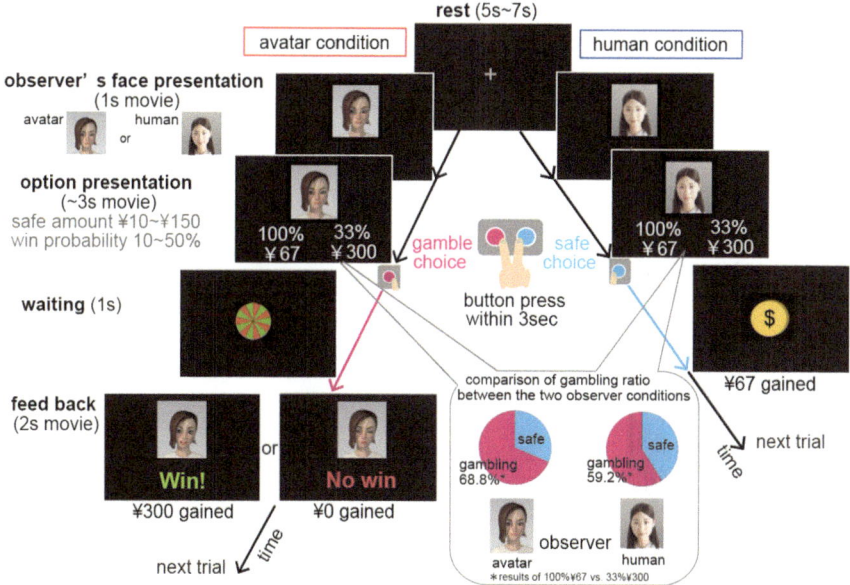

Fig. 7.8 Gambling task with observers

is, when people were unsure of their decision, the effect of the observer's appearance was more apparent. We then used a behavioral model to examine the factors that make a difference between the observer conditions. We assumed that the differences in gambling behavior between observers were because of differences in sensitivity to the social component of how they perceived feedback from their partner and introduced a feedback uncertainty term into the model. The full model includes a monetary reward-related term (the first term in Eq. 7.1), observer face-related terms (Eq. 7.2), and a loss aversion term (the third term in Eq. 7.1):

$$U_{\text{gamble}} = \beta_{\text{gamble}} * Wp * R^{\theta_1}_{\text{gamble}} + \text{FACE} + \beta_{\text{lossav}} * (1 - Wp) * R^{\theta_2}_{\text{safe}} \qquad (7.1)$$

$$\text{FACE} = \beta_{\text{win}} * Wp^{\lambda_1} - \beta_{\text{nowin}} * (1 - Wp)^{\lambda_2} + \beta_{\text{uncertainty}} * \text{entropy} \qquad (7.2)$$

$$U_{\text{safe}} = \beta_{\text{safe}} * R^{\theta_2}_{\text{safe}} \qquad (7.3)$$

The subjective reward for the gambling option was calculated using subjective probability (Wp; according to Kahneman and Tversky (1979)). In the equations, βs, p, γ, R_s, and θ_s indicated the weights for the terms, the win probability, the subjective distortion index of the win probability, the reward sizes for the options, and the subjective distortion index of the monetary rewards for the options, respectively. Following the AIC and BIC criteria, the best model is shown in Eq. 7.4, and the model with a separate estimation of only $\beta_{\text{uncertainty}}$ for humans and the avatar is the

Fig. 7.9 Difference in gambling rates

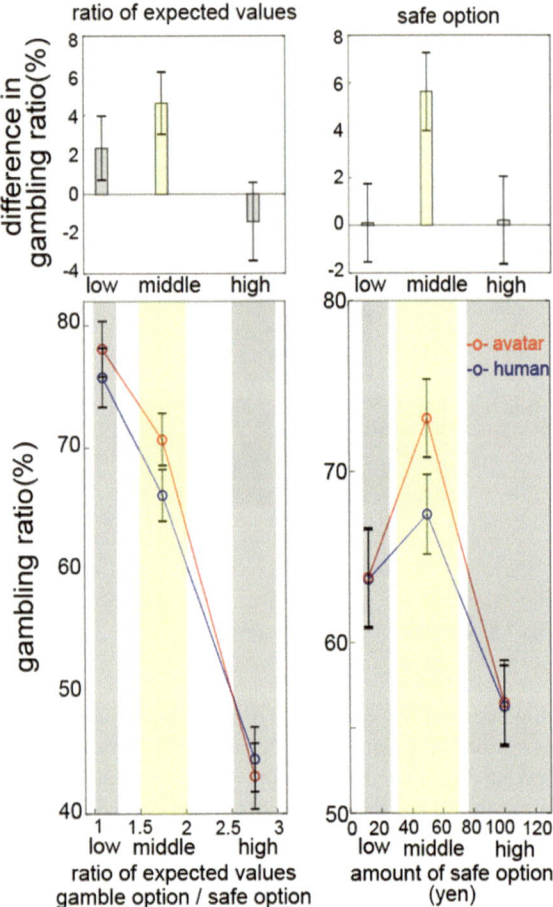

winning model:

$$U_{\text{gamble}} = \beta_{\text{gamble}} * Wp * R_{\text{gamble}}^{\theta_1} + \beta_{\text{uncertanty}} * \text{entropy} \qquad (7.4)$$

It has been suggested that sensitivity to uncertainty plays an important role in changes in gambling rates. We confirmed that the difference in $\beta_{\text{uncertainty}}$ could explain the behavioral changes ($p = 7.3 \times 10^{-11}$). To investigate the neural basis for the changes in gambling rates supported by the changes in the uncertainty sensitivity because of the avatar observer, we performed model-based general linear model (GLM) analyses of fMRI data using the coefficient for uncertainty ($\beta_{\text{uncertainty}}$). At the offer presentation timing, the most significant difference was found in the amygdala activity between the human and avatar conditions ($p = 8.5 \times 10^{-6}$, FWE corrected). To investigate the brain regions that regulated the differences in the uncertainty sensitivity, a weighting of the difference in the entropy contrasts by the changes in the entropy sensitivity (the difference in $\beta_{uncertainty}$) was applied. We found that participants with smaller $\beta_{uncertainty}$ values in the human observer condition had higher amygdala and striatum activities for entropy in the human observer condition (Fig. 7.10a, right; the p-values were 0.019 and 4.3×10^{-3}, FWE corrected, respectively).

The more gambling was promoted in the avatar condition, the lower the avatar observer's activity was. The amygdala is a hub region for social decision-making, including emotional processing (Haruno et al. 2014; Haruno and Frith 2010; Takami

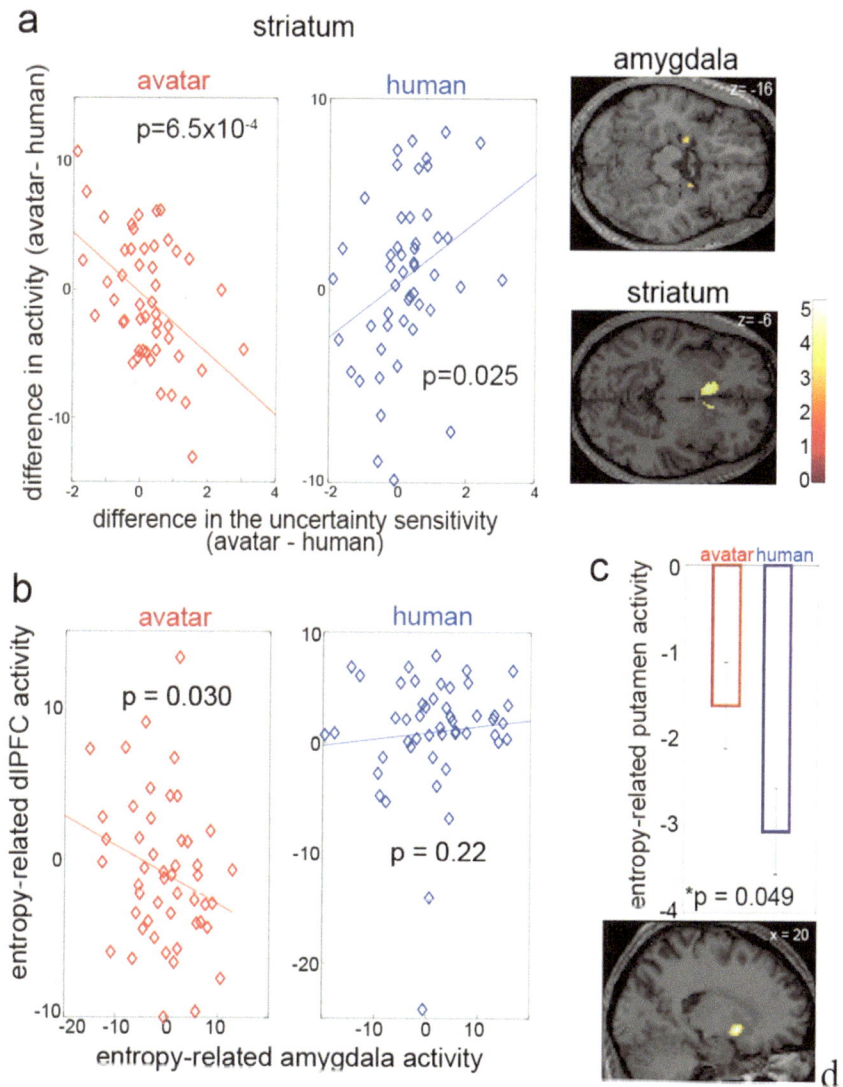

Fig. 7.10 Entropy-related brain activities

and Haruno 2019, 2020; Tanaka et al. 2017, 2022). Differences in amygdala activity may reflect individual differences in sensitivity to others. Furthermore, using a generalized psycho-physiological interaction analysis (McLaren et al. 2012), we suggested that the ventral striatum was involved in the emotional processing of uncertainty via interaction with the amygdala, and the effect was significantly larger in the human condition (the difference in slopes using the F distribution was significant; $p = 9.5 \times 10^{-3}$). These results suggest that the amygdala and ventral striatum work in coordination and are involved in controlling gambling choices altered by avatar use.

There are also reports indicating the emotional modulation of the ventral striatum by the amygdala (Watanabe et al. 2013), suggesting that the amygdala and striatum contribute to changes in decision-making driven by modulated emotions. As shown in Fig. 7.10b, amygdala activity was suppressed by the dorsolateral prefrontal cortex (dlPFC) of the avatar. Since it has been found that artificially stimulating the dlPFC selectively alters decision-making when caring about others' intentions (Nihonsugi et al. 2015), it is possible that changes in sensitivity to others by avatar observers may produce behavioral changes. Regardless of differences in uncertainty sensitivity, entropy-related putamen activity recovered in the avatar condition compared with the human condition (Fig. 7.10c). Since the putamen can be related to excitement regarding gambling (Takahashi et al. 2010; Joutsa et al. 2012; Linnet et al. 2011), it may be suggested that an avatar observer makes the task more enjoyable. This can be interpreted as a decrease in the fear of uncertain outcomes and an increase in excitement owing to the alteration of the avatar observer. Our results suggest that the balance of sensitivity to uncertainty, that is, whether gambling was anxious or enjoyable, was altered by the avatar observer and that the change in the balance of amygdala-striatum and dlPFC activities was related to behavioral changes during avatar use. We believe that our results have the potential to lead modern people, who tend to avoid challenges, toward embracing them through avatar use as a partner.

Our achievements are still only a small step, and basic scientific knowledge has not kept pace with the rapid technological and application advances. Undoubtedly, the use of avatars impacts brain function, as mentioned above. For innovation and healthcare, further investigation and understanding of the human brain's response to new, previously unexperienced situations while avoiding adverse effects are required.

7.5 Designing Experimental Systems for the Investigation of Psychobiological Responses to Cybernetic Avatars and Devices

New technologies can have both positive and negative impacts on society. Before CAs, which have the potential to change us significantly, become widespread in society, it is essential to investigate their positive and negative psychological and physiological effects for the effective use of CA. In this section, we point out that the use of CA involves two tasks: teleoperation of the avatar's body as an operator's proxy and communication with remote people. In previous studies with existing avatars, there was a lack of investigation of the physiological effects of the avatars in social settings, instead focusing on mental workload. To compare the psychological and physiological impacts of existing avatars and CAs and to characterize the effects of CAs, we introduce a simulator that we are currently developing to use avatars in social situations and present a preliminary study of the psychological and physiological impacts of CAs. The results show that several physiological features of the avatar operator, such as brain activity in some regions, are correlated with the mental workload in a social task. Our findings suggest that multimodal physiological data can help monitor or predict the mental state of an avatar operator.

7.5.1 Influence of Using New Technologies on Human

Advances in ICT, communications technology, and robotics continue to transform society, freeing us from the limitations of our minds and bodies. Telecommunication systems, such as videoconferencing systems, have made it possible to communicate with remote people in real time, and teleoperated robots can work in special places, such as disaster sites, where people cannot enter. VR allows for communication with other people by controlling virtual avatars as proxies in a simulated environment. This trend has been rapidly accelerated by the global lockdown policy during COVID-19, and we are now using these technologies in our daily lives to change the structure of society. Furthermore, the CA, which allows one person to control multiple avatars or multiple people to control a single avatar, offers uses beyond the conventional notions of the self and others (Fig. 7.11). In future, by integrating brain-machine interface (BMI) technology, which uses brain activities to control artificial systems, we expect that a new way of communication will be realized, in which avatars can be controlled simply by thinking, leading to significant changes in the concept of human beings.

Previous studies show the various effects that the introduction of new technologies will have on us; for example, the facilitation of disclosing personal information over a videoconferencing system rather than in person (Joinson 2001; Jiang et al. 2011), the phenomenon called the "illusion of body ownership" (Slater 2009), in which we feel as if the avatar's body is our own when we manipulate it, and the Proteus effect (Yee et al. 2009), in which we change our behavior when we operate an avatar whose appearance or gender is different from our own in VR. These effects are not only positive but also harmful, as are smartphone addiction (Haug et al. 2015), social media fatigue (Bright et al. 2015; Dhir et al. 2018), and cybersickness (Chang et al. 2020). We infer that using CA in our daily lives positively and negatively affects our minds and bodies. However, it is unclear how CA use in social situations affects psychological and physiological states. By addressing this question, we can develop an entire CA system that takes advantage of the positive impacts and controls the negative impacts in advance.

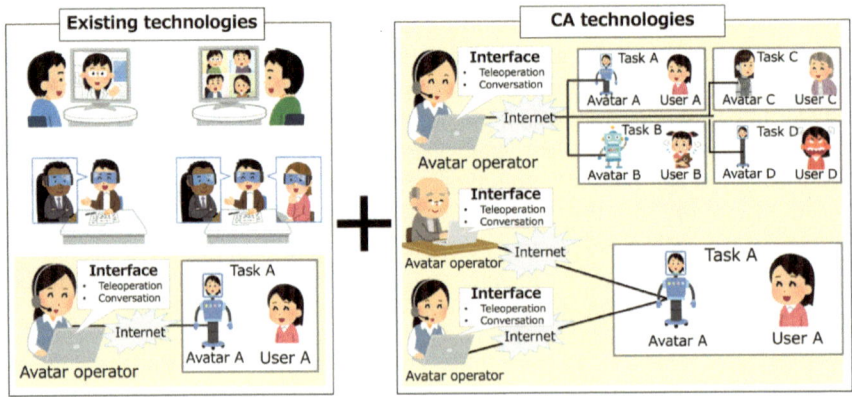

Fig. 7.11 Existing technologies (communication via videoconferencing, virtual environment, and an avatar (CG character or teleoperated robot)) and CA technologies, which allow one person to control multiple avatars or multiple people to control a single avatar. There is a lack of studies on the psychological and physiological impact of using not only CA but also an existing avatar on its operator in a social setting

In this section, we first provide an overview of the psychological and physiological impacts of using an avatar on the operator, focusing on the mental workload. Few studies using existing technologies have investigated the physiological and psychological effects of using an existing avatar on its operators in social situations. Next, we introduce our ongoing research on the relationship between the psychological and physiological effects of using an existing avatar in social situations using VR simulators as a comparison for future research to clarify the impact of CA use on its operator.

7.5.2 Mental Workload During Operating an Avatar

Using a CG character and teleoperated robot as an avatar is the current solution in several areas, such as communication, surgery, inspection, exploration, and rescue. While many studies have focused on teleoperating the avatar's movements only for manipulation and navigation in nonsocial tasks (Darvish et al. 2023), in a social context, using an avatar involves communicating with other people and teleoperating the avatar's social cues, such as emotional expressions and gestures. This multitasking nature increases the workload of the operator, which is linked to mental states such as attention and stress. Some intelligent control systems based on shared autonomy have been proposed to reduce workload (Glas et al. 2012). However, because they do not consider the physiological information of the operator, which reflects his/her mental and physical states, the system cannot adapt its support to the operator in consideration of his/her workload (Roy et al. 2020).

Few studies have proposed the monitoring of stress and attention levels in teleoperated nonsocial tasks using physiological signals. A physiologically attentive user interface was presented to improve teleoperated robot navigation by distinguishing between three induced mental states (rest, workload, and stress) using electroencephalography (EEG), electrocardiography (ECG), electrodermal activity (EDA), eye-tracking signals, and facial expressions from a camera (Singh et al. 2018). This dynamic interface enhanced the utility and ease of use, although it was tested on only six subjects. Another study addressed predicting a pilot's mental state during the operation of three unmanned aerial vehicles (UAVs) using behavioral and physiological measures (i.e., cerebral, cardiac, and oculomotor features). They reported that a higher intra-subject classification accuracy was achieved using ECG features alone or combined with EEG features and eye-tracking signals (Singh et al. 2021). Multiple physiological measures might help predict mental states because a study showed the difficulty in predicting emotional states with a single physiological measure during the teleoperation of a navigation robot (Yang and Dorneich 2017). These results indicate that some biomarkers can be used to monitor metrics related to teleoperation performance. However, no study has investigated physiological responses during communication with other people via an avatar.

Previous studies on communication have estimated mental stress during interviews or public speaking by processing multimodal signals, such as EEG, ECG, electromyogram, SCR, blood volume pulse (BVP), and temperature (Arza et al. 2019; Arsalan and Majid 2021). Other studies have used fNIRS data to estimate mental stress (Sumioka et al. 2019) and perceived difficulty (Keshmiri et al. 2019) during robot-mediated conversations. However, these studies do not include avatar controls. Therefore, it remains unclear how the use of an avatar in a social context affects psychological and physiological states and their relationships.

7.5.3 VR Simulator for Investigating the Psychological and Physiological Impact of an Avatar on Its Operator

Although CAs are expected to be used in social situations, few studies have investigated the psychological and physiological effects of avatar operators when existing avatars are used in such situations, as described above. Therefore, to characterize the impact of CA on operators during use, it is necessary to study CA use and the use of existing avatars in social situations. We realized this problem and developed a VR simulator for using avatars in social situations to investigate the psychological and physiological effects of using existing avatars. Here, we describe the system and briefly introduce our research on the mental workload.

The developed VR simulator is shown in Fig. 7.12. The simulator was displayed through a head-mounted display (HMD), HTC Vive Pro Eye (HTC Corporation), with an embedded eye tracker. The HMD had a refresh rate of 90 Hz and a field of view of 110°. The eye tracker had a frequency of 120 Hz, an accuracy between 0.5°

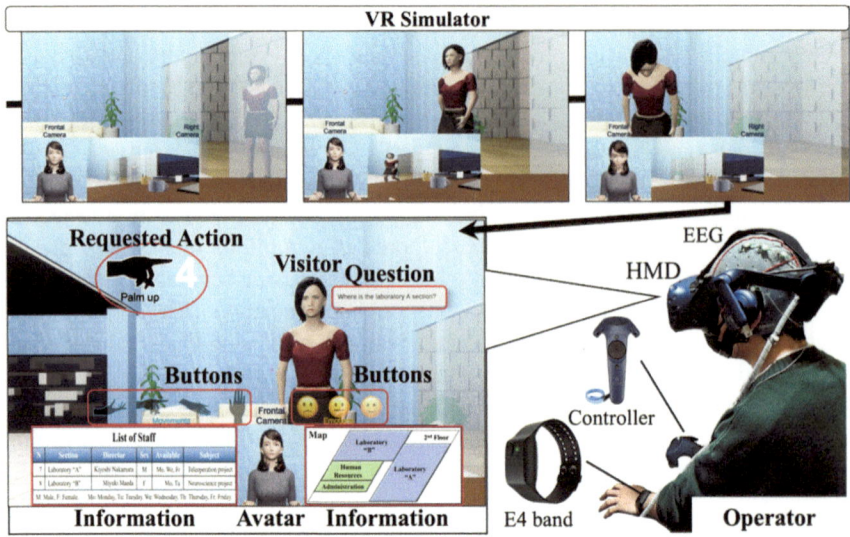

Fig. 7.12 VR simulator of teleoperated social tasks. Participants wore a head-mounted display (HMD) and used a controller to control the avatar's facial expressions and gestures while working as a receptionist in a virtual space. We measured participants' biosignals using Polymate Pro 6000 for EEG and an E4 band for such peripheral signals as SCR, temperature, BVP, and accelerometer

and 1.1°, and a field of view of 110°. Participants interacted with the simulator using a Vive controller. The simulator was developed using C# in Unity software (Unity Technologies). Signal processing, data recording, and VR simulation were performed using a PC with the following configuration: Windows 11 operating system, Intel Core i7-11,700 CPU up to 4.9 GHz, 32 GB RAM, and Nvidia GeForce RTX 3070 GPU.

The simulator was integrated with a physiological measurement system comprising the brain and peripheral signal sensors (Fig. 7.12). Brain signals were recorded using an EEG Polymate Pro 6000 with 13 active electrodes placed in the frontal, central, parietal, and occipital regions. In addition, an E4 wristband, which records peripheral signals such as EDA, temperature, BVP, and accelerometers, was placed on the non-dominant side of the avatar operator. Both brain and peripheral signals were collected based on the lab streaming layer (LSL) system, which allowed us to collect data with time synchronization.

The social task in the simulator was designed to induce a mental workload that influences the behavior of the avatar operator and is present in real applications (Rea et al. 2020). An avatar operator teleoperates a receptionist robot (avatar) that interacts with the visitor in a VR simulation by answering questions and clicking buttons corresponding to the avatar's actions using a Vive controller. The avatar was based on the humanoid robot ERICA (Glas et al. 2016). We designed the avatar's actions, including its facial expressions (sad, neutral, and happy) and hand movements (palm up, pointing, palm side, and raising hands), to be controlled by the operator. Not only can the operator freely choose the avatar's actions, but the system can also suggest actions as a "Requested action" with a time limit displayed to increase mental workload based on some cognitive tasks, such as a visual attention task during lectures (Ko et al. 2017) and the Stroop Color-Word Test (Ahn et al. 2019).

7.5.4 Investigation of Psychological and Physiological Influence During a Social Task

We conducted an experiment using the developed VR simulator to investigate the impact of avatar use on social task operators. In this section, we briefly describe preliminary experiments. Further details can be found in (Achanccaray and Sumioka 2023).

7.5.4.1 Participant

Twenty-five participants (13 males and 12 females) participated in the experiment. The patients were between 21 and 40 years of age (mean: 30.48, SD: 7.20). They had neither a history of neurological diseases nor experience with VR systems. Each participant signed an informed consent form before the experiment. This study was approved by the Ethics Committee of the Advanced Telecommunications Research Institute International, Japan. Each participant received a monetary incentive.

7.5.4.2 Experimental Procedure

An interaction with a simulated visitor lasted for 45 s (Fig. 7.12). First, the visitor entered a building and approached the avatar's desk. Then, the visitor and teleoperated avatar greeted each other (for 5 s). Next, the visitor asked the avatar to teleoperate the participant if a particular person was at work or his/her room was in the building. The participant had to respond to the visitor within 10 s, answering it verbally based on tables and maps displayed on the screen and selecting the avatar's action. The visitors were asked two questions. After being asked these two questions, the visitor would go into or left the building, depending on the participant's answers. The participants participated in all three conditions with different levels of mental workloads to select the avatar's actions: normal, attention, and stress conditions. Under normal conditions, the participants freely selected their avatar's actions. In the attention condition, participants had to choose the requested action suggested by the system. The stress condition was the same as the attention condition, except that a time limit was displayed. Each condition comprised 24 interactions. Before each condition, the participants performed some trials to practice and become accustomed to it. After each condition, the participants completed the NASA-TLX workload test (Hart and Staveland 1988), which rated six task load subscales (mental demand, physical demand, temporal demand, performance satisfaction, effort, and frustration) between 0 and 100 as well as provided an overall workload score.

7.5.4.3 Analysis

Several features were extracted from the collected physiological signals to explore their physiological influences on the avatar operator. Owing to the lack of multimodal data from 2 participants, we analyzed the physiological data from 23 participants. We calculated the EEG band powers (alpha, beta, theta, and delta bands) and computed features including the power asymmetry, theta/beta power ratio, and normalized power in some regions, such as the frontal, frontal-central, central, central-parietal, and occipital regions; these features are related to stress levels (Ahn et al. 2019) and attention levels (Putman et al. 2014; Ko et al. 2017). The mean, standard deviation (Std), number of peaks for the SCR, and mean temperature were calculated. The mean and gaze speed in the x-, y-, and z-axes were computed from the accelerometer and eye-tracking signals, respectively. In addition to these physiological features, we computed the total score on the NASA-TLX and its six subscales.

7.5.4.4 Results

The use of an avatar significantly affected the overall workload (the total NASA-TLX score). However, a post hoc comparison showed that the differences among the conditions were not substantial, although we did find a significant difference between the normal and stress conditions. When we analyzed the subscales, we found significant differences in the mental, physical, temporal, and effort demands. In particular, participants showed significant differences between the normal and attention conditions and between the normal and stress conditions in terms of mental and physical demands.

We found that several physiological features showed significantly different responses among the conditions. For example, there was a significantly different normalized alpha band power in the central-parietal region among the three conditions. Interestingly, we also found significant differences among the conditions in mean temperature, which is a peripheral signal. These results suggest that physiological signals have the potential to reflect mental workload better than subjective evaluations.

We also analyzed the correlations between physiological features and workload subscales. Most of the correlations were weak for all conditions. The normalized alpha band power in the central parietal region and mean temperature, which showed significant differences among conditions, did not show any correlations with the NASA-TLX subscales. However, we also found significant correlations with other features. These results imply that the mental workload perceived by an avatar operator reflects not a single physiological feature but multiple features.

7.5.5 Summary

In this section, we pointed out that previous studies on existing avatars have paid little attention to the psychological and physiological effects of avatars on avatar operators in social settings. To investigate the effects of CAs, we developed a simulator that enabled people to use avatars in social situations. As a preliminary investigation, we first examined the psychological and physiological effects of an existing avatar on its operator, focusing on the mental workload of using an avatar in a virtual environment. The results showed that the use of multiple physiological features may be essential for predicting an operator's mental workload. In future, we will investigate the psychological and physiological impacts of CAs and compare the future results with these current results to characterize the effects of CAs.

7.6 How to Create Health Indicators for Avatar Applications

7.6.1 What Is Health: Definition

The WHO Charter was signed in 1946 and became effective in 1948; it describes health as follows: "Health is not the absence of disease or infirmity, but a state of complete physical, mental, and social well-being" (Friends of WHO Japan 2023). Surprisingly, the definition of health included the words "mental and social health" when infectious diseases were the leading cause of death.

Here, we discuss health according to the WHO Charter and the Ministry of Health, Labour and Welfare (MHLW) of Japan, dividing it into three categories: physical health, mental health, and social health.

7.6.1.1 Physical Health

Physical health refers not only to the health of individual organs but also to the proper functioning of the body as a functional system. For example, the liver is responsible for normal detoxification, and the kidneys excrete excess water as urine. The heart maintains a sufficient cardiac output to distribute blood to all parts of the body, and the lungs have sufficient functionality to absorb the necessary oxygen and exchange carbon dioxide.

7.6.1.2 Mental Health

According to the World Health Organization (WHO), mental health is "a state of well-being in which an individual realizes his or her own abilities, can cope with the normal stresses of life, can work productively and is able to make a contribution to his or her community" (WHO 2022). Mental health is an integral and essential component of health and is a basic human right (WHO 2022). Mental health is more than the absence of mental disorders or disabilities, and it is determined by a range of socioeconomic, biological, and environmental factors (WHO 2022, 2023). The WHO also states that cost-effective public health and intersectoral strategies as well as interventions exist to promote, protect, and restore mental health (WHO 2023). Mental health is fundamental to our collective and individual abilities as humans to think, emote, interact with each other, earn a living, and enjoy life (WHO 2023).

In Japan, the MHLW defines mental health as "an important condition for leading an active life in one's own way." Specifically, it refers to being able to recognize and express one's feelings (emotional health), being able to think appropriately in response to situations and solve practical problems (intellectual health), and being able to establish constructive and good relationships with others and society (social health). Finding purpose and meaning in life and making independent life choices (human health) are also important factors, and mental health has a significant impact on "quality of life" (Ministry of Health, Labour and Welfare of Japan 2000).

7.6.1.3   Social Health

Social health refers to one's ability to form and maintain relationships, cope with social situations, and participate in social activities that enhance well-being. Social health is an integral part of health, as it influences and is influenced by other dimensions of health, such as physical, mental, and environmental factors.

Many factors affect social health, including socioeconomic status, education, culture, gender, social norms, social policies, and social support. These factors can create social inequalities and inequities in health outcomes among different groups. Therefore, addressing the social determinants of health is essential to promote social health and reduce health disparities.

In Japan, social health is defined by the MHLW as "the ability to have constructive and good relationships with others and society" (Ministry of Health, Labour and Welfare of Japan 2000), as mentioned previously. In other words, human beings are social beings, and they should be in situations in which one is able to maintain good relationships with society, the aggregate of others, and specific others.

7.6.2   Evaluation of Health

Identifying how the use of avatars affects health is important for creating health standards. A possible point of reference when considering the impact of an avatar on health is to consider how the three aforementioned health factors are evaluated.

In Japan, there is a historical background in which these three types of health have been evaluated in accordance with changes in social and working conditions. This section summarizes the results.

7.6.2.1 Historical Background of Sanitary Administration

Health evaluation in Japan began with the evaluation of physical health to improve sanitary standards. After the government announced its policy of adopting Western medicine in the late eighteenth century, the sanitary administration in Japan began to take off. In the beginning, the control of infectious diseases was an important issue, and the focus was on measures to prevent acute infectious diseases, such as cholera, and chronic infectious diseases, such as tuberculosis, by improving the sanitary environment. It was necessary to evaluate the effectiveness of these measures. As the number of deaths from infectious diseases declined because of changes in disease structure, including the effects of widespread immunization after World War II, lifestyle-related diseases accounted for a large proportion of deaths. This resulting increase in life expectancy has led to an increased focus on lifestyle-related diseases and cancer control (Ministry of Health, Labour and Welfare of Japan 2014).

7.6.2.2 Historical Background of Occupational Health

Subsequently, with the enactment of the Labor Standards Law in 1947, the evaluation of workers' health was promoted in Japan. Since the enactment of the Occupational Safety and Health Law in 1972, employers have been responsible for developing measures to prevent health problems in response to changes in the nature of work and to create a more comfortable working environment.

In addition, since the 1980s, the aging of the workforce because of increased life expectancy and rapid changes in work styles due to technological innovation have necessitated mental health measures in response to increased stress among workers, making it necessary to focus on mental and physical health. In other words, for the first time in Japan, the WHO came up with the idea of mandating mental health measures from an occupational health perspective. In 1988, the Occupational Health and Safety Law was revised to require employers to take the necessary measures to maintain and promote the health of their employees on a continuous and systematic basis. Specifically, occupational physicians qualified as medical doctors will take the lead in measuring the health of individual workers, and, according to the results, staff with sufficient knowledge and skills in their respective fields will provide "exercise guidance," "mental health care," "nutrition guidance," and "health guidance" to assess the health of workers from both physical and mental aspects. The new law requires companies to assess and maintain the physical and mental health of their employees (Ministry of Health, Labour and Welfare of Japan 2014).

7.6.2.3   Historical Background of Occupational Health in Information Technology Work

With the proliferation of computers, an evaluation of their impact on health in the workplace has been historically incorporated into medical examinations in Japan.

To answer the question of how to evaluate health in response to the spread of avatars, it is important to carefully examine how this indicator was created.

Technological innovations centered on computers and information processing have led to rapid advances in office automation in various industries. The widespread use of visual display terminals (VDTs) has resulted in eye, body, and mental symptoms caused by long hours of VDT work. Thus, VDT syndrome has become a problem. In 1985, the "Occupational Health Guidelines for VDT Work" were issued, and health examinations for VDT workers, called VDT health examinations, began to be conducted (although they were not mandatory). Subsequently, "Guidelines for Occupational Health Management in VDT Work" were issued in 2002 in light of the fact that VDT work is now widely performed in workplaces other than offices. Furthermore, "Guidelines for Occupational Health Management in Information Equipment Work" were issued in 2019 (and partially revised in 2021) to address the diversification of work forms.

7.6.2.4   Historical Background for Understanding the Health of the Elderly

In terms of human health, we believe that different responses are required from the elderly. This is because their physical functions have declined over time, and they are considered more vulnerable to the effects of environmental changes. This subsection discusses how Japan addresses the needs of the elderly.

In the immediate postwar period, only a small number of low-income households were covered, as the average life expectancy was around 50 years, and 3 households lived together. However, with an increase in the number of elderly people and changes in the industrial structure, nuclear families have become more common, and the environment for the elderly has changed. In 1963, the Elderly Welfare Law was enacted, making the welfare of the elderly the responsibility of national and local governments. In 1973, free medical care for the elderly began, the problem of social hospitalization for the elderly in need of care arose, and a long-term care insurance system was established to solve this problem. The government aimed to control spending and improve the overall health of its population. In addition, in response to the decline in the labor force, measures such as extending the retirement age and raising the age at which pension benefits begin have been implemented, with the expectation that the elderly will function as workers. Health check-ups known as "specified medical check-ups" are mandatory up to the age of 74 under the Law on Ensuring Medical Care for the Elderly, but priority is given to those who are required to receive occupational health check-ups under the Occupational Safety and Health

Law, and health management is expected at the workplace (Ministry of Health Labour and Welfare of Japan 2000).

7.6.2.5   Importance of Health Evaluation

In the process of creating and disseminating new things in the environment and culture, it can be said that country's policy has been to evaluate and promote the maintenance of health in response to health hazards that have occurred, as inferred from the historical background mentioned above. However, it seems that the rapid expansion of information equipment has been carried out without sufficient evaluation. The same danger may arise in future when the needs to use avatars and improve work efficiency increase rapidly in the industrial world. Therefore, it is important to develop avatars from an occupational health perspective as much as possible to ensure their healthy use. In addition, attention should be paid to what the government is thinking about and implementing in terms of understanding the characteristics of the elderly and creating products that are friendly for them.

7.6.2.6   Summary

This section discusses the historical background of health administration with a special focus on occupational health and health assessment in working life. This discussion is followed by a discussion on older people. As society ages and other social conditions change, the issues requiring special attention change, and it is necessary to keep these changes in mind as we move forward. In the next section, we will focus specifically on VDT health screenings and discuss how we have considered what to look for when using new products in the work environment and how we have used this framework as a guideline.

7.6.3   Issues to Consider When Assessing the Health of Avatar Operators

To answer the question of what to consider when evaluating the health of avatar operators, we discuss various guidelines that have been published to date.

As mentioned in the previous section, the workplace is obligated to provide a safe and healthy environment. The VDT work referred to here was classified as a special type of work when computers first entered the work environment; therefore, it was given special attention as a separate examination item in accordance with these guidelines. However, it cannot be denied that it has become a skeleton in accordance with the rapid spread of PCs, tablet terminals, and smartphones today. However, it is necessary to fully discuss how to define the new avatars to be developed. If we can grasp the perspective of safety and health management required at workplaces and identify the points that can be improved in advance, we may be able to make them more user-friendly.

7.6.3.1 Issues to Consider Regarding Avatar Characteristics

Avatar operations are not only limited to computer-mediated tasks such as working on a display screen but also include factors such as the use of an HMD, the combined use of VR or AR, the complexity of the equipment used to perform the operations, and the restrictive nature of the operating environment. This subsection describes the issues that must be considered based on these characteristics.

7.6.3.2 Head-Mounted Display (HMD) Applications and Considerations for Physical Health

An HMD is a device worn on the head to view images in three dimensions and can be a means of VR, AR, etc. VR chatting, in which avatars talk to each other in a VR space, is widely used.

HMDs can provide a powerful and realistic visual experience with a display close to the eyes; however, HMDs already have known side effects.

VR sickness is a major complication of using HMDs. VR sickness is a symptom of discomfort, nausea, eye strain, and dizziness associated with viewing VR content using an HMD (Adhanom et al. 2022).

The main cause of VR sickness is thought to be sensory confusion between the visual and vestibular systems; however, many other factors are involved, including light stimulation, spatial information, workload, and individual differences. Subjective questionnaires and scales are commonly used to assess VR sickness (Sevinc and Berkman 2020); however, objective biometric signals and behavioral indicators are also used (Jang et al. 2022). Countermeasures for VR sickness have been proposed, including the design of VR content and presentation methods, user adaptation and training, and the use of medications and devices.

VR sickness has become an important issue with the spread of VR and the expansion of its use, and theoretical clarification and the development of effective countermeasures are needed.

7.6.3.3 Possible Influence of Using Avatars on Mental Health and Social Health

Operational complexity is inevitable in a world with multiple avatars. Because health management precautions for workers who perform complex operations are positioned as work with a higher mental load (Ministry of Health, Labour and Welfare of Japan, 2021, 2023; Tokyo Labor Bureau 2018), the condition of workers should be monitored via periodic health checks and, if necessary, consultation with a doctor. It is important to ensure that the working environment is carefully managed and that sufficient breaks are taken during and after work.

Highly restrictive work can create a strong sense of anxiety in some individuals; therefore, the appropriateness of such work must be carefully assessed. If different types of avatars are to be developed in future, each example should be evaluated based on its advantages and disadvantages.

Particularly, avatar addiction may be the worst issue for users who focus on mental and social health. Avatar addiction is a term used to describe the excessive and problematic use of online games involving avatars, such as massive multiplayer online role-playing games. Avatar addiction can lead to IGD, which is a condition characterized by impaired control over gaming, increased priority given to gaming over other activities, and the continuation or escalation of gaming, despite its negative consequences (WHO 2020). Avatar addiction may be influenced by several factors, including identification with the avatar, an experience of flow, and a clarity of self-concept (Zhang et al. 2022). Some studies have suggested that avatar identification, or the degree to which players perceive their avatars as extensions of themselves (Casale et al. 2023), may influence IGD through the mediating roles of flow and self-concept clarity. Flow is an optimal experience that involves high levels of concentration, enjoyment, and intrinsic motivation. Self-concept clarity is the degree to which individuals have a clear and consistent sense of self. Avatar identification may increase the likelihood of experiencing flow, which may decrease self-concept clarity and increase IGD symptoms (Stavropoulos et al. 2022). Thus, avatar addiction may reflect a complex interaction between psychological processes and features of online gaming.

The ICD-11 and DSM-5 have already included the concept of IGD as an important disease (Carlisle 2021; WHO 2020). Not only during gaming but also during work, using avatars may become an issue that is worthy of consideration for maintaining good health among the general population.

7.6.4 How to Confirm the Positive Effects of Using Avatars

We have discussed the possibility of using avatars not voluntarily but for work. Here, we discuss the method used to confirm the biological effects of avatars, including their positive effects.

Thus far, we have focused on the beguiling aspect of avatars as a cautionary point, assuming that avatars will enter the industrial world from the same perspective as industrial robots; however, avatars are reported to have positive effects because of their characteristics of being virtual entities that serve as one's alter ego. Participation in activities such as avatar-based psychological counseling (Coleman et al. 2019), treatment for chronic auditory hallucinations (Stefaniak et al. 2019), smoking cessation, avatar-led acceptance, and commitment therapy has already been reported with positive results (Karekla et al. 2020).

Table 7.1 Expectations for avatar symbiosis in an aging society

	Current issues	Expectations of the avatar society
Physical health	Sarcopenia and frailty Lack of physical activity because of a decrease in physical work	Improvement of physical functions Creating a lifestyle habit that can increase metabolism
Mental health	Decreased motivation Depression, anxiety Cognitive decline, dementia	Opportunities for physical activity Targets that one can get absorbed in A safe place for communication
Social health	Social isolation	Creating a system to prevent isolation Social participation through avatars

Avatars have the potential to realize a society free from the constraints of body, brain, space, and time; however, at the same time, there are various challenging issues that need to be resolved before the health benefits of avatars can be enjoyed.

7.6.5 Summary

As long as sanitation is good and access to goods and medical care is adequate in modern society, except in special cases of nutrition, population aging will not stop. As the elderly constitute a large proportion of the population in developed countries, users will age rapidly, and society as a whole will have to focus on thinking about the aging of operators in the process of maintaining health and accepting new and different technologies. To maintain health, we have organized past policies and literature on what we should pay attention to at this time, what we should consider, and to what we should refer to in order to generate methods to evaluate health. It is important to be able to consider the aging operators and how to maintain health by referring to past examples as well as good effects, and it is also important to be able to examine the effects of the system, especially on the aging population.

A current administrative challenge in Japan is bridging the gap between life expectancy in general and life expectancy for social participation.

To this end:

1. What is required to maintain physical health?
2. What is required to maintain mental health?
3. What can be done to improve social health?

We must create a society that coexists with avatars while considering these questions. It is necessary to assess the influences of avatars on humans to improve human life. As mentioned in Table 7.1, by living in harmony with an avatar, a society in which people can maintain better health can be expected.

References

Achanccaray D, Sumioka H (2023) Analysis of physiological response of attention and stress states in teleoperation performance of social tasks. In: The 45th annual international conference of the IEEE Engineering in Medicine and Biology society (EMBC)

Adamovich SV, August K, Merians A, Tunik E (2009) A virtual reality-based system integrated with fmri to study neural mechanisms of action observation-execution: a proof of concept study. Restor Neurol Neurosci 27(3):209–223. https://doi.org/10.3233/RNN-2009-0471

Adhanom I, Halow S, Folmer E, MacNeilage P (2022) VR sickness adaptation with ramped optic flow transfers from abstract to realistic environments. Front Virtual Reality 3. https://doi.org/10.3389/frvir.2022.848001

Ahn JW, Ku Y, Kim HC (2019) A novel wearable EEG and ECG recording system for stress assessment. Sensors 19(9). https://doi.org/10.3390/s19091991

Ahn SJG, Bostick J, Ogle E, Nowak KL, McGillicuddy KT, Bailenson JN (2016) Experiencing nature: embodying animals in immersive virtual environments increases inclusion of nature in self and involvement with nature. J Comput-Mediat Commun 21(6):399–419. https://doi.org/10.1111/jcc4.12173

Álvarez-Pérez Y, Rivero F, Herrero M, Viña C, Fumero A, Betancort M, Peñate W (2021) Changes in brain activation through cognitive-behavioral therapy with exposure to virtual reality: a neuroimaging study of specific phobia. J Clin Med 10(16):3505. https://doi.org/10.3390/jcm10163505

American Diabetes Association (2006) Diagnosis and classification of diabetes mellitus. Diab Care 29(suppl_1):s43–s48. https://doi.org/10.2337/diacare.29.s1.06.s43

Arsalan A, Majid M (2021) Human stress classification during public speaking using physiological signals. Comput Biol Med 133:104377. https://doi.org/10.1016/j.compbiomed.2021.104377

Arza A, Garzón-Rey JM, Lázaro J, Gil E, Lopez-Anton R, de la Camara C, Laguna P, Bailon R, Aguiló J (2019) Measuring acute stress response through physiological signals: towards a quantitative assessment of stress. Med Biol Eng Compu 57(1):271–287

Banakou D, Groten R, Slater M (2013) Illusory ownership of a virtual child body causes overestimation of object sizes and implicit attitude changes. Proc Natl Acad Sci 110(31):12846–12851. https://doi.org/10.1073/pnas.1306779110

Banakou D, Kishore S, Slater M (2018) Virtually being Einstein results in an improvement in cognitive task performance and a decrease in age bias. Front Psychol 9(JUN). https://doi.org/10.3389/fpsyg.2018.00917

Beaudoin M, Barra J, Dupraz L, Mollier-Sabet P, Guerraz M (2020) The impact of embodying an "elderly" body avatar on motor imagery. Exp Brain Res 238(6):1467–1478. https://doi.org/10.1007/s00221-020-05828-5

Beebe K, Kennedy AD (2016) Sharpening precision medicine by a thorough interrogation of metabolic individuality. Comput Struct Biotechnol J 14:97–105. https://doi.org/10.1016/j.csbj.2016.01.001

Bright LF, Kleiser SB, Grau SL (2015) Too much facebook? An exploratory examination of social media fatigue. Comput Hum Behav 44:148–155. https://doi.org/10.1016/j.chb.2014.11.048

Carlisle K (2021) Utility of DSM-5 criteria for internet gaming disorder. Psychol Rep 124(6):2613–2632. https://doi.org/10.1177/0033294120965476

Casale S, Musicò A, Gualtieri N, Fioravanti G (2023) Developing an intense player-avatar relationship and feeling disconnected by the physical body: a pathway towards internet gaming disorder for people reporting empty feelings? Curr Psychol 42(24):20748–20756. https://doi.org/10.1007/s12144-022-03186-9

Charbonneau P, Dallaire-Cote M, Cote SS-P, Labbe DR, Mezghani N, Shahnewaz S, Arafat I, Irfan T, Samaraweera G, Quarles J (2017) Gaitzilla: exploring the effect of embodying a giant monster on lower limb kinematics and time perception. In: 2017 International conference on virtual rehabilitation (ICVR), 1–8 June 2017. https://doi.org/10.1109/ICVR.2017.8007535

Chang E, Kim HT, Yoo B (2020) Virtual reality sickness: a review of causes and measurements. Int J Hum-Comput Inter 36(17):1658–1682. https://doi.org/10.1080/10447318.2020.1778351

Cho YU, Lee D, Lee J-E, Kim KH, Lee DY, Jung Y-C (2017) Exploratory metabolomics of biomarker identification for the internet gaming disorder in young Korean males. J Chromatogr B 1057:24–31. https://doi.org/10.1016/j.jchromb.2017.04.046

Coleman D, Black N, Ng J, Blumenthal E (2019) Kognito's avatar-based suicide prevention training for college students: results of a randomized controlled trial and a naturalistic evaluation. Suicide Life-Threat Behav 49(6):1735–1745. https://doi.org/10.1111/sltb.12550

Darvish K, Penco L, Ramos J, Cisneros R, Pratt J, Yoshida E, Ivaldi S, Pucci D (2023) Teleoperation of humanoid robots: a survey. IEEE Trans Robot 39(3):1706–1727. https://doi.org/10.1109/TRO.2023.3236952

Davies SK, Ang JE, Revell VL, Holmes B, Mann A, Robertson FP, Cui N, Middleton B, Ackermann K, Kayser M, Thumser AE, Raynaud FI, Skene DJ (2014) Effect of sleep deprivation on the human metabolome. Proc Natl Acad Sci 111(29):10761–10766. https://doi.org/10.1073/pnas.1402663111

Dhir A, Yossatorn Y, Kaur P, Chen S (2018) Online social media fatigue and psychological well-being—a study of compulsive use, fear of missing out, fatigue, anxiety, and depression. Int J Inf Manage 40:141–152. https://doi.org/10.1016/j.ijinfomgt.2018.01.012

Donath MY, Dinarello CA, Mandrup-Poulsen T (2019) Targeting innate immune mediators in type 1 and type 2 diabetes. Nat Rev Immunol 19(12):734–746. https://doi.org/10.1038/s41577-019-0213-9

Fiehn O (2002) Metabolomics—the link between genotypes and phenotypes. Plant Mol Biol 48(1–2). https://doi.org/10.1023/A:1013713905833

Fiehn O, Kopka J, Dörmann P, Altmann T, Trethewey RN, Willmitzer L (2000) Metabolite profiling for plant functional genomics. Nat Biotechnol 18(11):1157–1161. https://doi.org/10.1038/81137

Friends of WHO JAPAN (2023) What is the WHO charter (Japanese). https://japan-who.or.jp/about/who-what/charter/

Fukusaki E, Kobayashi A (2005) Plant metabolomics: potential for practical operation. J Biosci Bioeng 100(4):347–354. https://doi.org/10.1263/jbb.100.347

Glas DF, Kanda T, Ishiguro H, Hagita N (2012) Teleoperation of multiple social robots. IEEE Trans Syst Man Cybern Part A: Syst Hum 42(3):530–544. https://doi.org/10.1109/TSMCA.2011.2164243

Glas DF, Minato T, Ishi CT, Kawahara T, Ishiguro H (2016) ERICA: the ERATO intelligent conversational android. In: 2016 25th IEEE international symposium on robot and human interactive communication (RO-MAN), pp 22–29

Harada K, Ohyama Y, Tabushi T, Kobayashi A, Fukusaki E (2008) Quantitative analysis of anionic metabolites for *Catharanthus roseus* by capillary electrophoresis using sulfonated capillary coupled with electrospray ionization-tandem mass spectrometry. J Biosci Bioeng 105(3):249–260. https://doi.org/10.1263/jbb.105.249

Hart SG, Staveland LE (1988) Development of NASA-TLX (task load index): results of empirical and theoretical research. In: Hancock PA, Meshkati N (eds) Human mental workload. North-Holland, pp 139–183

Haruno M, Frith CD (2010) Activity in the amygdala elicited by unfair divisions predicts social value orientation. Nat Neurosci 13(2):160–161. https://doi.org/10.1038/nn.2468

Haruno M, Kimura M, Frith CD (2014) Activity in the nucleus accumbens and amygdala underlies individual differences in prosocial and individualistic economic choices. J Cogn Neurosci 26(8):1861–1870. https://doi.org/10.1162/jocn_a_00589

Hasin Y, Seldin M, Lusis A (2017) Multi-omics approaches to disease. Genome Biol 18(1):83. https://doi.org/10.1186/s13059-017-1215-1

Haug S, Castro RP, Kwon M, Filler A, Kowatsch T, Schaub MP (2015) Smartphone use and smartphone addiction among young people in Switzerland. J Behav Addict 4(4):299–307. https://doi.org/10.1556/2006.4.2015.037

Takahashi H, Matsui H, Camerer C, Takano H, Kodaka F (2010) Dopamine D_1 receptors and nonlinear probability weighting in risky choice. J Neurosci 30(49):16567–16572

Ille AM, Lamont H, Mathews MB (2022) The central dogma revisited: insights from protein synthesis, CRISPR, and beyond. WIREs RNA 13(5). https://doi.org/10.1002/wrna.1718

Iwasaki M, Yokota Y, Naruse Y (2023) The proteus effect in virtual space modulated early stage of neural processing. J Hum Interface Soc 25(3):273–282

Izumi Y, Matsuda F, Hirayama A, Ikeda K, Kita Y, Horie K, Saigusa D, Saito K, Sawada Y, Nakanishi H, Okahashi N, Takahashi M, Nakao M, Hata K, Hoshi Y, Morihara M, Tanabe K, Bamba T, Oda Y (2019) Inter-laboratory comparison of metabolite measurements for metabolomics data integration. Metabolites 9(11):257. https://doi.org/10.3390/metabo9110257

Jang K-M, Kwon M, Nam SG, Kim D, Lim HK (2022) Estimating objective (EEG) and subjective (SSQ) cybersickness in people with susceptibility to motion sickness. Appl Ergon 102:103731. https://doi.org/10.1016/j.apergo.2022.103731

Jiang LC, Bazarova NN, Hancock JT (2011) The disclosure-intimacy link in computer-mediated communication: an attributional extension of the hyperpersonal model. Hum Commun Res 37(1):58–77. https://doi.org/10.1111/j.1468-2958.2010.01393.x

Joinson AN (2001) Self-disclosure in computer-mediated communication: the role of self-awareness and visual anonymity. Eur J Soc Psychol 31(2):177–192. https://doi.org/10.1002/ejsp.36

Joutsa J, Johansson J, Niemelä S, Ollikainen A, Hirvonen MM, Piepponen P, Arponen E, Alho H, Voon V, Rinne JO, Hietala J, Kaasinen V (2012) Mesolimbic dopamine release is linked to symptom severity in pathological gambling. Neuroimage 60(4):1992–1999. https://doi.org/10.1016/j.neuroimage.2012.02.006

Trivedi DK, Hollywood KA, Goodacre R (2017) Metabolomics for the masses: the future of metabolomics in a personalized world. Eur J Mol Clin Med 3(6):294. https://doi.org/10.1016/j.nhtm.2017.06.001

Kahneman D, Tversky A (1979) Prospect theory: an analysis of decision under risk. Econometrika 47(2):263–291

Karekla M, Savvides SN, Gloster A (2020) An avatar-led intervention promotes smoking cessation in young adults: a pilot randomized clinical trial. Ann Behav Med 54(10):747–760. https://doi.org/10.1093/abm/kaaa013

Kato H, Izumi Y, Hasunuma T, Matsuda F, Kondo A (2012) Widely targeted metabolic profiling analysis of yeast central metabolites. J Biosci Bioeng 113(5):665–673. https://doi.org/10.1016/j.jbiosc.2011.12.013

Keshmiri S, Sumioka H, Yamazaki R, Ishiguro H (2019) Decoding the perceived difficulty of communicated contents by older people: toward conversational robot-assistive elderly care. IEEE Robot Autom Lett 4(4):3263–3269. https://doi.org/10.1109/LRA.2019.2925732

Kilteni K, Bergstrom I, Slater M (2013) Drumming in immersive virtual reality: the body shapes the way we play. IEEE Trans Visual Comput Graphics 19(4):597–605. https://doi.org/10.1109/TVCG.2013.29

Kilteni K, Groten R, Slater M (2012) The sense of embodiment in virtual reality. Presence: Teleoperators Virtual Environ 21(4):373–387. https://doi.org/10.1162/PRES_a_00124

Ko L-W, Komarov O, Hairston WD, Jung T-P, Lin C-T (2017) Sustained attention in real classroom settings: an EEG study. Front Hum Neurosci 11:388. https://doi.org/10.3389/fnhum.2017.00388

Kodama M, Oshikawa K, Shimizu H, Yoshioka S, Takahashi M, Izumi Y, Bamba T, Tateishi C, Tomonaga T, Matsumoto M, Nakayama KI (2020) A shift in glutamine nitrogen metabolism contributes to the malignant progression of cancer. Nat Commun 11(1):1320. https://doi.org/10.1038/s41467-020-15136-9

Krekhov A, Cmentowski S, Krüger J (2018) VR animals: surreal body ownership in virtual reality games. In: Proceedings of the 2018 annual symposium on computer-human interaction in play companion extended abstracts, pp 503–511. https://doi.org/10.1145/3270316.3271531

Lange M, Fedorova M (2020) Evaluation of lipid quantification accuracy using HILIC and RPLC MS on the example of NIST® SRM® 1950 metabolites in human plasma. Anal Bioanal Chem 412(15):3573–3584. https://doi.org/10.1007/s00216-020-02576-x

Linnet J, Møller A, Peterson E, Gjedde A, Doudet D (2011) Dopamine release in ventral striatum during Iowa Gambling task performance is associated with increased excitement levels in pathological gambling. Addiction 106(2):383–390. https://doi.org/10.1111/j.1360-0443.2010.03126.x

Longo SK, Guo MG, Ji AL, Khavari PA (2021) Integrating single-cell and spatial transcriptomics to elucidate intercellular tissue dynamics. Nat Rev Genet 22(10):627–644. https://doi.org/10.1038/s41576-021-00370-8

McLaren DG, Ries ML, Xu G, Johnson SC (2012) A generalized form of context-dependent psychophysiological interactions (gPPI): a comparison to standard approaches. Neuroimage 61(4):1277–1286. https://doi.org/10.1016/j.neuroimage.2012.03.068

Ministry of Health Labour and Welfare of Japan (2000) Rest and mental health. https://www.mhlw.go.jp/www1/topics/kenko21_11/b3.html

Ministry of Health Labour and Welfare of Japan (2014) Changes in policies concerning health in Japan. https://www.mhlw.go.jp/wp/hakusyo/kousei/14/dl/1-01.pdf

Ministry of Health Labour and Welfare of Japan (2021) A guide to mental and physical health promotion in the workplace. https://www.mhlw.go.jp/stf/seisakunitsuite/bunya/0000055195_00012.html

Ministry of Health Labour and Welfare of Japan (2023) Occupational health measures in the workplace. https://www.mhlw.go.jp/stf/seisakunitsuite/bunya/koyou_roudou/roudoukijun/anzen/anzeneisei02.html

Morville T, Sahl RE, Moritz T, Helge JW, Clemmensen C (2020) Plasma metabolome profiling of resistance exercise and endurance exercise in humans. Cell Rep 33(13):108554. https://doi.org/10.1016/j.celrep.2020.108554

Nakatani K, Izumi Y, Takahashi M, Bamba T (2022) Unified-hydrophilic-interaction/anion-exchange liquid chromatography mass spectrometry (Unified-HILIC/AEX/MS): a single-run method for comprehensive and simultaneous analysis of polar metabolome. Anal Chem 94(48):16877–16886. https://doi.org/10.1021/acs.analchem.2c03986

Nihonsugi T, Ihara A, Haruno M (2015) Selective increase of intention-based economic decisions by noninvasive brain stimulation to the dorsolateral prefrontal cortex. J Neurosci 35(8):3412–3419. https://doi.org/10.1523/JNEUROSCI.3885-14.2015

Nishiumi S, Izumi Y, Hirayama A, Takahashi M, Nakao M, Hata K, Saigusa D, Hishinuma E, Matsukawa N, Tokuoka SM, Kita Y, Hamano F, Okahashi N, Ikeda K, Nakanishi H, Saito K, Hirai MY, Yoshida M, Oda Y, Matsuda F, Bamba T (2022) Comparative evaluation of plasma metabolomic data from multiple laboratories. Metabolites 12(2):135. https://doi.org/10.3390/metabo12020135

Nishiumi S, Kobayashi T, Ikeda A, Yoshie T, Kibi M, Izumi Y, Okuno T, Hayashi N, Kawano S, Takenawa T, Azuma T, Yoshida M (2012) A novel serum metabolomics-based diagnostic approach for colorectal cancer. PLoS ONE 7(7):e40459. https://doi.org/10.1371/journal.pone.0040459

Oyanagi A, Narumi T, Ohmura R (2021) An Avatar that is used daily in the social VR contents enhances the sense of embodiment. TVRSJ 25(1):50–59

Oyanagi A, Ohmura R (2017) Investigating the sense of body ownership over a bird avatar for enhancing immersion in flying experience. Trans Virtual Reality Soc Jpn 22(4):513–522

Peck TC, Seinfeld S, Aglioti SM, Slater M (2013) Putting yourself in the skin of a black avatar reduces implicit racial bias. Conscious Cogn 22(3):779–787. https://doi.org/10.1016/j.concog.2013.04.016

Putman P, Verkuil B, Arias-Garcia E, Pantazi I, van Schie C (2014) EEG theta/beta ratio as a potential biomarker for attentional control and resilience against deleterious effects of stress on attention. Cogn Affect Behav Neurosci 14(2):782–791

Rea DJ, Seo SH, Young JE (2020) Social robotics for nonsocial teleoperation: leveraging social techniques to impact teleoperator performance and experience. Curr Robot Rep 1(4):287–295. https://doi.org/10.1007/s43154-020-00020-7

Ries B, Interrante V, Kaeding M, Anderson L (2008) The effect of self-embodiment on distance perception in immersive virtual environments. In: Proceedings of the 2008 ACM symposium on virtual reality software and technology, pp 167–170. https://doi.org/10.1145/1450579.1450614

Röder PV, Wu B, Liu Y, Han W (2016) Pancreatic regulation of glucose homeostasis. Exp Mol Med 48(3):e219–e219. https://doi.org/10.1038/emm.2016.6

Rosenberg RS, Baughman SL, Bailenson JN (2013) Virtual superheroes: using superpowers in virtual reality to encourage prosocial behavior. PLoS ONE 8(1):e55003. https://doi.org/10.1371/journal.pone.0055003

Rosendo-Rios V, Trott S, Shukla P (2022) Systematic literature review online gaming addiction among children and young adults: a framework and research agenda. Addict Behav 129:107238. https://doi.org/10.1016/j.addbeh.2022.107238

Roy RN, Drougard N, Gateau T, Dehais F, Chanel CPC (2020) How can physiological computing benefit human-robot interaction? Robotics 9(4):100. https://doi.org/10.3390/robotics9040100

Russell G, Lightman S (2019) The human stress response. Nat Rev Endocrinol 15(9):525–534. https://doi.org/10.1038/s41574-019-0228-0

Safaryan K, Mehta MR (2021) Enhanced hippocampal theta rhythmicity and emergence of eta oscillation in virtual reality. Nat Neurosci 24(8):1065–1070. https://doi.org/10.1038/s41593-021-00871-z

Setoyama D, Yoshino A, Takamura M, Okada G, Iwata M, Tsunetomi K, Ohgidani M, Kuwano N, Yoshimoto J, Okamoto Y, Yamawaki S, Kanba S, Kang D, Kato TA (2021) Personality classification enhances blood metabolome analysis and biotyping for major depressive disorders: two-species investigation. J Affect Disord 279:20–30. https://doi.org/10.1016/j.jad.2020.09.118

Sevinc V, Berkman MI (2020) Psychometric evaluation of simulator sickness questionnaire and its variants as a measure of cybersickness in consumer virtual environments. Appl Ergon 82:102958. https://doi.org/10.1016/j.apergo.2019.102958

Singh G, Bermúdez i Badia S, Ventura R, Silva JL (2018) Physiologically attentive user interface for robot teleoperation: real time emotional state estimation and interface modification using physiology, facial expressions and eye movements. In: 11th International joint conference on biomedical engineering systems and technologies. SCITEPRESS-Science and Technology Publications, pp 294–302

Singh G, Chanel CPC, Roy RN (2021) Mental workload estimation based on physiological features for pilot-UAV teaming applications. Front Hum Neurosci 15

Slater M (2009) Inducing illusory ownership of a virtual body. Front Neurosci 3(2):214–220. https://doi.org/10.3389/neuro.01.029.2009

Sreekumar A, Poisson LM, Rajendiran TM, Khan AP, Cao Q, Yu J, Laxman B, Mehra R, Lonigro RJ, Li Y, Nyati MK, Ahsan A, Kalyana-Sundaram S, Han B, Cao X, Byun J, Omenn GS, Ghosh D, Pennathur S, Alexander DC, Berger A, Shuster JR, Wei JT, Varambally S, Chinnaiyan AM (2009) Metabolomic profiles delineate potential role for sarcosine in prostate cancer progression. Nature 457(7231):910–914. https://doi.org/10.1038/nature07762

Stavropoulos V, Dumble E, Cokorilo S, Griffiths M, Pontes H (2022) The physical, emotional, and identity user-avatar association with disordered gaming: a pilot study. Int J Ment Health Addiction 20:183–195

Stefaniak I, Sorokosz K, Janicki A, Wciórka J (2019) Therapy based on avatar-therapist synergy for patients with chronic auditory hallucinations: a pilot study. Schizophr Res 211:115–117. https://doi.org/10.1016/j.schres.2019.05.036

Strimbu K, Tavel JA (2010) What are biomarkers? Curr Opin HIV AIDS 5(6):463–466. https://doi.org/10.1097/COH.0b013e32833ed177

Sumioka H, Nakae A, Kanai R, Ishiguro H (2013) Huggable communication medium decreases cortisol levels. Sci Rep 3(1):3034. https://doi.org/10.1038/srep03034

Sumioka H, Keshmiri S, Ishiguro H (2019) Information-theoretic investigation of impact of huggable communication medium on prefrontal brain activation. Adv Robot 33(19):1019–1029. https://doi.org/10.1080/01691864.2019.1652114

Takami K, Haruno M (2019) Behavioral and functional connectivity basis for peer-influenced bystander participation in bullying. Soc Cogn Affect Neurosci 14(1):23–33. https://doi.org/10.1093/scan/nsy109

Takami K, Haruno M (2020) Dissociable behavioral and neural correlates for target-changing and conforming behaviors in interpersonal aggression. Eneuro 7(3), ENEURO.0273-19.2020. https://doi.org/10.1523/ENEURO.0273-19.2020

Takeda H, Izumi Y, Takahashi M, Paxton T, Tamura S, Koike T, Yu Y, Kato N, Nagase K, Shiomi M, Bamba T (2018) Widely-targeted quantitative lipidomics method by supercritical fluid chromatography triple quadrupole mass spectrometry. J Lipid Res 59(7):1283–1293. https://doi.org/10.1194/jlr.D083014

Tanaka T, Okamoto N, Kida I, Haruno M (2022) The initial decrease in 7T-BOLD signals detected by hyperalignment contains information to decode facial expressions. Neuroimage 262:119537. https://doi.org/10.1016/j.neuroimage.2022.119537

Tanaka T, Yamamoto T, Haruno M (2017) Brain response patterns to economic inequity predict present and future depression indices. Nat Hum Behav 1(10):748–756. https://doi.org/10.1038/s41562-017-0207-1

Tokyo Labor Bureau (2018) Key points of the revised occupational health and safety law. https://jsite.mhlw.go.jp/tokyo-roudoukyoku/content/contents/000372681.pdf

Vandenbon A, Mizuno R, Konishi R, Onishi M, Masuda K, Kobayashi Y, Kawamoto H, Suzuki A, He C, Nakamura Y, Kawaguchi K, Toi M, Shimizu M, Tanaka Y, Suzuki Y, Kawaoka S (2023) Murine breast cancers disorganize the liver transcriptome in a zonated manner. Commun Biol 6(1):97. https://doi.org/10.1038/s42003-023-04479-w

Watanabe N, Sakagami M, Haruno M (2013) Reward prediction error signal enhanced by striatum-amygdala interaction explains the acceleration of probabilistic reward learning by emotion. J Neurosci 33(10):4487–4493. https://doi.org/10.1523/JNEUROSCI.3400-12.2013

WHO (2020) Addictive behaviours: gaming disorder. https://www.who.int/news-room/questions-and-answers/item/addictive-behaviours-gaming-disorder

WHO (2022) Mental health: strengthening our response. https://www.who.int/news-room/fact-sheets/detail/mental-health-strengthening-our-response

WHO (2023) Mental health. https://www.who.int/data/gho/data/themes/theme-details/GHO/mental-health

Xia J, Broadhurst DI, Wilson M, Wishart DS (2013) Translational biomarker discovery in clinical metabolomics: an introductory tutorial. Metabolomics 9(2):280–299. https://doi.org/10.1007/s11306-012-0482-9

Yang E, Dorneich MC (2017) The emotional, cognitive, physiological, and performance effects of variable time delay in robotic teleoperation. Int J Soc Robot 9(4):491–508

Yee N, Bailenson J (2007) The proteus effect: the effect of transformed self-representation on behavior. Hum Commun Res 33(3):271–290. https://doi.org/10.1111/j.1468-2958.2007.00299.x

Yee N, Bailenson JN, Ducheneaut N (2009) The proteus effect: implications of transformed digital self-representation on online and offline behavior. Commun Res 36(2):285–312. https://doi.org/10.1177/0093650208330254

Yoshida M, Hatano N, Nishiumi S, Irino Y, Izumi Y, Takenawa T, Azuma T (2012) Diagnosis of gastroenterological diseases by metabolome analysis using gas chromatography–mass spectrometry. J Gastroenterol 47(1):9–20. https://doi.org/10.1007/s00535-011-0493-8

Zhang D, Cao M, Tian Y (2022) Avatar identification and internet gaming disorder among Chinese middle school students: the serial mediating roles of flow and self-concept clarity. Int J Mental Health Addict. https://doi.org/10.1007/s11469-022-00923-w

Chapter 8
Field Experiments in the Real World

Takahiro Miyashita, Hirokazu Kumazaki, Shuichi Nishio, Takashi Hirano, Shinichi Arakawa, and Yoshinao Sodeyama

Abstract This chapter explores social-field experiments of cybernetic avatars (CAs). Previous chapters introduced the research and development initiatives necessary to achieve a society where individuals can actively participate remotely by operating CAs. To understand the feasibility of social participation through CA utilization, one approach is to conduct experiments in real-world scenarios and evaluate their effectiveness. In the Avatar Symbiotic Society Project, a variety of real-world validation experiments are being conducted across different countries, industries, sectors, teleoperators, and user demographics. This chapter outlines the essential elements for conducting social–field experiments and highlights those done in the fields of nursing care, mental health, and elderly support.

T. Miyashita (✉)
Advanced Telecommunications Research Institute International, Seika-cho, Soraku-gun, Kyoto, Japan
e-mail: miyasita@atr.jp

H. Kumazaki
Nagasaki University, Nagasaki, Nagasaki, Japan
e-mail: kumazaki@tiara.ocn.ne.jp

S. Nishio · T. Hirano
Osaka University, Toyonaka, Osaka, Japan
e-mail: nishio@botransfer.org

T. Hirano
e-mail: hirano@irl.sys.es.osaka-u.ac.jp

S. Arakawa
Osaka University, Suita, Osaka, Japan
e-mail: arakawa@ist.osaka-u.ac.jp

Y. Sodeyama
Sony Group Corporation, Shinagawa, Tokyo, Japan
e-mail: Yoshinao.Sodeyama@sony.com

H. Ishiguro et al. (eds.), *Cybernetic Avatar*,
https://doi.org/10.1007/978-981-97-3752-9_8

8.1   Introduction

The Avatar Symbiosis Society Project has set two primary objectives: (1) creating a society where anyone, including seniors and the physically/mentally challenged, can freely participate in various activities using multiple CAs, and (2) everyone can work and study anytime, anywhere, with minimized commuting time to a company or a school and have plenty of free time. Social-field experiments are one method for assessing the progress toward these goals in the research and development of cybernetic avatars (CAs).

Social-field experiments, unlike laboratory experiments, are set in actual society where the target population (ordinary citizens) serves as participants. The technology under research and development is refined to a level where participants can experience it, and its effects are verified. In the context of CA research and development, real-world experiments are conducted in society using ordinary citizens who are aligned with the attributes of CA teleoperators and the end users of CA services. These experiments aim to provide CA services in real-world scenarios by validating whether the anticipated effects of research and development are achieved.

Although technical validation experiments can sometimes be conducted in simulated environments within laboratories or computer simulations, such aspects as the appropriateness of CA services and the societal acceptance of CAs are often best understood when tested with ordinary people in real-world situations. Social-field experiments have proven to be an effective method for comprehensive verification because they encompass human perceptions and associated behaviors.

This chapter introduces the equipment, communication technologies, and corporate consortium essential for the social-field experiments of cybernetic avatars. The project has focused on conducting such experiments in particularly challenging areas for commercialization, such as nursing care, mental health, and elderly support. Specific experiments will be discussed in this chapter.

8.2   Field Experiment Platform for Cybernetic Avatars

8.2.1   Requirements for Social-Field Experiments of CA

To conduct social-field experiments, several essential elements must be addressed, which differ from experiments in laboratories or computer simulations.

1. Create research and development outcomes that can be experienced by participants:

    In social-field experiments of CA services, it is imperative that individuals can experience these services. If a CA service does not exist as a viable resource, participants cannot contemplate its acceptance. For experiments assessing whether CA services can effect societal changes, it is essential to introduce them into society and conduct experiments over months or years to evaluate their societal impact. To achieve this, participants, not just engineers, must appropriately experience the technology.

2. Establish an experimental environment (communication and safety):

   When activating CAs, a wireless communication environment must be established. In the social-field experiments of CA services, as experiments unfold in our everyday living spaces, ensuring the safety of participants and those around them is paramount. A deep understanding of how the location is used is required, and the experimental environment must be configured in such a way that it does not obstruct or endanger others.

3. Ensure participant safety through ethical review and risk assessment:

   Following the preparation of experimental environments, a safe experimental plan is devised, which must be thoroughly appraised by ethics review committees comprised of such experts as physicians and lawyers. Such committees evaluate the plans of experiments from various perspectives to ensure safety and avoid any potential violation of participant rights. Identified issues are broached, improvements are made, and approval is obtained before an experiment can move forward.

   Without scrutinizing these three points, the execution of experiments is impossible. Additionally, aspects related to the external design of CAs are significant. For instance, if the experiment facility is a commercial establishment, the CA's external design must align with the acceptance criteria of that facility before the experiment can proceed.

8.2.2 CA's Social-Field Experiment Platform

To facilitate the social-field experiments of CA services within the Avatar Symbiotic Society Project, a foundational infrastructure known as the CA social-field experiment platform has been established. This infrastructure is comprised a CA platform (middleware), CAs, and sensor systems. In the course of infrastructure development, such components are incrementally constructed as part of other research and development projects within this initiative and extracted, refined, and made publicly available on an annual basis. Then they are utilized, customized, and maintained for use by member companies of the Corporate Consortium for Avatar Symbiotic Society (C-CAS2), among others. The results of social-field experiments with this platform are reflected in the construction of CA platform and international standardization activities, as described in Chap. 5. Below is a summary of the elements that comprise the CA social-field experiment platform:

CA Platform (Middleware, Public Release Version): As introduced in Chap. 5, the CA platform serves as the information and communication infrastructure connecting teleoperators to CAs for teleoperations. Although as of 2024 the CA platform is in its research and development stage, a version capable of connecting three CAs to two teleoperators (Fig. 8.1) has been made publicly available to C-CAS2 members free of charge for non-commercial use. This opportunity enables researchers and non-developers alike to conduct social-field experiments with CA services.

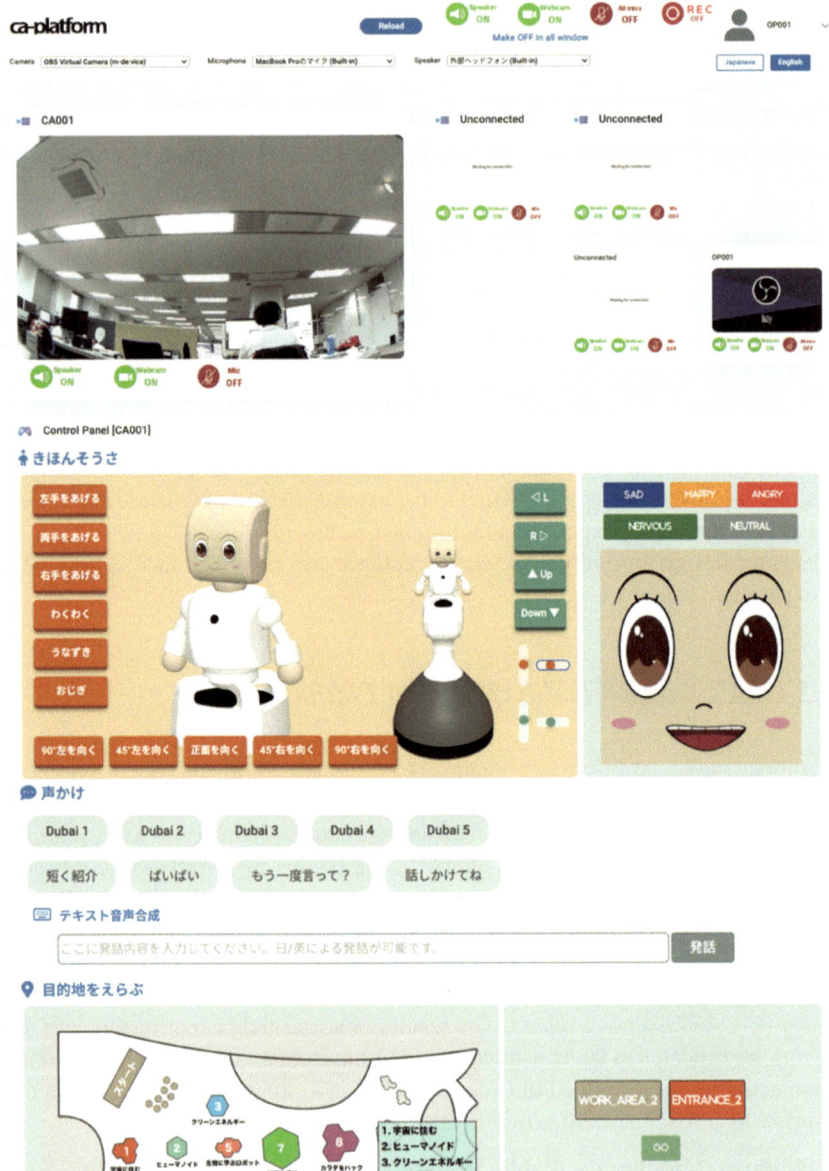

Fig. 8.1 Teleoperation UI of CA platform

CAs: CAs include movable CAs (Teleco), location-fixed CAs (ERICA, an android type), Sota and CommU (small-size humanoid types), and digital signages for CG-CAs (Fig. 8.2), all of which are available for use in social-field experiments.

Sensor Systems: These systems include human position measurement systems utilizing 2D and 3D LiDARs (Glas et al. 2014, 2015), network camera systems,

(a) Teleco (movable CA)   (b) ERICA (location-fixed CA)

(c) Sota and CommU (location-fixed CAs)   (d) CG-CAs in digital signage

Fig. 8.2   CAs in social-field experiment platforms

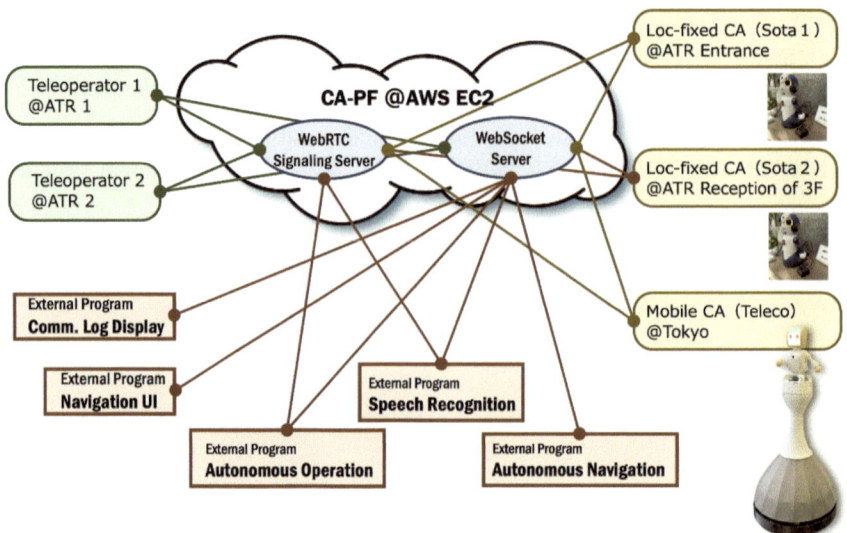

Fig. 8.3 Experimental system example utilizing social-field experiment platform of CA

and other systems that measure human behaviors. These systems are deployable in social-field experiments within the project.

Wireless Communication Devices: Specifically developed for the stable operation of movable CAs, we deployed an MD-WLAN developed by ATR (Kondo et al. 2023) as a wireless communication device to ensure stable communication during physical movements.

Leveraging these social-field experiment platforms, an example of the assembled experimental system is shown in Fig. 8.3. Two teleoperators can control two location-fixed CAs and one movable CA, all located remotely, simplifying the configuration of such systems.

8.2.3 Experiments that Utilize CA's Social-Field Experiment Platform

In this section, we introduce an experiment conducted using the social-field experiment platform of CA. If diverse social activities such as work, education, health care, and daily life can be achieved through avatars, thereby overcoming the constraints of time, space, and physical limitations, people's lifestyles will significantly change. How will the general public perceive such changes? To explore this idea, the Avatar Symbiotic Society Project used the previously introduced social-field experiment platform of CA to conduct a societal validation experiment called the Avatar 100 Experiment (or the Avatar Festival) from July 11 to 20, 2023, within commercial facilities and actual living spaces with restricted conditions. Figure 8.4 shows an advertisement for the event aimed at the general public. The aim was to partially create an avatar symbiotic society, allowing the general public to experience participating in it. Through this experience, participants were encouraged to reflect on the ideal form of an avatar symbiotic society and the necessary research and development to achieve such a futuristic society. This planning approach, starting from a pseudo-realized future and contemplating the current society and research and development, is known as backcasting.

In this experiment, 104 multiple heterogeneous CAs were placed throughout a commercial complex called the Asia–Pacific Trade Center (ATC), in Osaka, Japan. These avatars provided information about the facility, guided visitors to shops and events, introduced the project, and performed such tasks as customer service and security. Operated remotely from multiple teleoperation rooms, these avatars were controlled by individuals who physically visited the ATC building or other teleoperation rooms. Scenes of the experiment are shown in Fig. 8.5.

Key findings from a survey conducted during the experiment (645 valid responses) included the revelation that many participants, even though they seemed familiar

Fig. 8.4 Advertisement for avatar 100 experiment (avatar festival)

Fig. 8.5 Scenes of avatar 100 experiment (avatar festival)

with the term "avatar," were unaware that it refers to CG agents or robots that can be teleoperated as surrogates. About 74.4% reported a clarity in their awareness and knowledge of avatars through the experiment. We gathered such sentiments as "Using avatars expands communication and human connections" (37.2%) and "I learned that avatars are operated by people" (32.7%).

Moving forward, we plan to conduct more backcasting-type social-field experiments in the Avatar Symbiotic Society Project to increase the general public's awareness of potentially new ways of working and participating in society through CAs. By gathering candid opinions based on personal experiences, we will contemplate the future directions of research and development for an avatar symbiotic society, where diverse social activities will become possible that exceed the constraints of time, space, and physical limitations in work, education, health care, and daily life.

8.2.4   Collaborative Efforts for Social Implementation Through Corporate Partnerships

The effective utilization of cybernetic avatars (CAs) varies across industries and sectors. To ensure that the considerations extend beyond CA researchers and engineers, in August 2021 the Avatar Symbiotic Society Project established the Corporate Consortium for Avatar Symbiotic Society (C-CAS2) to actively involve such interests in its actual operation (123 corporate members, as of February 2024).

The consortium's membership is divided into two types: Information Members, who share research and development outcomes related to CAs, and Subcommittee

Members, who specifically explore CA utilization in their respective business operations. In the consortium, CA applications to various industries and sectors are referred to as Avatar Transformation (AX) in alignment with Digital Transformation (DX). To date, four subcommittees (health care and medical, educational support, IT infrastructure, and town development) have been established by consortium members and researchers, each progressing in its exploration of AX. As the examination of AX deepens and demonstration projects are organized, collaboration between research institutions and companies will be fostered to conduct social-field experiments.

8.2.5 Discussion

This section introduced a social-field experiment platform designed to facilitate real-world social experiments using cybernetic avatars (CAs) and presented the Avatar 100 Experiment (Avatar Festival) as an illustrative example of a social-field experiment. We discussed corporate collaborations toward the social implementation of CAs, including Avatar Transformation (AX) and the Corporate Consortium for Avatar Symbiotic Society (C-CAS2). Through numerous social-field experiments, the Avatar Symbiotic Society Project emphasizes that both the research and the development of CA technology and social-field experiments play a vital role for achieving an avatar symbiotic society in which everyone can actively participate. Our goal is to create an avatar symbiotic society that people are eager to accept, based on CA's social acceptability determined through social-field experiments.

8.3 5G Communication Systems for Cybernetic Avatars

8.3.1 Motivation and Scope

Communication systems are essential to enjoy a cybernetic avatar (CA) life. Starting in 2020, 5th generation cellular networks (5Gs) are eventually deployed throughout the world (Atat et al. 2017). In 5G, more radio frequency bands are assigned for telecommunication. A channel bandwidth allocated to 5G communication systems can be 100 MHz (except the milli-wave system), which is five times larger than 4G/LTE (Long-Term Evolution). The CAs connected with 5G will extend their applications by transferring various kinds of information, not only compressed video streams like Zoom or YouTube but also such multidimensional information streams as virtual reality or 3D-point cloud information. All of these applications will enhance our daily lives.

5G communication systems have been standardized by both network carriers and vendors. Many protocols have been defined, each of which describes a procedure through which information is exchanged between such user equipment (UE) as smartphones and such carrier equipment as a base station. With standardization, users can choose their favorite brand/type of smartphone and network carriers (in terms of prices or connectivity) for 5G communication.

In addition to upgrading radio frequencies and channel bandwidth through 5G, the most notable change of 5G communication systems is that a few bands of radio frequencies are offered to non-network carriers, allowing a company or a university to operate its own private 5G communication system. Japan licenses the radio frequency from 4.6 to 4.9 GHz for such purposes, and our research group obtained a license in December 2021 to investigate the applicability of 5G systems for CA lives.

From a CA perspective, the emergence of private 5G communication systems increments the candidates of communication systems. One of our focuses includes the deployment of private 5G communication systems to CA. In this section, we present some results of field experiments on private 5G communication systems. They include the characteristics of communication areas that are superior to WiFi systems and the service qualities for interactive VR services.

Another focus is enhancing 5G communication systems for CA. Note that such enhancement does not include modifications to 5G standards; it does include developments of algorithms or methodologies which are outside of the standards. Due to space limitations, our recent progress on this aspect is not included. For further information, refer to the following article (Eum et al. 2022).

8.3.2 Field Experiments on Private 5G Communication Systems

8.3.2.1 Radio Coverage

Our first result was derived from a field experiment at a shopping mall in Osaka. In July 2023, many project groups demonstrated their research results to visitors. Our research group also set up a private 5G communication system and helped one of the demonstrations achieve higher 5G communication coverage.

Figure 8.6 shows a teleoperated robot at the demonstration, developed by another project group (Hu et al. 2023). Our 5G UE is attached via Ethernet. A teleoperator, located at the same shopping mall in this demonstration, receives a video stream, map information, and a control state from teleoperated robots and sends commands for actuators or voice message to interact with visitors. Figure 8.7 illustrates our demonstrated system with our private 5G systems for which we obtained another license for this field experiment. Two base stations and six UEs were operated during the field experiment. Four UEs (not shown) were used for signal strength monitoring and throughput measurements. Since the bitrate of the teleoperation's video stream topped out at 30 Mbps, accommodating traffic by a private 5G communication system was relatively simple.

Fig. 8.6 Teleoperated robots

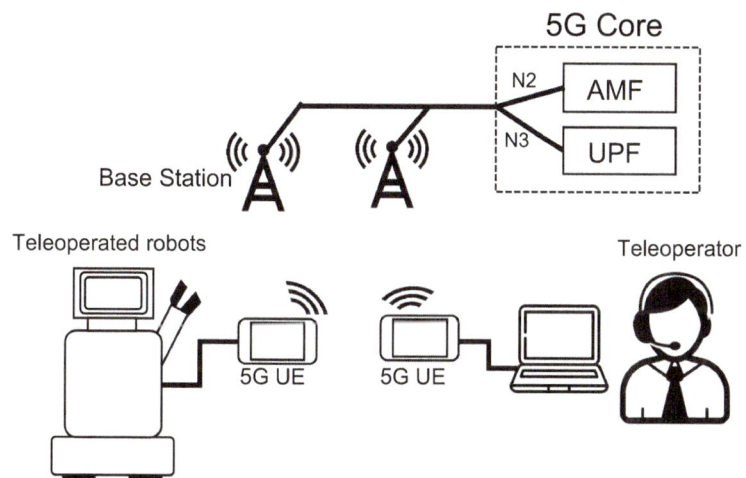

Fig. 8.7 Illustration of demonstrated system

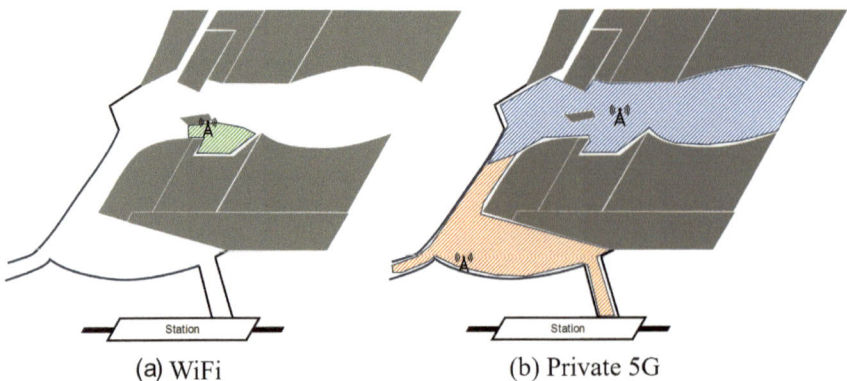

(a) WiFi   (b) Private 5G

Fig. 8.8 Operating area where teleoperated robots can communicate

Figure 8.8 shows the operating area when a private 5G communication system or a WiFi system is deployed. The most significant advantage of our private 5G communication system is manifested in the operating area where teleoperation can be performed. The colored region represents the region where the teleoperator can control the teleoperated robot; outside of it, the teleoperated robot stops moving due to a lack of reachability. Note that the region does not directly represent the physical (PHY)-level signal strength with enough signal-to-noise ratio (SNR). The region represents the area the teleoperator regards as operatable. The teleoperator is near the WiFi's access point.

Comparing Fig. 8.8a (WiFi) and b (private 5G), the private 5G communication system has a significant advantage in the operating area. Almost half of the floor can be covered by a single base station. We were dismayed by the sparsity of the WiFi system's operating area because it should have included around a half of the area of the private 5G communication system based on the specifications of transmission power. The WiFi system's poor coverage was caused by radio interference from other access points. Since WiFi uses an unlicensed frequency band, anyone in the shopping mall can freely set up WiFi systems. The strength of the radio signals of our WiFi system was strongest among other WiFi systems in the operating region. However, as the distance from the access point increased, the strength of the radio signals of our WiFi system was weakened, and so radio signals were affected by the other WiFi systems, lowering the SNR.

We may select WiFi systems for their obvious advantage: no required paperwork as long as the system is certified by government regulations. However, because of this advantage, CA systems may be affected by other systems. We believe that 5G communication systems are a viable candidate to operate CAs outside of laboratories.

8.3.2.2 Quality of Experience on Interactive, Cooperative VR Services

In this subsection, we describe the impact of shortening communication delays by 5G communication systems on the service quality of applications. We focus on VR applications that involve interactions among users.

With the improvement of virtual reality (VR) technology, a variety of new VR services have appeared. VR, which has been used for a wide range of societal purposes and functions from remote medical operations to human interactions in a metaverse space, has the potential to significantly change our daily lives (Hu et al. 2020). Applications that incorporate interaction in such VR spaces require real-time synchronization of information, and so such synchronization must be sped up to improve the quality of user experiences. In 5G, resources are allocated in both frequency and time domains. A time domain is slotted. For example, there are 20 slots within 10 ms at 30-kHz subcarrier spacing (SCS). Low-latency communications are achieved by utilizing the bandwidth available in 5G standards and shortening the slots (Elbamby et al. 2018).

The effect of the amount of delay on users in shooting games (Quax et al. 2004) and certain VR applications (Brunnström et al. 2018) has also been studied, although there remains insufficient literature on VR applications with interactions. We conducted experiments where participants experienced VR applications in our private 5G environment and measured the quality of their experiences based on subjective evaluations. For evaluation purposes, we developed a simple VR application where two or more users interact in VR space.

Figure 8.9 shows a screenshot of our application. Two users enter the VR space, and one user (sender) takes a can off the table and puts it on a conveyor belt, which starts to move when VR objects (cans) are placed on it. The movement's speed is constant. The sender places the cans on the conveyor belt and waits until they are grabbed by another user (receiver). After the sender realizes that the receiver successfully took a can, the sender immediately takes another from the table and places it on the conveyor belt. The receiver observes the sender's behavior. When the can starts its journey, the receiver grabs it at an appropriate timing to avoid falling and moves it to another table.

Figure 8.10 shows our experimental environment. The 5G environment in our laboratory is licensed and operated at a 4.85 GHz center frequency with 100 MHz bandwidth. We use two Oculus Quests as a head-mounted display (HMD). With this device and a dedicated controller, we track the user's hands and 6 DoF movements. The VR application adopts offload rendering where graphic rendering is performed on PCs connected to HMD devices. The VR application running on each PC synchronizes the position information on the VR objects through the application server.

To see the impact of communication delay more clearly, we also prepared wired environments where PCs are connected to the application server via Ethernet and injected a delay simulator just before it. By changing the simulator's setting, packets from/to the application server experience arbitrary delays or jitter. Based on the average delay and its variance observed in 5G communication environments, we

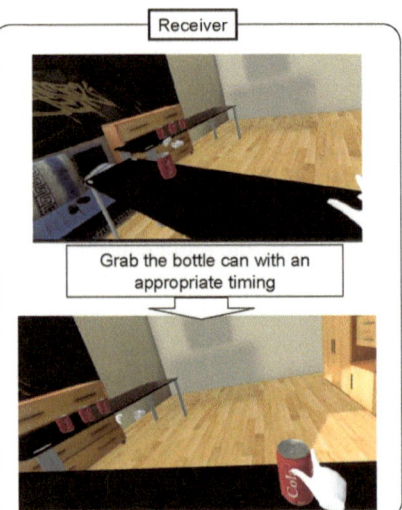

Fig. 8.9 VR application

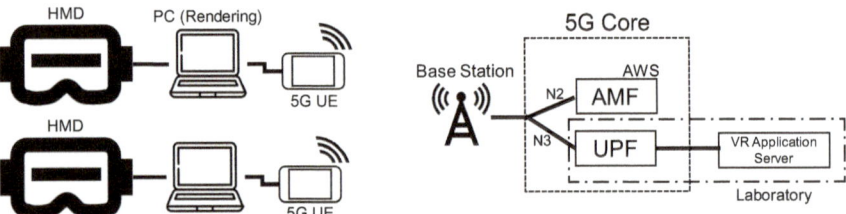

Fig. 8.10 Experimental environment

added the following delay patterns to evaluate the service quality: pattern A where the average delay is close to 5G but has higher variance (jitter); pattern B where the jitter is low, but the average delay is high. The actual delay/jitter measured in our experiment environment is summarized in Table 8.1.

Note that such an intentional injection of delay/jitter can never happen in actual communication systems. However, obtaining quality service in various environments is useful from a networking research perspective. Since a communication system is

Table 8.1 Delay patterns

	Min. (ms)	Max. (ms)	Average (ms)	Variance
5G	10.8	43.6	19.9	24.7
Pattern A (wired)	5.6	55	22	61
Pattern B (wired)	37	45	41	4.1

shared by many users, perhaps it cannot always assign enough resources to guarantee optimal service quality for every user. Thus, a communication system, in our current case, a base station, must prioritize UEs for resource assignment. Such an eventuality happens in commercial 5G communication systems where many of users are attached to a limited resource. In general, commercial communication systems struggle to provide "fair" throughput in a best-effort manner regardless of the perceived service quality. Our challenge is to reveal the appropriate priorities on UEs from the CA perspective, for example, giving greater priorities to sensitive tasks or lower priorities to those that are less sensitive against delay/jitters and introducing resource allocation mechanisms for CA in commercial 5G communication systems. The following experimental results with our private 5G communication system are the first step of our challenge.

The following is the procedure with which we obtained subjective evaluations. For each pair of participants, we first explained the sender/receiver operations in VR space. Participants wore HMDs and held a controller to learn how to manipulate objects, how to change the viewpoint, and how to move in VR space. Then the user's role (sender or receiver) was decided, followed by a five-minute practice for both roles. Before they started to experience the VR application, we explained our questionnaires and asked if they understood them. After the VR application was finished, the participants answered the questionnaires. This process was repeated three times to experience the VR application under 5G and delay patterns A and B.

The questionnaires completed by our participants were prepared to discern how they felt about the communication or service quality in VR space based on whether they knew/recognized the exact position of the other user's hand and the object's position. We asked the following five questions: (1) Did you understand the position of the receiver's hand? (2) Did you feel any discomfort from the receiver's hand movement? (3) Did you understand the object's position? (4) Did you feel any discomfort from its movement? (5) Did you recognize the receiver's action when grabbing the object? Following the Likert scale, each question was graded from 1 (worst) to 5 (best). The questionnaires for the receivers were identical except that they were asked about the sender's behavior.

Figures 8.11 and 8.12 show the results of the scores averaged over three pairs of participants. In our private 5G communication system, the scores are high for all items, which means that the 5G communication system provided good service quality in the VR applications. Looking at the results of delay patterns A and B, the 4th question (about the object's movement) and the 5th question (about the recognition of actions by the other side) severely decreased, indicating that both delay and jitter should be kept low to maintain the service quality of VR applications. We conducted additional experiments and found that when the jitter is large, a discrepancy occurs when updating the coordinates of objects. This flaw causes variations in the speed at which objects move, making them appear to move erratically. Lowering the delay is important; when the average delay is under a certain level, lowering the jitter becomes more important. Our results suggest that even in situations where allocating adequate resources to UEs is difficult, if we reduce the jitter, i.e., variations of delays through

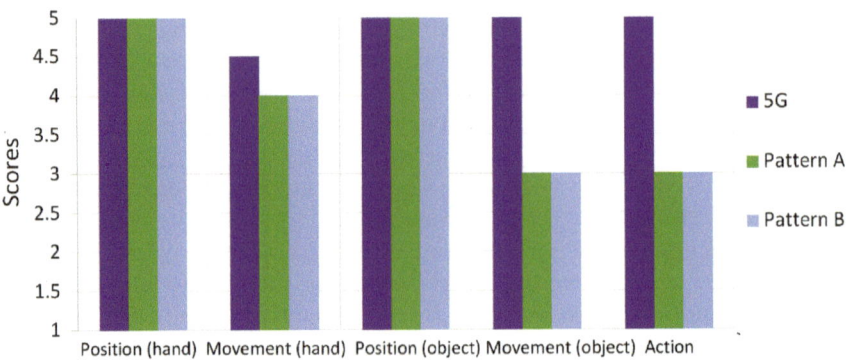

Fig. 8.11 Evaluation scores (sender)

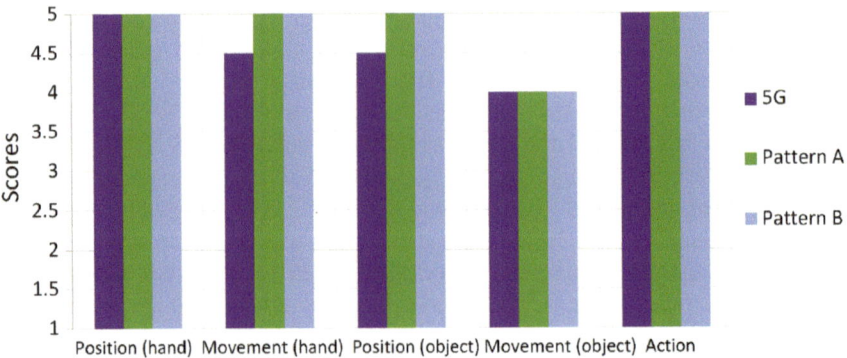

Fig. 8.12 Evaluation scores (receiver)

an appropriate scheduling of resource assignment, we can improve the quality of user experiences in VR applications.

8.4 CA Systems in Field of Nursing Care

In this section, we introduce a cybernetic avatar (CA) system called HANAMOFLOR (Fig. 8.13). Its nickname is Hana-chan. It is a voiced sound such as "HA" and "NA" which are generated by opening the mouth wide, and it is easy for the elderly to speak and hear. The name HANAMOFLOR suggests activities for the goddess of blooming flowers and conveys activities that fulfill dreams. It is a child-type robot that was designed for the field of nursing care. It moves autonomously within shared living spaces in elderly care facilities and spends time with residents to combat psychological instability. For each resident, individual recreational activities are discussed: taking one's temperature, singing together, and assisting with phone calls

Fig. 8.13 HANAMOFLOR

from family. Such activities are carried out to contribute to the psychological stability and improvement of the quality of life of senior residents.

8.4.1   Background

Japan has already become a super-aged society (Oshima 2015). Its working-age population is decreasing and is expected to continue on such a trajectory in the future, threatening traditional support for this burgeoning number of seniors. Japan's national government is exploring support through new approaches and technologies related to elderly care in such a context (Hirukawa 2016; The Ministry of Health 2021). With this background in mind, based on an actual needs assessments of caregiving facilities, we conducted a concept design for the necessary technology and systems and have been progressing with the research and development of this CA system.

We conducted a task analysis survey that targeted the most common form of elderly care facilities in Japan: private-room, unit-type nursing homes (Toyama 2002). In these facilities, a single unit accommodates up to ten residents, and caregiving is holistically provided. Each resident has his/her own individual room; a shared living space surrounds these rooms. Residents have their meals, watch TV, read newspapers, and spend their daily lives as they please in this shared living area. We conducted a time study on the caregiving tasks of professionals in such environments. Our most significant insight from the results concerned the issue of insufficient monitoring of residents in the living areas due to a shortage of workers (Fig. 8.14). Situations often arise where a single caregiver is responsible for ten users in one unit. When this

Fig. 8.14 Nursing care facility issues and needs

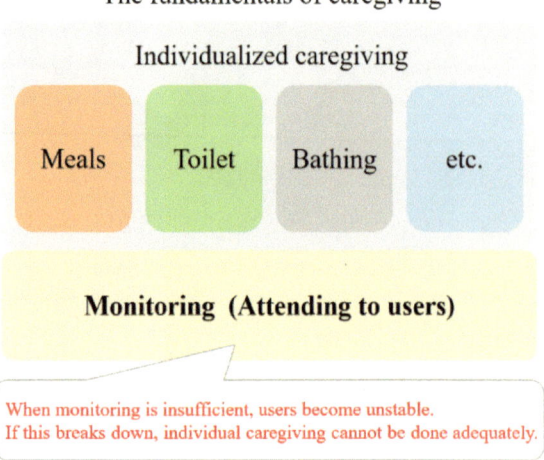

The fundamentals of caregiving

Individualized caregiving

| Meals | Toilet | Bathing | etc. |

Monitoring (Attending to users)

When monitoring is insufficient, users become unstable.
If this breaks down, individual caregiving cannot be done adequately.

single caregiver is providing assistance in individual rooms, such as with toilet assistance, the living area structurally becomes unattended. Some residents feel uneasy and unsettled when no staff members are present. Many such seniors also have dementia. A peripheral symptom of dementia is BPSD, which is characterized by easily becoming anxious and afraid, etc. When staff members are absent, users may cry, become angry, or attempt to stand even though they are unable to walk. Monitoring, which is the foundation of such basic caregiving responsibilities as meal assistance, toilet assistance, and bathing assistance, allows each of these tasks to be performed while maintaining the residents' psychological stability. Providing monitoring of all seniors as a baseline task is essential for ensuring efficient caregiving and maintaining psychological stability.

8.4.2 Design

We designed HANAMOFLOR as a concept to respond to the problem of lack of monitoring due to shortage of manpower. This child type of robot moves autonomously within the living area and leads individual recreational activities to prevent psychological instability. One distinctive feature is its ability to interact with seniors with dementia without causing fear and with clear and understandable interactions. Even in the presence of memory impairment, its design is based on the insight that "important episodic memories are not lost (Tulving 1985)" and invokes a design that is reminiscent of a child or grandchild. HANAMOFLOR's life-like size and physical parameters are designed to resemble to a 2-year-old child. At 83 cm, its height allows those in wheelchairs to look slightly down at it, facilitating its direct interaction with seniors while it moves autonomously. It shifts its head and eyes through face-tracking control and makes eye contact during conversations.

HANAMOFLOR also has a stable design with a flexion axis at the waist. Through this, it can bow, maintain eye contact, and continue conversations while moving its face close to users. Elucidation is enhanced by such movements. For seniors with diminished visual, auditory, and cognitive functions, it is essential to clarify the source of communication and identify the speaker. A gradual approach is emphasized to control the progression of scenarios. The robot first speaks from a slight distance, observing the response, and slowly approaches. Its software also gradually transitions from distant topics to the main topic in its control of the progression of scenarios. Our aim is to achieve a high degree of receptivity from dementia sufferers and advance easy-to-understand conversations by adhering to the interaction methods implemented by caregivers as methods of care (Gagnon et al. 2009; Gineste and Marescotti 2010; Sato 2014; Honda et al. 2016; Nakazawa et al. 2020).

8.4.3 Real-World Proof-of-Concept Testing

8.4.3.1 Proof-of-Concept Testing on Elderly Dementia Patients at a Caregiving Facility

We have been conducting regular proof-of-concept testing at actual caregiving facilities every few months since November 2021. Individual recreational activities were done for seniors one by one in an actual environment of a shared living area (Fig. 8.15). We used a HANAMOFLOR to engage in interaction that included conversation about doing various activities together. Up to this point, we had repeatedly conducted the tests with 18 participants. Since we did these tests during the COVID-19 pandemic, they were conducted after thoroughly weighing measures to prevent infection. We conducted investigations through interviews with the caregivers on a 5-point scale At-At evaluation sheet (Danish Technological Institute and Yamaguchi 2019). Our summary of the results identified no concerns about receptivity, and no participant was frightened. Participants with moderate dementia symptoms gave particularly high ratings. The system's system was deemed to be positive, contributing to an improvement in the participants' quality of life.

A notable benefit of conducting tests multiple times is that those individuals with moderate dementia symptoms and short-term memory impairment began to remember HANAMOFLOR even though the tests were conducted over intervals spanning several months. The remarks made by the participants during the tests were also positive, as shown in the following examples: "You're so cute, Hana-chan," "I'm happy to see you again," "I'm so happy that I feel like crying," "When I go to heaven, I'll tell the angels that you were cute," etc. Participants usually have difficulty conversing with each other due to compatibility issues, and staff members also lack the time for extensive interaction. The COVID-19 pandemic restricted family visits and limited conversation time. The staff members also observed a significant increase in the conversational activities of their residents compared to without the CA.

Fig. 8.15 Proof-of-concept test at a nursing care facility

We also found cases of disuse syndrome accompanying dementia as well as instances where participants who rarely spoke with staff members were singing with this system. Caregivers observed that the participants cared for the CA as if it was a grandchild and sang with it as if recovering a type of maternal/paternal instinct.

Quantitative data collection and evaluation will be conducted in the future, although based on the results of previous implementations, clearly positive stimulation is being imparted, based on the interview results. The CA system shows potential for reducing the progression of dementia and improving the quality of life of those suffering from this disease.

8.4.3.2   Proof-of-Concept Testing on General Public at an Avatar Festival

When implementing a CA system in caregiving facilities, its reception by the general public must also be addressed. Systems are needed that will be accepted by societies in which various systems of values coexist. We conducted proof-of-concept testing on members of the general public at the Avatar Festival (see 8.2.3) at a shopping mall in Osaka to gauge their degree of receptivity. Visitors observed demonstrations of our CA system and watched videos depicting its use in caregiving facilities, followed by interviews and surveys (Fig. 8.16). Two types of operation, autonomous and teleoperation, were conducted during the demonstrations. In the latter, remote control was done using a dialogue-operation module to select the CA's speech content and recreational activities to be carried out. The remote operator observed the speech and the reactions of the visitors through microphones and cameras installed at the venue

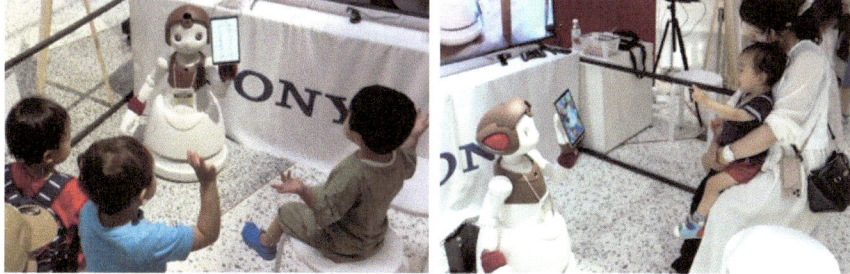

Fig. 8.16 Proof-of-concept testing at the avatar festival

and executed appropriate motions. We conducted a stable demonstration in real time through a remote operation between Tokyo and Osaka.

A summary of the results of the 10-day test with 243 responses shows that our CA system was positively received by the general public. Ninety % of the respondents had a positive attitude toward using it in caregiving facilities and supported its implementation it in a facility in which a family member resides. We conducted an investigation of adjective pairs using the semantic differential (Snider and Osgood 1969) to evaluate the impressions received by individuals. The results identified a particularly favorable impression with such opinions as cute, kind, and likable. The word "cute" was mentioned most frequently in the survey's free comment section, especially in the context of Hanachan's voice. The high evaluations pertain not only to the hardware design but also the results of conversational interactions. The reactions and evaluations from children and families were particularly positive. The system especially successfully captured the attention of pre-school-age toddlers, and during recreational singing, multiple children clapped their hands and sang with smiles, a frequently occurring scene.

8.4.4   Conclusion and Outlook

In this section, we introduced a CA system (a child-type monitoring and care-robot named HANAMOFLOR) designed for use in the caregiving field along with background information, a concept design, and examples of proof-of-concept testing. The primary focus in this CA system's concept design is enhancing its receptivity by elderly dementia patients. We aimed for a design that is easily accepted by individuals whose average age is 90 as well as those suffering from reduced visual, aural, and cognitive functions. We successfully achieved both receptivity and value for this target audience. As a result, our system design is a prime example of an inclusive design, which was positively received as cute/adorable by individuals of all age groups. In particular, the system showed high receptivity among children aged two and three, suggesting a broad range of potential applications, such as monitoring children of this age group and assisting pre-school teachers, etc.

In the future, we will continue to promote activities to quantitatively demonstrate the effects described above with an aim toward CA's real-world applications and commercialization. We will continue to explore consumer opportunities in fields outside of elderly caregiving, for example, children care, education and hospital, etc. Our CA system is a life-sized humanoid robot with dual-arm mobile manipulators and modules for mobility, dialogue, observation, and operation. We will further research and develop these modules in the future for operating them more efficiently as a skill-specialized remote-control system. Our goal is to achieve a perfect CA system that exceeds the limits of autonomous functionality and operator efficiency through remote control. By implementing such a system, we can create a CA system with an even higher level of hospitality. We hope to provide value to a broader range of targets.

8.5   Social-Field Experiments for Individuals with Developmental Disorders and Depressive Disorders

The core symptoms of autism spectrum disorder (ASD) are impaired social communication and restricted repetitive behaviors (American Psychiatric Association 2013). Individuals with ASD experience marked difficulties in social functioning (Bishop-Fitzpatrick et al. 2017). It is estimated that approximately one in 36 American children suffers from ASD, based on the Autism and Developmental Disabilities Monitoring (ADDM) Network (Maenner et al. 2021). In the USA, the medical costs of the social problem of ASD are estimated to be $3 million per person (Cakir et al. 2020).

Several interventions are available for patients with autism spectrum conditions (ASC). Evidence-based studies have demonstrated that developmental interventions are partially effective in ASC individuals. However, many such individuals cannot easily maintain high motivation or focus on human interventions (Warren et al. 2015).

Intensive sensory processing in ASC individuals may be influenced by dynamic facial features and expressions, which likely induce sensory and emotional overstimulation and distraction (Johnson and Myers 2007). Consequently, they tend to actively avoid sensory stimuli and concentrate on predictable rudimentary features, which can hinder learning. Therefore, novel and effective supportive measures and therapeutic intervention strategies must be specifically established for patients with ASC.

Anecdotal evidence shows that individuals with ASC may have a unique opportunity to use cybernetic avatars (CAs), such as robots. Robots allow them to control and recreate situations with smooth and precise conversations despite their reactions, thus contributing to a more structured and standardized intervention. Unlike humans, robots that operate within a predictable and lawful system provide ASC individuals a highly structured learning environment that facilitates their focus on relevant stimuli. Structured interactions with humanoid robots are likely to form standardized social situations in which specific social behaviors can occur.

In this section, we introduce many examples of using CAs in the psychiatric field.

8.5.1 Our System: Supporters Operate a Robot

Previous studies have demonstrated the effectiveness of robotic interventions in ASC. A typical form of support for ASC individuals with a robot involves a therapist who operates the robot to interact with another person.

Since ASC individuals generally struggle to write personal narratives, it is difficult for ASC individuals and their supporters to understand their inner psychological lives. The Sentence Completion Test (SCT) is a semi-structured projection widely used by clinicians and psychologists to explore needs, inner conflicts, fantasies, attitudes, desires, adjustment disorders as well as the possibility of sexual abuse. Even with SCT, providing sentence starters is insufficient to compensate for atypicalities in creativity and imagination, and self-disclosures are difficult for ASC patients.

We developed a novel projection system using an android robot because many ASC individuals often achieve higher task engagement through interaction with robots, and robotic systems may be useful for eliciting and facilitating social communication, such as self-disclosure (Kumazaki et al. 2022a). An android's appearance often evokes a specific personality, making it easy for some ASC individuals to imagine their own personality. Therefore, we investigated whether the addition of an android robot exemplification to SCT encouraged self-disclosure in ASC patients. We compared the differences in disclosure statements and subjective feelings on test forms between SCTs with the addition of model answers by an android robot and from a control group (human interviewers). For comparison, we also assessed disclosure statements and subjective feelings in the SCT using model answers written on the test form.

Our results found that quantitative data suggest that model answers by android robots promote more self-disclosures, especially on negative topics, compared with those by human interviewers and the model answers on the test form. Participants' embarrassment with the android robot's model answers appeared lower than in the human interviewee condition. Eliciting self-disclosure is an urgent issue in the assessment and support of ASC individuals. Our results suggest that this system may be useful for eliciting self-disclosure in ASC individuals. However, an android robot is not necessarily better at eliciting self-disclosure. For some individuals with ASC, humans elicit self-disclosure better than android robots. We must elucidate the strengths and weaknesses of individuals and robots that elicit self-disclosure.

8.5.2 Our System: Patients Operate a Robot

Another form of support for ASC sufferers with a robot involves the individual who is operating the robot to interact with a therapist.

Education in communication skills is essential for ASC individuals to attain their full potential. To provide communication education while maintaining social distancing, we developed a communication training system with a teleoperated robot (Kumazaki et al. 2021). In this system, each participant was provided with a PC and robot. The participants worked in pairs and communicated by a teleoperated robot. The objectives were to verify whether this system continue to motivate ASC individuals for training. We found that it was useful for improving their communication skills. The participants were randomly assigned to one of two groups: teacher-only (TCT) or robot-mediated communication training (RMC). Participants in the former group received lessons on communication skills from their teachers. Participants in the latter group communicated through a teleoperated robot once a week for four weeks (five times per week) in pairs in addition to teacher-led lessons. Twenty ASC patients were participated in this study. We identified the following significant greater gains in self-assessment, "I am good at explaining my thoughts to others," and in self-assessment and teacher assessment, "I am good at listening to others' thoughts and feelings." As expected, motivation for training with this system was maintained throughout the sessions. In summary, this system effectively improved communication skills (e.g., listening to others' thoughts and feelings). Some ASC individuals can deftly operate keyboards, although they struggle with face-to-face communication. Therefore, our proposed system is suitable for such patients. However, various methods are available for remote-control robots. If ASD individuals struggle to operate a keyboard, there are many other ways to remotely control robots. We must customize how ASC sufferers operate a keyboard for each individual.

Social skills training (SST) helps ASD patients better understand others' perspectives and social interactions, develop empathy skills, and learn how to engage socially with others. We developed a social skills training program (STUH) using several humanoid robots, including an android robot, to familiarize ASC patients with the perspectives of others and improve their socialization and empathy skills (Noguchi et al. 2023). This study investigated the efficacy of STUH in such patients. In STUH, we prepared 50 social exercises consisting of conversations and behavioral interactions between an android robot and a simple humanoid robot. We prepared another humanoid robot that had a cartoon-like mechanical design that acted as a host. Therefore, this study investigated the efficacy of STUH in this population. Fifty social exercises were prepared for the STUH consisting of conversations and behavioral interactions between an android robot and a simple humanoid robot. In the first half of the STUH, participants engaged in exercises from the perspective of outsiders, whereas in the second half, participants engaged in simulated experiences using robots as avatars. Interventions related to STUH lasted for five days. Fourteen ASC individuals were participated. All sociability index items improved between the pre-intervention and follow-up sessions. Our program enabled the participants to become familiar with the perspectives of others and improved their sociability. Android robots exhibit various expressions. Although their expressions might be complex, they are obviously much simpler than those of humans. Humanoid robots are critical for designing tools that are efficacious for assisting ASC individuals. The ultimate goal of robot-assisted ASC therapy is to generalize the social skills obtained during robotic sessions to subsequent interactions with humans. For this purpose, more humanlike robots are probably advantageous than mechanical robots. The optimal appearance of robots used for ASD therapy should be located at some point on the humanoid-non-humanoid spectrum, and varying this point based on the severity of ASC might be beneficial.

8.5.3   Importance of Considering Individuality

Previous studies have suggested that preferences and reactions to robot interactions vary widely among individuals with ASC. Given that most previous studies experimented with robots in the same non-verbal environment and that ASC individuals have strong likes and dislikes, ensuring not only an optimal appearance of the robot but also optimal movement is important for smooth interactions and potential robot interventions. We investigated whether ASC patients are more likely to talk to an android robot that barely moves (only opening/closing its mouth when speaking) or to one that moves frequently (not only opening/closing its mouth when speaking but also moving its eyes left to right, up and down, blinking, taking deep breaths, rotating its neck and body, and making random movements).

   We investigated which of the two types of individuals was more likely to talk to each other (Kumazaki et al. 2022b). This was a crossover study in which 25 ASC individuals experienced a simulated interview with an android robot that made many spontaneous facial and body movements and another that made almost none. We compared the demographic information of the participants who indicated that they were more comfortable when speaking with an android robot that made many movements than with the one that made almost no movements and those who indicated the opposite results. Furthermore, we examined how each demographic data item is related to the participants' sense of ease in the context of an interview with an android robot. Fourteen participants indicated that an android robot that made fewer movements was easier to talk to than one that made many movements; eleven indicated the opposite. Significant differences were also found between the two groups in sensory sensitivity scores, which reflect a tendency toward lower neurological thresholds. We also found a correlation between sensation-seeking scores, which reflect a tendency toward higher neurological thresholds, and self-assessments of comfort in each condition.

These results provide preliminary support for the importance of establishing the android robot's behavior by considering the sensory characteristics of ASC individuals. Adaptation, which is a central feature of the nervous system, is defined as a short-term decrease in the responsiveness or sensitivity of neurons after prolonged exposure to a particular stimulus (or attribute) to which the neuron is sensitive (Lawson et al. 2018). This process, which is an important aspect for adapting to environmental changes (Störtkuhl et al. 1999), has been classified as abnormal in ASCs compared to that in controls (Lawson et al. 2014). The traits of ASC are gradually being considered in the motions of android robots. It is estimated that participants prefer the android robot's motion after a long interaction time.

8.5.4 Intervention Using Online Training

The demand for online interviews has been increasing rapidly. Since many ASC individuals are uncomfortable with online or in-person interviews, such online interactions present a hurdle to social participation. Training with computer graphics offers several advantages, including active participation rather than passive observation, a unique training experience, low cost, and accessibility.

We developed a group-based online interview training program using a virtual robot (GOT) in which the interviewer and interviewee were projected onto the screen as virtual robots, and five participants were grouped together as interviewers, interviewees, and evaluators (Kumazaki et al. 2022c). The participants randomly played every role. Each session was comprised a primary interview, feedback, and a secondary interview. Participants underwent 25 sessions. Before and after GOT sessions, they underwent a mock online interview with a professional interviewer (MOH) to assess GOT's effectiveness. Fifteen ASC individuals were participated in the study, which improved their confidence, their motivation, their understanding of others' perspectives, their verbal and non-verbal abilities, and their interview test performances. Their perceptions of the importance of the interviewer's or evaluator's viewpoint also increased significantly after the second MOH compared to the first one. Realizing the importance of interview skills using VR robots and experiencing a different perspective from the interviewer's or evaluator's viewpoint may have helped maintain the motivation and confidence of the ASC individuals. In another study (Yoshikawa et al. 2023), we found that a CG-based program improved real-world interview skills (verbal, non-verbal, and interview performance).

Unfortunately, the cost of preparing robots is high. Engineers are scarce in clinical settings, and preparing and maintaining robotic systems are difficult, particularly during system malfunctions. Since preparing and maintaining a CG system are easier than using a robotic system, CG systems are expected to be used in clinical practice in the near future.

8.5.5 Applications to Other Psychiatric New Diseases

The number of patients worldwide with psychiatric disorders has been rapidly increasing. For example, according to a cohort study in New Zealand, lifetime prevalence of mental disorders is over 70% (Caspi et al. 2020). We previously reported not only in the field of ASC but also in others, such as social anxiety, schizophrenia, and social withdrawal. Social anxiety disorder (SAD) generally co-occurs with ASC (Spain et al. 2018). Some individuals struggle to speak in front of others, a problem linked to social restrictions (Muris and Ollendick 2021). We previously described a case of an individual with comorbid SAD and ASC who could not speak in public and used a humanoid robot as his avatar (Yoshida et al. 2022). For patients with SAD/ASC comorbidities who cannot speak in public, the intervention of a teleoperated robot as an avatar might be very beneficial, since it allows them to speak and respond to others through a robot called CommU in the presence of others while avoiding eye contact by concentrating on PC operations. This study's results suggest that PC manipulation is beneficial. A robot-based intervention was deemed especially helpful for getting patients with comorbid SAD and ASC to understand that the interlocutor's response to their speech was less negative than they interpreted. This realization reduced social anxiety and increased their confidence in their speech. Although not limited to ASC and SAD individuals, many lack self-confidence in their public speech. Since increasing confidence is one crucial factor for more self-confidence, the appropriate use of robots may be linked to active social participation.

However, few studies have examined the use of robots for patients with schizophrenia. These patients have difficulty interpreting facial expressions and gaze directions. The interpretation of both facial expression forms is significantly related to social competence. Interventions are required to improve these interpretations in schizophrenia patients. We showed that an android robot's intervention for schizophrenic patients with comorbid ASC might improve their interpretations of facial emotions and gaze direction (Kumazaki et al. 2023). Thus, a 3D learning environment, which includes interaction with an android robot (avatars are also used by schizophrenia patients), may be more effective for such patients. Some ASC patients with schizophrenia and comorbid are suspicious of others, an aspect that obviously also hinders communication rehabilitation. However, some patients with schizophrenia believe that robots will not betray them, a belief that strengthens their sense of security. The relationship between ASC and schizophrenia has received considerable attention in recent years (Zheng et al. 2018). Treating patients with both conditions is especially difficult, partly because of their negative attitudes toward others. Contrastingly, those with ASC often achieve higher task engagement in robotic interventions. Our results suggest that robotic interventions may be effective for patients with schizophrenia who are complicated by ASC. The cognitive and social rehabilitation of patients with schizophrenia has many challenges. For those who are distrustful, the time spent with robots can be important. Clinicians must constantly attempt to treat these patients.

8.5.6 Summary

Although the number of individuals with psychiatric disorders is increasing worldwide, various challenges must be conquered before such psychiatric patients can receive support. The lives of psychiatric patients might be enriched through the appropriate placement of robots in society. Unfortunately, progress in robot-mediated interventions remains modest, and few advances have been made regarding their clinical applicability (Begum et al. 2016).

The potential role of robots in interventions with ASC individuals is currently not recognized by the potential end users of this technology, such as ASC individuals, their caregivers, and clinicians. Collaborations between technologists and psychiatrists are necessary to advance this field.

8.6 Social-Field Experiments for Older Adults

In Japan, life expectancy continues to increase, its birth rate is decreasing, and its shrinking workforce and the need to create a meaningful life for seniors have become critical social issues. In particular, the COVID-19 pandemic created a challenging social environment for seniors. Although widespread vaccinations curtailed the most recent spread of COVID-19, new variants might spawn future outbreaks. Simultaneously, online work spread rapidly during the pandemic, and various issues might emerge when seniors cannot cope with such changes. However, other pertinent issues linked to the work environment and type of work are inevitable. Many seniors continue to harbor an intense desire to work, although many probably seek shorter daily work periods and fewer hours. They want to participate in society while working close to their homes. Additionally, many seniors want work that resembles what they did before retiring to utilize their existing knowledge and experience. Unfortunately, society might no longer require the same services from them, and starting over from scratch at a new job may be highly difficult for many seniors.

One way to resolve this problem is to perform work and related tasks by remotely controlling or teleoperating CAs, such as robots and similar devices. An image of such teleoperation is shown in Fig. 8.17. By working through teleoperated avatars, the problem of working location is resolved, and people might be allowed to work for shorter periods or hours. Furthermore, teleoperation makes it possible to systematically intervene in the content of operations (semi-automated teleoperation) and

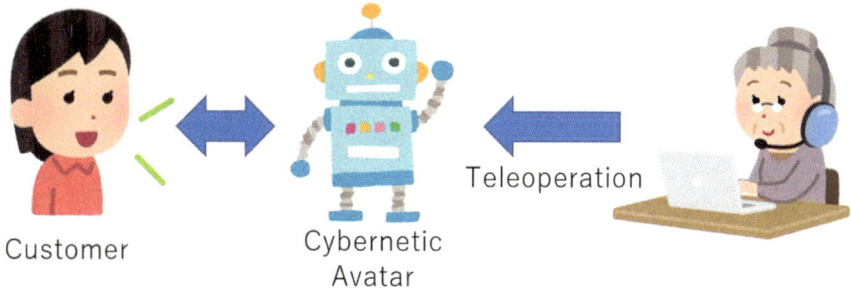

Fig. 8.17 Teleoperation flow

provide on-screen and voice instructions to assist operators while they perform their tasks.

In this study, we developed a prototype robotic teleoperation system that older adults can readily use and tested it in a real-world setting to evaluate its applicability, including whether seniors can accept teleoperation. In this section, we reported its preliminary results. In this study, we used an interactive robot as a teleoperated avatar, and the scope of its application was limited to such conversational tasks as facility guidance.

8.6.1 Equipment

8.6.1.1 Robot

We used a small interactive robot called RoBoHoN, manufactured by the Sharp Corporation (Fig. 8.18).

RoBoHoN is a small, 20-cm tall puppet-like robot that weighs ~ 400 g. It includes a built-in camera, and a microphone is installed in its head. It has built-in motors in its neck and arms, allowing it to gesture and speak using speech synthesis. In the present study, in collaboration with the Sharp Corporation, we modified the firmware to enable voice recording and control servo motors and developed an embedded teleoperation application for which the details are provided below.

8.6.1.2 Teleoperation System

Our teleoperation system was developed using WebRTC with an interface that runs on a web browser. We also developed an application for the robot using a native library and connected it to a commercial signaling server. We generated robot speech using the following three methods:

1. Direct transmission of the operator's voice.

Fig. 8.18 RoBoHoN

2. Speech synthesis produced by pressing buttons on a screen.
3. Conversation based on speech synthesis after speech recognition of the operator's speech.

Here, since a voice synthesis mechanism which was built into the robot was used for Steps 2 and 3, the robot's voice is identical regardless of the operator. In Step 1, the operator's voice is transmitted, and so the voice heard remotely is different, depending on the operator. Figure 8.19 shows how the teleoperation was performed.

The screen's top left shows the image captured by the robot's camera. The operator can recognize the other person and determine, for example, whether he/she is in front of the robot. In this area, if the mouse is moved over the left, center, and right halves of the screen, buttons are revealed that can change the angle of the robot's neck to −

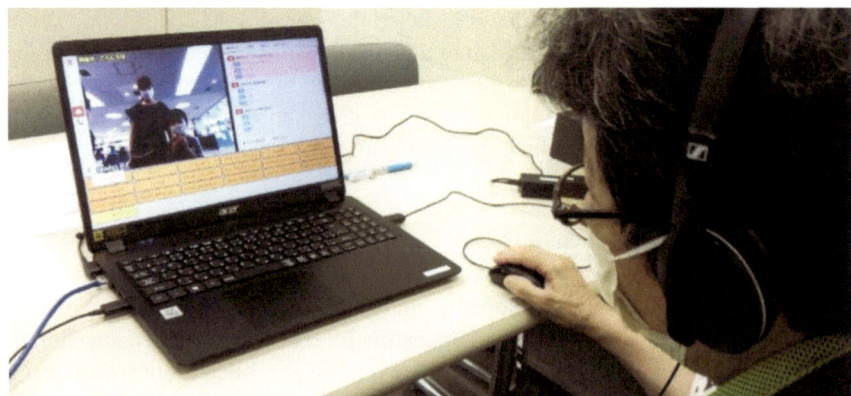

Fig. 8.19 Teleoperation performed by a senior

45° (left), 0° (forward), and 45° (right). If the mouse is moved to the top right-hand corner, a communication-disconnect button appears, which can end the teleoperation. The left-hand side of the screen is comprised of buttons that cancel the voice content, switch from the robot's voice to the operator's voice by voice recognition, and mute.

On the screen's right side, there is an area to set the speech content based on the task. In this experiment, we conducted a quiz and operated a facility guide. In the quiz, the chosen category is played automatically, and the operator determines whether the answer is right or wrong. A voice is played based on the answer. In the facility guide, the selected voice is played for each question.

8.6.2 Methods

8.6.2.1 Experimental Procedure

In this experiment, with the cooperation of Sakai City in Japan, three older adults tele-operated robots at a local governmental ward office, and seven teleoperated those at a large children's center. The experiment was conducted with the approval of the Ethics Committee on Research Involving Human Subjects from the School of Engineering Science, Osaka University (approval number R2-32–2). Written informed consent was obtained from the operators. Explanations of the experiment were provided to visitors to the facility, and consent was obtained on an opt-out basis. The details of the experiments conducted at each facility are described below.

In the experiment conducted at the city's ward office, three seniors operated the robot for 90 min over three days. The robot was located on the first floor of the ward office, where pamphlets and other materials are placed; the robot greeted visitors to attract their interest and encourage dialogues. When visitors approached it, the robot was controlled remotely from another location within the facility, allowing it to engage in dialogues and conduct quizzes about the district. The operator also freely controlled the robot before and after the quizzes. A speaker with a microphone connected via Bluetooth was placed near the robot to simplify hearing the other person's voice and to make the robot's voice louder.

In the experiment, conducted at a large children's center, seven seniors operated the robot for two days. On the first day, it was placed in a hallway on the second floor, and two participants took turns operating it for three 30-min periods (with a 30 min break between each 30-min period), for a total of one hour and 30 min. A speaker with a Bluetooth-connected microphone function was placed near the robot to simplify hearing the other person's voice and to make the robot's voice louder. On the second day, the robot was set up in the exhibition hall on the fourth floor, and three 30-min periods were allocated for operating the robot (with a 15-min break between each 30-min period), and two seniors in the morning and three in the afternoon alternately operated it. Since our experiment took place in a children's center and during an extended holiday period, the facility had many visitors; no special attempts were made to attract visitors. For the visitors who approached the robot, the operator freely interacted with them by remotely controlling the robot from another location within the facility to guide visitors through the facility. On the second day, some people complained that the microphone speaker's echo cancelation function made it challenging to know when to speak. Unfortunately, given the relatively quiet exhibition room, no speaker with a microphone function was installed.

8.6.2.2 Questionnaire Interviews

Following the experiment's completion, each operator filled out a questionnaire and was interviewed. We asked about the following situations:

- Work situation/desire to work.
- Anxiety regarding COVID-19 and its impact on their lifestyles.
- Effects of aging, anxiety about starting this experiment, or a new job/activity.
- Opportunities to meet with children, expectations for and anxieties regarding said interactions.
- Frequency of smartphone/PC use.
- Impressions of robots and their remote operation.
- Level of interest in joining future experiments.

8.6.3 Results

Ten of the senior operators in this experiment were not currently working. Among them, including those who are currently volunteering, six said that they were content just volunteering, two wanted paid employment, and two said that they did not want to work. Six admitted to feeling anxious about COVID-19, and all ten said that the pandemic disrupted their lifestyles. These findings suggest that a demand exists for elderly people to work/volunteer by remotely operating and guiding robots.

At the same time, eight of the seniors expressed concerns about starting something new, such as a job. They admitted that their concentration, physical strength, and memory were declining due to age, which we inferred to be fueling such worries. However, when asked if they were anxious before operating the robot, eight felt no anxiety, suggesting that even seniors can operate a robot with such an interface without excessive anxiety.

In terms of their impressions of the children, they expressed relatively little resistance, although note that this may have been because seven of the seniors who took part in the experiment were volunteers at the Children's Center, a fact that undoubtedly influenced their answers.

In interviews, nine respondents enjoyed using the robot and found it interesting. When asked if they would like to participate again, nine of ten said yes. One respondent on the fence about a subsequent participation admitted that she/he had difficulty leaving the house. However, since she/he added that operating the robot from home would be possible if she/he had an internet connection and a PC, his/her interest in teleoperations can be inferred. Respondents said that they liked how visitors responded pleasantly to them and enjoyed talking to them from the perspective of a robot. This idea indicates that interaction by teleoperating the robot was generally well received. Of course, since this was the first experiment for every participant, the results might change if the experiment was repeated several times due to habituation or boredom.

Concerning the positive aspects of the robot's teleoperated dialogues, by acting in place of the robot or pretending to be one, they felt that it was easier to engage in dialogues, describing a phenomenon that resembled a masking effect. They did experience some disadvantages to standing in for a robot.

The fact that millions and millions of people use computers and smartphones regularly suggests that operators, as a whole, do not have strong resistance to the manipulation of electronic devices. In terms of handling such devices, four operators stated that the time lag caused by the voice recognition complicated its uses. In this experiment, when speaking with the robot's speech synthesis, a time lag occurred because the voice input was converted into text through speech recognition, after which the robot talked using speech synthesis from text. Our system allows operators to speak in their own voice, similar to making a telephone call, which causes almost no delay. However, we explained this idea to the operators and advised them to speak in their own voice. However, they described the experience of hearing their own voice coming out of the robot as strange and embarrassing and refused to use it. Perhaps, the operators were overly aware that they were speaking on behalf of the robot. Based on this result, one possible solution is real-time voice transformation, perhaps with a near real-time voice changer. In recent years, although voice changers that utilize deep learning have been developed, most struggle regarding voice quality after conversion. Further research and development are expected in the future.

Another problem observed during the operation was the difficulty knowing what the robot was saying at any given moment. This happened because, during the experiments, the robot and the microphone speaker were connected via Bluetooth. The microphone used by the speaker had a higher maximum volume than the robot's on-board speaker, and the microphone had a better pick-up performance. By using a speaker with a microphone function, it was easier for the visitors to hear the robot as well as the voices of the visitors. The speaker with a microphone function also had an echo cancelation function so that the voice spoken by the robot was not heard on the operator's side. When the first experiment was carried out under these conditions, approximately half of the operators reported that they couldn't hear themselves and couldn't tell when they were speaking.

Therefore, in the second experiment, we used a standard microphone and a speaker on the robot instead of connecting a speaker with a microphone function. For the experiment carried out under this condition, some participants commented that it was difficult to hear the visitors' voices. The robot in this experiment was equipped with a microphone sufficient for autonomous conversation in a quiet room, although its performance was inadequate as a device for dialogue in a large facility with noise. Based on these considerations, the microphone used in the robot requires a good sound-collection performance and must provide some kind of feedback, such as a no echo cancelation function.

For the operating screen, one person thought that the "button positions are difficult to recall," an opinion that reflects that too many buttons were prepared. Consideration was given to operators who could acclimate to button positioning after continuously operating the robot for a certain amount of time; however, at the same time, the size, categorization, and arrangement of the characters must be addressed to facilitate operation especially among inexperienced users.

Furthermore, through interviews, another drawback surfaced: the inability to deal with problems on one's own when they occurred. When implementing this system in the real world, it must stably and easily recover during malfunctions.

8.6.4   Discussion and Future Issues

From the two experiments described in this section, we found that seniors easily operated the robot remotely and enjoyed using it during interactions with children. Although the remote operation system used in the experiments leaves much room for improvement, even seniors remotely operating the robot for the first time could do so and engage in dialogues without any difficulty. Furthermore, many seniors were happy to talk to children, entertain them, and help them and their parents. Although such sentiments reflect desire for approval and sociability, since many questionnaire respondents said that they were "currently not working," "would like to volunteer," and "anxious about a new job," this suggests that they are reluctant to participate in social activities because of unfamiliar tasks and responsibilities and their own mental and physical deterioration. Although the seniors who participated in this experiment seem actively involved in such social activities as volunteering, many are willing to help but have few opportunities to interact with others, such as the increasing number of the elderly living alone in recent years. Many seniors might benefit from effective remote interaction.

In addition to solving the above problems, mechanisms are required to support seniors. For example, for tasks that are difficult to remember and respond to, such as those involving information about large buildings or commercial products, or tasks that require multi-step procedures, such as how to use equipment, support needs to be based on the context of what is being said at the time and the information that must be conveyed. Although given recent advances in automated dialogue generation technology, many conversational tasks can be replaced by automated systems. However, it remains difficult for automated dialogue systems to assimilate content that users cannot explain well or to express emotional and sensory dialogue, such as communication that provides feelings of security. The advantage of teleoperation is that human communication can be conducted by a person, while information retrieval and presentation can be performed automatically. Therefore, the research and development of such semi-automatic remote operation mechanisms must be promoted, particularly types that can be easily operated by seniors to provide them with a genuine sense of satisfaction.

References

American Psychiatric Association (2013) Diagnostic and statistical manual of mental disorders. American Psychiatric Association

Atat R, Liu L, Chen H et al (2017) Enabling cyber-physical communication in 5G cellular networks: challenges, spatial spectrum sensing, and cyber-security. IET Cyber Phys Syst Theory Appl 2:49–54. https://doi.org/10.1049/iet-cps.2017.0010

Begum M, Serna RW, Yanco HA (2016) Are robots ready to deliver autism interventions? A comprehensive review. Int J Soc Robot 8:157–181. https://doi.org/10.1007/s12369-016-0346-y

Bishop-Fitzpatrick L, Mazefsky CA, Eack SM, Minshew NJ (2017) Correlates of social functioning in autism spectrum disorder: the role of social cognition. Res Autism Spectr Disord 35:25–34. https://doi.org/10.1016/j.rasd.2016.11.013

Brunnström K, Sjöström M, Imran M et al (2018) Quality of experience for a virtual reality simulator. Electron Imaging 30:1–9. https://doi.org/10.2352/ISSN.2470-1173.2018.14.HVEI-526

Cakir J, Frye RE, Walker SJ (2020) The lifetime social cost of autism: 1990–2029. Res Autism Spectr Disord 72:101502. https://doi.org/10.1016/j.rasd.2019.101502

Caspi A, Houts RM, Ambler A et al (2020) Longitudinal assessment of mental health disorders and comorbidities across 4 decades among participants in the Dunedin birth cohort study. JAMA Netw Open 3:e203221. https://doi.org/10.1001/jamanetworkopen.2020.3221

Danish Technological Institute, Yamaguchi J (2019) ATAT: assistive technology assessment tool (in Japanese). https://unit.aist.go.jp/harc/arrt/ATAT_eval_ver_1_0.pdf. Accessed 8 Dec 2023

Elbamby MS, Perfecto C, Bennis M, Doppler K (2018) Toward low-latency and ultra-reliable virtual reality. IEEE Netw 32:78–84. https://doi.org/10.1109/MNET.2018.1700268

Eum S, Arakawa S, Murata M (2022) A probabilistic grant free scheduling model to allocate resources for eXtreme URLLC applications. In: 2022 IEEE Latin-American conference on communications (LATINCOM). IEEE, pp 1–6

Gagnon L, Peretz I, Fülöp T (2009) Musical structural determinants of emotional judgments in dementia of the Alzheimer type. Neuropsychology 23:90–97. https://doi.org/10.1037/a0013790

Glas DF, Ferreri F, Miyashita T et al (2014) Automatic calibration of laser range finder positions for pedestrian tracking based on social group detections. Adv Robot 28(9):573–588. https://doi.org/10.1080/01691864.2013.879272

Glas DF, Brscic D, Miyashita T et al (2015) SNAPCAT-3D: calibrating networks of 3D range sensors for pedestrian tracking. In: 2015 IEEE international conference on robotics and automation (ICRA), pp 712–719. https://doi.org/10.1109/ICRA.2015.7139257

Gineste Y, Marescotti R (2010) Interest of the philosophy of Humanitude in caring for patients with Alzheimer's disease. Soins Gerontol (85):26-277

Hirukawa H (2016) Robotic devices for nursing care project (in Japanese). J Robot Soc Jpn 34:228–231. https://doi.org/10.7210/jrsj.34.228

Honda M, Ito M, Ishikawa S et al (2016) Reduction of behavioral psychological symptoms of dementia by multimodal comprehensive care for vulnerable geriatric patients in an acute care hospital: a case series. Case Rep Med 2016:1–4. https://doi.org/10.1155/2016/4813196

Hu F, Deng Y, Saad W et al (2020) Cellular-connected wireless virtual reality: requirements, challenges, and solutions. IEEE Commun Mag 58:105–111. https://doi.org/10.1109/MCOM.001.1900511

Hu H, Nozawa R, Iwata K et al (2023) Investigating non-verbal communication in human-multiple teleoperated robots interaction. In: Proceedings of ICRA2023 workshop on Avatar-Symbiotic Society

Johnson CP, Myers SM (2007) Identification and evaluation of children with autism spectrum disorders. Pediatrics 120:1183–1215. https://doi.org/10.1542/peds.2007-2361

Kondo Y, Yomo H, Nishimura S et al (2023) A practical implementation of multi-radio Wi-Fi for teleoperated mobile robots. In: 2023 IEEE international conference on Omni-layer intelligent systems (COINS). https://doi.org/10.1109/COINS57856.2023.10189247.

Kumazaki H, Muramatsu T, Yoshikawa Y et al (2022) Android robot promotes disclosure of negative narratives by individuals with autism spectrum disorders. Front Psychiatry 13. https://doi.org/10.3389/fpsyt.2022.899664

Kumazaki H, Muramatsu T, Yoshikawa Y et al (2021) Enhancing communication skills of individuals with autism spectrum disorders while maintaining social distancing using two tele-operated robots. Front Psychiatry 11. https://doi.org/10.3389/fpsyt.2020.598688

Kumazaki H, Muramatsu T, Yoshikawa Y et al (2022) Differences in the optimal motion of android robots for the ease of communications among individuals with autism spectrum disorders. Front Psychiatry 13. https://doi.org/10.3389/fpsyt.2022.883371

Kumazaki H, Muramatsu T, Yoshikawa Y et al (2023) Android robot was beneficial for communication rehabilitation in a patient with schizophrenia comorbid with autism spectrum disorders. Schizophr Res 254:116–117. https://doi.org/10.1016/j.schres.2023.02.009

Kumazaki H, Yoshikawa Y, Muramatsu T et al (2022) Group-based online job interview training program using virtual robot for individuals with autism spectrum disorders. Front Psychiatry 12. https://doi.org/10.3389/fpsyt.2021.704564

Lawson RP, Aylward J, Roiser JP, Rees G (2018) Adaptation of social and non-social cues to direction in adults with autism spectrum disorder and neurotypical adults with autistic traits. Dev Cogn Neurosci 29:108–116. https://doi.org/10.1016/j.dcn.2017.05.001

Lawson RP, Rees G, Friston KJ (2014) An aberrant precision account of autism. Front Hum Neurosci 8. https://doi.org/10.3389/fnhum.2014.00302

Maenner MJ, Shaw KA, Bakian AV et al (2021) Prevalence and characteristics of autism spectrum disorder among children aged 8 years—autism and developmental disabilities monitoring network, 11 sites, United States, 2018. MMWR Surveill Summ 70:1–16. https://doi.org/10.15585/mmwr.ss7011a1

Muris P, Ollendick TH (2021) Selective mutism and its relations to social anxiety disorder and autism spectrum disorder. Clin Child Fam Psychol Rev 24:294–325. https://doi.org/10.1007/s10567-020-00342-0

Nakazawa A, Mitsuzumi Y, Watanabe Y et al (2020) First-person video analysis for evaluating skill level in the humanitude tender-care technique. J Intell Robot Syst 98:103–118. https://doi.org/10.1007/s10846-019-01052-8

Noguchi Y, Kamide H, Tanaka F (2023) How should a social mediator robot convey messages about the self-disclosures of elderly people to recipients? Int J Soc Robot 15:1079–1099. https://doi.org/10.1007/s12369-023-01016-x

Oshima S (2015) Medical and nursing care in a Super-Aged Society. Med Soc Aff 25:49–57

Quax P, Monsieurs P, Lamotte W et al (2004) Objective and subjective evaluation of the influence of small amounts of delay and jitter on a recent first person shooter game. In: Proceedings of ACM SIGCOMM 2004 workshops on NetGames '04 Network and system support for games—SIGCOMM 2004 workshops. ACM Press, New York, USA, pp 152–156

Sato M (2014) Current status and future of non-drug therapy for dementia (in Japanese)). Cogn Neurosci 15:207–213

Snider JG, Osgood CE (1969) Semantic differential technique: a sourcebook. Aldine, Chicago

Spain D, Sin J, Linder KB et al (2018) Social anxiety in autism spectrum disorder: a systematic review. Res Autism Spectr Disord 52:51–68. https://doi.org/10.1016/j.rasd.2018.04.007

Störtkuhl KF, Hovemann BT, Carlson JR (1999) Olfactory adaptation depends on the Trp Ca^{2+} channel in drosophila. J Neurosci 19:4839–4846. https://doi.org/10.1523/JNEUROSCI.19-12-04839.1999

The Ministry of Health L and WJ (2021) Overview of the 2020 survey of long-term care service facilities and business places: status of facilities and business places. https://www.mhlw.go.jp/toukei/saikin/hw/kaigo/service20/index.html. Accessed 8 Dec 2023

Toyama T (2002) A study on the introduction of private rooms and small scale units at long-term care insurance facilities. Health Econ Res 11:63–89

Tulving E (1985) Elements of episodic memory: educational publication

Warren ZE, Zheng Z, Swanson AR et al (2015) Can robotic interaction improve joint attention skills? J Autism Dev Disord 45:3726–3734. https://doi.org/10.1007/s10803-013-1918-4

Yoshida A, Kumazaki H, Muramatsu T et al (2022) Intervention with a humanoid robot avatar for individuals with social anxiety disorders comorbid with autism spectrum disorders. Asian J Psychiatr 78:103315. https://doi.org/10.1016/j.ajp.2022.103315

Yoshikawa Y, Muramatsu T, Sakai K et al (2023) A new group-based online job interview training program using computer graphics robots for individuals with autism spectrum disorders. Front Psychiatry 14. https://doi.org/10.3389/fpsyt.2023.1198433

Zheng Z, Zheng P, Zou X (2018) Association between schizophrenia and autism spectrum disorder: a systematic review and meta-analysis. Autism Res 11:1110–1119. https://doi.org/10.1002/aur.1977

Chapter 9
Cybernetic Avatars and Society

Yukiko Nakano, Takayuki Kanda, Jani Even, Alberto Sanfeliu, Anais Garrell, Minao Kukita, Shun Tsugita, Fumio Shimpo and Harumichi Yuasa

Abstract Toward a future symbiotic society with Cybernetic Avatars (CAs), it is crucial to develop socially well-accepted CAs and to discuss legal, ethical, and socioeconomic issues to update social rules and norms. This chapter provides interdisciplinary discussions for these issues from the perspectives of technological and social sciences. First, we propose avatar social implementation guidelines

Y. Nakano (✉)
Seikei University, Musashino, Tokyo, Japan
e-mail: y.nakano@st.seikei.ac.jp

T. Kanda · J. Even
Kyoto University, Kyoto, Kyoto, Japan
e-mail: kanda@i.kyoto-u.ac.jp

J. Even
e-mail: even.jani.5x@kyoto-u.ac.jp

A. Sanfeliu · A. Garrell
Universitat Politècnica de Catalunya, Barcelona, Spain
e-mail: alberto.sanfeliu@upc.edu

A. Garrell
e-mail: anais.garrell@upc.edu

M. Kukita
Nagoya University, Nagoya, Aichi, Japan
e-mail: minao.kukita@gmail.com minao.kukita@is.nagoya-u.ac.jp

F. Shimpo
Keio University, Fujisawa, Kanagawa, Japan
e-mail: shimpo@sfc.keio.ac.jp

H. Yuasa
Meiji University, Chiyoda, Tokyo, Japan
e-mail: yuasa@meiji.ac.jp

S. Tsugita
University of Toyama, Toyama-shi, Toyama, Japan
e-mail: tsugita@hmt.u-toyama.ac.jp

© The Author(s) 2025
H. Ishiguro et al. (eds.), *Cybernetic Avatar*,
https://doi.org/10.1007/978-981-97-3752-9_9

313

and present studies that contribute to the development of socially well-accepted CAs. The second part of this chapter addresses the ethical and legal issues in installing CAs in society and discusses solutions for them.

9.1 Introduction

Cybernetic Avatars (CAs) are an innovative technology that enables people to extend their physical, cognitive, and perceptual capabilities and become free from the constraints of time and space. Therefore, CA technologies may affect human life, including workstyles and communication with others. Toward such a future in a symbiotic society with CAs, it is crucial to study design methods for socially well-accepted CAs. In addition to technological issues, it is indispensable to discuss legal, ethical, and socioeconomic issues to update social rules and norms in order to realize a future society with CAs. To envision the implementation of CAs in society, this chapter provides interdisciplinary discussions on these issues from technological and social science perspectives.

Section 9.2 discusses the aspects of CA ethics that need to be considered when installing CAs in society, based on which, we propose avatar social implementation guidelines that consider concern and trust issues in the development of CAs, followed by a description of a use case and a discussion of the steps to disseminate CA guidelines in society.

Section 9.3 focuses on the technologies for socially well-accepted CAs by overviewing the moral problems specifically occurring in communications through CAs and proposes "moral computing" research that challenges to solve these problems. As examples of moral interaction research with CA, we propose a CA guardsman and CA cashier that can change inappropriate language from remote workers working as guardsmen or cashiers, to appropriate and polite language. As another aspect of social acceptance of CAs, we review the literature on cultural differences in human communication and discuss culturally adaptive social robots. As the third topic of this section, we propose a SOCIAL-PIA model consisting of environmental perception, intention inference, and sharing cooperative plans. A handover task between a human and a robot is presented as a use case for the SOCIAL-PIA model.

Section 9.4 discusses the ethical issues of CAs by focusing on two aspects: the impact of CAs on people's work and life and the ethical issues of gendering CAs. We then discuss how human society addresses ethical issues when encountering new technology.

Section 9.5 discusses the legal issues for CAs by defining the legal status of CAs and discussing corporate CAs separately from individual ones. We then show that not only ELSI issues but also Ethics, Law, Society, and Economics (ELSE) issues should be considered when installing CAs in society.

Section 9.6 discusses the use of CAs in election campaigning. In terms of using new technologies in elections, past cases of Internet use are presented.

9.2 Implementing Cybernetic Avatars in a Society

In human history, the introduction of new technology that significantly impacts people's lives has necessitated discussions on their application, use, and the need to update social rules and ethics to ensure their acceptance and dissemination in society. Regarding misuse, even if the technology is used in a way that the developer did not intend, it may cause social problems and lead to social condemnation of the developers. Therefore, defining and sharing social norms with people in society are indispensable for implementing new technologies. Given the potential of Cybernetic Avatar (CA) technology to revolutionize people's lives and work, it is imperative to establish ethical guidelines for CA. However, there are no regulations regarding the development and use of CA. In this section, we present a comprehensive discussion on the ethical considerations and factors that should be contemplated when implementing CA in society, along with guidelines for doing so. We also discuss further actions to disseminate these guidelines to the public.

9.2.1 Legal, Ethical, and Socioeconomic Principles for CA

As CA technology is deeply related to and developed based on Artificial Intelligence (AI) and robotics, which have recently been changing human lives, we begin by reviewing the legal and ethical principles of AI and robotics. In 2019, OECD and partner countries adopted policy guidelines on AI (OECD 2019) that aim to ensure robust, safe, fair, and trustworthy AI systems by upholding international standards. In 2019, the European Commission's expert group proposed ethical principles for trustworthy AI, including respect for human autonomy, prevention of harm, fairness, and explicability (European Commission 2019). Through this preparation, in 2021, the European Commission proposed minimum requirements to address the risks and problems associated with AI (EUR-Lex 2021). In 2021, UNESCO proposed a recommendation for AI ethics, which consists of four primary values for AI systems: AI systems should work for the good of humanity, individuals, societies, and the environment (UNESCO 2021).

Among AI principles proposed by academia, one of the earliest proposals is the Asilomar AI principles (Asilomar Conference 2017), which were formulated by over 100 researchers at the Asilomar conference organized by the Future of Life Institute. The principles consist of three parts: research issues, ethics and values, and long-term issues. As a proposal from an academic association, the IEEE, a professional association for electronic and electronics engineers, proposed General Principles of Ethical Autonomous and Intelligent Systems (IEEE SA 2019).

For the ethics on robotics, the European Parliament proposed civil law and ethical aspects of robotics, which were updated in 2017 (European Parliament 2017). The code of conduct for robotics engineers includes the principles of beneficence, non-maleficence, autonomy, and justice.

Despite the numerous proposals for AI ethics, articles that have reviewed AI principles have found a remarkable degree of coherence and overlap among them (Floridi et al. 2018; Morley et al. 2020). Floridi et al. (2018) proposed five common principles: beneficence, non-maleficence, autonomy, justice, and explicability. Beneficence refers to the principle of promoting well-being, preserving dignity, and sustaining the planet. Non-maleficence is the principle of avoiding negative consequences, particularly for privacy and security. Autonomy is the principle of an individual's right to make decisions, whether to decide by ourselves or delegate the decision to AI agents. Justice is the principle of promoting prosperity and preserving solidarity, while seeking to eliminate discrimination. Explicability is a principle of intelligibility and accountability, which can be expressed as transparency. Except for explicability, four principles of AI ethics are commonly proposed in robotics ethics (European Parliament 2017). Therefore, we adopted these five principles as the fundamental principles of CA ethics and added instrumental principles that protect and promote the core principles (Canca 2020). Table 9.1 list these principles.

Some instrumental principles contribute to the multiple core principles. For example, good behavior change can be advantageous for humans, while bad behavior change should be discouraged. Diversity should be considered in defining social justice and in decision-making.

9.2.2   Avatar Social Implementation Guidelines

Based on the discussion for principles of CA ethics in Sect. 9.2.1, we proposed guidelines for avatar social implementation. Although not all the principles are mentioned in the guidelines, this is the first draft of the CA ethics guidelines, which we designed to be as concise as possible.

<Preamble>
Innovation can make significant strides toward diversity and inclusion. The Moonshot R&D program aims to create a world where people can operate freely through avatar-related innovations. These guidelines are aimed at developers of

Table 9.1 Principles of CA ethics

Core principles	Instrumental principles
Beneficence	Human rights, dignity, sustainability and inclusive, well-being, trust and trustworthy, behavior change
Non-maleficence	Privacy, safety, security, anonymity, trust and trustworthy, disruptive behavior and harassment, psychological impacts, behavior change, error
Autonomy	Identity, diversity
Justice	Discrimination and disparity, rule of law, trust and trustworthy, democracy, literacy, morality, diversity
Explicability	Responsibility and accountability, trust and trustworthy

Cybernetic Avatars (CAs) and summarize the main aspects to be considered when implementing avatar-related innovations in society.

Due to the rapid pace of technological progress, new technologies often lack usage guidelines. However, they should continue to progress while maintaining a balance between R&D and social implementation, with the primary goal of enhancing social well-being. It is imperative to address issues of concern and trust, including legal and ethical considerations to ensure widespread adoption of new technologies.

<Solicitude>

In addition to general considerations for ensuring the reliability of information systems, the following can be considered as ways of addressing issues related to concern and trust in the development of CAs:

1. **Solicitude for Operators**

   (Safety) It is recommended to provide instructions to users to prevent them from using the CA inappropriately or equip the CA with functions that inhibit such behaviors.

   (Secure) The operator's consent should be obtained for the use of the operator's operating record.

2. **Solicitude for Users**

   (Safety) It is recommended to provide instructions to operators to prevent inappropriate use of CA, or equip the CA with a function that prevents such behavior.

   (Secure) CA operations may be performed autonomously by AI; it is desirable to indicate this fact to users while the AI is in operation.

<Dissemination>

Goals to be aimed at the spread of CA to society include the following;

- Anyone should have equal opportunities to use CAs. Therefore, when developing CAs, accessibility considerations are necessary to ensure that everyone (e.g., children, people with disabilities, and the elderly) can use CAs.
- To improve productivity, it is recommended that an individual utilizes multiple CAs.
- It is recommended to improve the service quality by using CAs.
- CAs must be acceptable to users, operators, and other relevant people. Therefore, it is recommended to consider the purpose and environment of CA usage in designing CA.
- It is recommended to contribute to a sustainable society by reducing the emissions of greenhouse gases, such as CO_2, using CAs.
- It is recommended to provide literacy education to properly use CAs in society.

9.2.3   Use Case

In this subsection, we present a possible use case and demonstrate how the guidelines proposed in Sect. 9.2.2 can be applied to the service or application of the use case.

Suppose that Mr. A works remotely as a shop clerk at a convenience store. Mr. A operates his CA from his home, far from the store. To prevent customer harassment, it is desirable to provide instructions to the store customers not to abuse CA clerks. It is also useful to provide CA functions that prevent customer harassment behaviors. One possibility is to automatically detect and admonish inappropriate behavior. To ensure the security of the CA operator when recording the operating log, it is necessary to obtain consent from Mr. A.

For convenience store customers, it is desirable to instruct CA operators not to serve customers improperly, for instance, crudely providing customer service. It is also recommended to provide functions to prevent CA clerks from performing inappropriate operations, such as automatically changing the clerks' tone of voice. To ensure customer security when the clerkship avatar is AI-controlled, it is recommended to provide customer notification.

9.2.4   Steps Toward Disseminating CA Guidelines in Society

As reviewed in Sect. 9.2.1, the principles are listed in AI ethics. However, as Prem (2023) suggests, these principles are defined at very high levels and do not specify how to realize them or translate them into operationalizable actions. Moreover, ethical standards may differ depending on the culture or community. Therefore, to put principles into practice, it is crucial to specify system requirements and develop tools and technologies to concretely address ethical issues. In this process, communication between developers and potential end-users is indispensable. Moreover, in the final step of the CA ethics practice, assessing the ethical aspects of the developed CAs is necessary. At this stage, the development of assessment tools is necessary.

9.3   Technologies for Implementing Socially Well-Accepted CAs

This section presents studies contributing to the development of socially well-accepted CAs. To this end, the following subsection focuses on the following three topics: moral computing, culturally adaptive design, and cooperative human–robot interaction.

9.3.1 Moral Interaction with Cybernetic Avatars

This subsection provides an overview of the "moral problems" that arise when moving from face-to-face to interactions using Cybernetic Avatars (CAs). We introduce "moral computing for CAs" and explain why achieving a harmonious cohabitation between the people and CAs is essential. We provide information on the current state of moral computing, outline our plans, and discuss the likely development of moral computing.

In many developed countries, including Japan, serious problems related to aging populations and the subsequent decline in the working population are emerging. As such, there are high expectations for robots and AI regarding their replacement or, at least, assisting human workers by taking on roles such as security guards, cashiers, clerks, receptionists, delivery personnel, attendants, and cleaners. Recent advances in robotics have demonstrated the feasibility of automating specific tasks, including parcel delivery. However, fully autonomous robots are not ready to perform most tasks requiring human-level communication skills. This space is designated for the deployment of CAs as "avatar workers" who interact with people in the service industry. We focus on CAs that engage in interactions with people to facilitate discussion and illustrate our presentation. Given the first potential for large scale use of CAs in this field, it is imperative to address the issue urgently. To facilitate clarity, we will henceforth refer to the worker operating the CA as the "remote worker" and the person interacting with the CA as the "customer" in the remainder of this subsection.

First, we evaluate the feasibility of substituting in-person workers with remote workers who perform the same tasks using CAs. Potential difficulties become apparent when the specifics of in-person workers are considered more closely. For example, fast-food cashiers greet customers in a friendly manner by politely taking and delivering orders. However, at a less perceptible level, cashiers also observe the environment from a human perspective. This aspect is crucial as the presence of "human eyes" can prevent some unscrupulous people from disturbing others or deteriorating the environment. A sense of order and security can be established by deterring low-moral behavior through the observation of others.

Concurrently, cashiers are themselves subject to the presence of "human eyes." They work in an environment in which people can see all their actions. Particularly, they face customers in person in an environment that encourages them to do their best and provide services with great professionalism. Specifically, being physically in the environment makes it difficult to ignore customers' requests and to behave in a rude manner.

Concerning a CA cashier, the "human eyes" of an in-person worker are replaced by those of a remote worker. To ensure environmental compliance, we must examine whether a remote worker can exert comparable peer pressure through the CA. In an adjacent context, surveillance cameras may already provide such functionalities, as one could argue. Certainly, surveillance cameras can inhibit serious crimes but cannot prevent low-moral behavior. Recently, the problem of low-moral behavior has become more serious, even in cities where large networks of surveillance cameras have been installed (Velastin 2005; Gayet-Viaud 2017). Simply watching people who engage in low-moral behavior is not enough; interactions with them are necessary. If CAs can perform their duties in a friendly and accommodating manner while simultaneously monitoring the environment, similar to in-person cashiers, they may be accepted by society without raising ethical, legal, or societal concerns that could lead to fear of becoming a surveillance society.

When CAs are used to provide services, remote workers are not subject to direct pressure from customers. Thus, we anticipate a novel phenomenon—a substantial increase in unethical conduct aimed at customers. For example, when working in person, ignoring a customer is very difficult, whereas it is relatively easy behind the screen. Similarly, confronting a customer or starting an argument is less worrying behind the screen. We can draw an analogy between online and offline behaviors, where people tend to exhibit more negative behavior when interacting on the Internet than in face-to-face interactions. Of course, this issue depends on the remote workers' level of accountability and anonymity. However, we can imagine situations easily getting out of hand when remote workers are tired, stressed, or upset. If we do not want people to judge that the services provided by CAs are of inferior quality or even worse, plain rude, we must address this issue by preparing safeguards to handle remote worker issues.

Customers interacting with CAs may experience reduced social pressure to behave politely, leading to increased incidence of low-moral behavior directed toward remote workers. This phenomenon is not novel, as customers have been observed to exhibit low-moral behavior toward in-person workers. In contrast, remote workers may be less affected by such low-moral behaviors as they do not interact in-person with customers. However, we hypothesize that remote workers may be less susceptible to the negative impact of justified customer complaints and may be more resilient in their response.

In addition, immoral behavior induces immoral behavior in others (for instance, the "broken windows" theory (Wilson and Kelling 1982)). If people see others engaging in low-moral behaviors, such as abandoning bicycles, littering, spray-painting graffiti, making fun of others, making noise, running around (e.g., in libraries), eating and drinking in places where such activities are prohibited, urinating outdoors, and mismanaging dog feces, they are more likely to do so themselves. Then, we posit that CAs ought to adhere to "moral rules" to curb the propagation of immoral behaviors at an early stage and prevent further degradation.

Is it possible for CAs to play meaningful roles in moral interactions? Can they offer similar functionalities to "human vision" to reduce immoral or morally questionable behaviors while maintaining the professionalism that in-person workers naturally exhibit? The answer is unclear, and this remains a pressing academic challenge in human–robot interaction (HRI) research on CAs. We focus on the use of moral computing to address these challenges. Moral computing is an interdisciplinary field that combines computer and social sciences. The objective is to integrate human morals and values into the designing, operation, and management of computer systems, particularly regarding human–robot interactions. As illustrated by our cashier example, we believe that moral computing applies to human-CA interactions and should allow for more harmonious and socially acceptable interactions.

9.3.1.1 Moral Interaction for Cybernetic Avatars

Morality is a fundamental element of a symbiotic society. Because most people are morally equipped, they tend to respect each other and engage in prosocial behaviors (e.g., helping one another). "Moral interaction" in the context of HRI is defined as an interaction in which a robot encourages people to respect it as a peer and moral recipient and brings a sense of security to the environment. In short, moral interactions with robots elicit both "peer pressure" and "peer respect."

Contrary to robots, CAs are "projections" of the remote workers. Then, during an interaction with a CA, we can assume that this occurs between morally equipped people (the customer and remote workers). However, "peer pressure" and "peer respect" are both diminished by the lack of physical proximity and the resulting communication barriers. Moral interaction for CAs should ensure that both "peer pressure" and "peer respect" are strong enough or propose workarounds to enable harmonious interactions. However, achieving this goal remains a largely unexplored issue. Research into moral interaction should ideally strive to elucidate the following question: How does using CA induce or promote moral behavior in both customers and remote workers?

9.3.1.2 Examples of Moral Interaction Research with Cybernetic Avatar

Two examples of research aimed at addressing moral interaction and enabling CAs to provide services in harmony with people are presented.

Study 1: Cybernetic Avatar as a Guardsman
The first study (Daneshmand et al. 2023) predicts that CA technology will enable remote work opportunities for individuals who are typically excluded from the workforce (Takeuchi et al. 2020). The working style of these newcomers is envisioned as a gig economy, where remote workers have the freedom to frequently switch between different CA-enabled jobs. In the service industry, inexperienced remote workers can manage CAs to perform brief work assignments and address the labor shortage.

This study considers using CAs to enable remote workers to fill in for missing guardsmen in shopping malls. In Japan, guardsmen are expected to remain polite and talk appropriately under all circumstances. Achieving this component requires experience and skills that novice workers typically do not acquire. One difficulty is that consistently talking appropriately and politely, as guardsmen would do to customers who are not always well-behaved, imposes a severe mental burden on such novice workers. These novice workers are likely to control CAs in a familiar environment where the pressure to act professionally is relatively low compared to working in person in a shopping mall. Then, with fatigue and stress, it is very likely that they let some inappropriate utterances slip, and the quality of the service may suffer.

To solve this issue, we propose a support system that allows novice remote workers to talk freely without considering appropriateness and politeness, while maintaining the quality of the service. The support system functions as a "moral safeguard" that oversees the maintenance of appropriateness and politeness during interactions. This method aims to improve the performance of CAs by providing polite and friendly interactions, even when faced with challenges, such as remote workers' frustration and inexperience. A proposed system enables remote workers to express their intentions verbally, and an intent recognition pipeline determines the appropriate wording (Huggins et al. 2021). The remote worker and support system collaborate to enhance customer experience. While the remote worker observes customer behavior and communicates freely, the system identifies the probable purpose of the communication and uses appropriate and polite language to express it. As illustrated in Fig. 9.1, the inappropriate language of an angry remote worker is not conveyed to customers by the CA. This safeguarding system ensures that the CA behaves as expected.

Figure 9.1 illustrates the pipeline. First, the remote worker's spoken words are transcribed into text using automatic speech recognition (ASR). Next, the intent recognition module categorizes the text according to the intent it expresses. Finally, the speech-generation module produces a polite and appropriate response that reflects the recognized intent. The CA then communicates the appropriate utterance to the customer.

To develop the intent recognition module, we obtained a dataset covering nine tasks that guardsmen are expected to perform (including greeting customers, thanking customers, and admonishing customers who smoke or litter). For each task, we gathered approximately 40 utterances that expressed intent. Half of these were utterances appropriate for a guardsman, and the other half were rude or used broken language. The intent classifier was then created by fine-tuning a large language model (BERT trained on the Japanese version of Wikipedia (Suzuki and Takahashi 2020)) using this dataset.

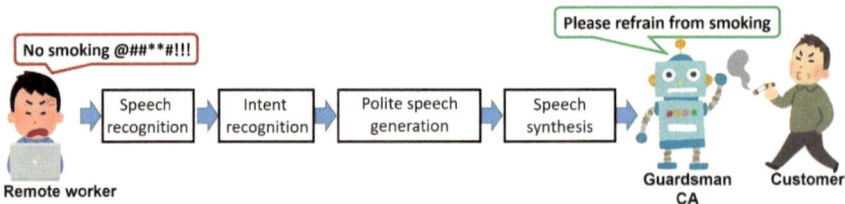

Fig. 9.1 Intent recognition pipeline for generating appropriate utterances from expressed intent

We conducted a user study with 23 participants who acted as novice remote workers controlling a guardsman's CA responsible for monitoring customer behaviors. The confederates who acted as customers were directed to engage in low-moral behaviors, such as smoking or littering, and to initially disregard or respond to the robot when admonished before either exiting the scene or complying and leaving the scene. This is illustrated in Fig. 9.2.

Each participant controlled the CA under two conditions: (1) with the help of the proposed support system and (2) without the help of the proposed support system. For each condition, we measured the workload using the NASA task load index (Hart and Staveland 1988; Hart 2006) and rated the politeness of the CA utterances. Participants were interviewed and debriefed after experiencing both conditions.

The support system demonstrated an expected mean classification accuracy of 96% across all participants. Specifically, the system accurately estimated the intent of participants, even when they spoke freely, and reformulated it appropriately.

The workload was significantly lower ($p<0.001$, Cohen's $d=1.23$) when using the proposed support system ($M=46.07$, $SD=14.36$) than when not using it ($M=62.74$, $SD=12.70$). There was no significant difference in the perceived politeness of the CA with and without support.

The analysis of post-experiment interviews showed that the reduction in workload was attributed to the system alleviating the pressure to speak politely or generate on-the-spot appropriate responses. Some participants regarded the system as a backup to correct mistakes. In addition, participants felt that the support system better protected the remote worker from negative customer behavior aimed at the CA.

The proposed support system is a "moral safeguard" that corrects utterances when the loss of peer pressure causes remote workers to act unprofessionally. On the customer side, the CA seems to adhere to the "moral rules" dictated by peer pressure, and harmonious interactions are possible. Simultaneously, the support system can potentially improve the welfare and mental health of remote workers by reducing their workload and cognitive pressure. However, it also raises concerns about worker autonomy and employer control, as it converts casual statements into polite and preset expressions, thus limiting the operators' ability to choose words and express themselves freely. The system limits operators to a

Fig. 9.2 Remote worker admonishing a confederate playing a smoking customer with and without support

set of predefined actions that can augment the employer's authority over worker behavior. This approach may result in interactions becoming more scripted and less personalized, which could lead to decreased engagement and a suboptimal user experience. Further developments should address these concerns by seeking a balance between support and employee autonomy. Involving stakeholders, such as users, CA designers, and ethicists, in discussions can help minimize potential negative impacts and ensure the ethical and responsible use of technology.

Study 2: Cybernetic Avatar as a Cashier

This second study (Yamada et al. 2023) investigated another support system designed to assist CA operators in customer service settings. The assumption is that most remote workers providing services using CAs will act professionally, with only a minority displaying subpar performance. Service providers should enable proficient operators to engage in unrestricted communication with customers, which leads to the delivery of high-quality services. However, there is a need to assist less-competent remote workers in achieving a decent quality of service. Then, we understand the need for a system that can classify remote workers as "competent workers" or "subpar workers."

To develop such a system, we selected a CA taking orders at a fast-food restaurant as an application scenario. Similar to an in-person worker, the CA must take orders from customers and answer more general questions. This second point is important because the quality of the service, and especially how hospitable it is, is often related to the ability of workers to perform their primary tasks. We judge a cashier as hospitable, which, in addition to taking our order, could indicate local tourist attractions when prompted. Regarding a CA cashier, competent and professional remote workers should be able to provide such a hospitable service, which we cannot expect from remote subpar workers.

In the proposed system, a remote worker is classified as competent if the estimated probability of using appropriate utterances exceeds 0.9. In contrast, if the estimated probability of using inappropriate utterances exceeds 0.06, the remote worker is classified as a subpar worker. These thresholds were determined based on preliminary experiments. Probabilities were estimated from the number of appropriate and inappropriate utterances made by remote workers. The appropriateness of an utterance was determined using an accurate utterance classifier obtained by fine-tuning a large language model (BERT trained on the Japanese version of Wikipedia (Suzuki and Takahashi 2020)) with a dataset of appropriate and inappropriate utterances. These utterances are used by competent and subpar remote workers when controlling the CA to greet customers and take orders.

In practice, the results from the remote worker classifier were used to alternate between the two operating modes of the cashier CA.

1. Free Mode: This mode enables competent remote workers to interact freely with customers.
2. Support Mode: This mode supports subpar remote workers by replacing their utterances with preset appropriate ones.

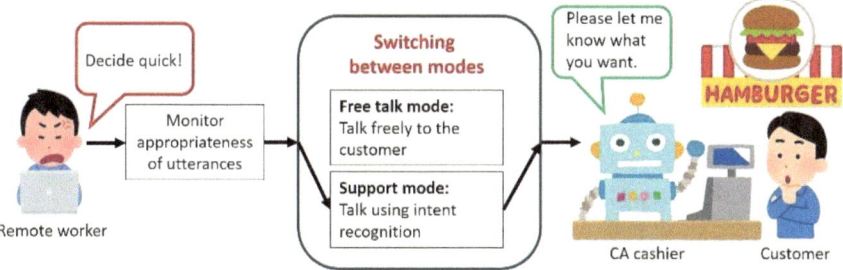

Fig. 9.3 System is switched to support mode and assist subpar remote worker to act professionally

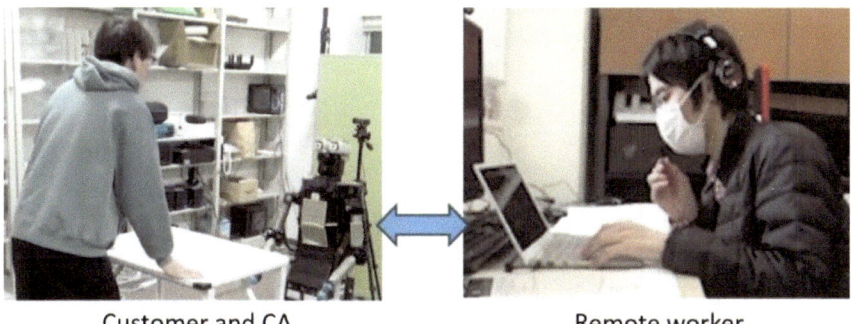

Fig. 9.4 Participant taking order from another participant using the cashier CA

The support mode (see Fig. 9.3) is based on an intent recognition (Takeuchi et al. 2020) pipeline to that used in the first study (Daneshmand et al. 2023). The intent recognition module was trained to classify utterances into classes that represented the different steps necessary for greeting customers and taking orders at a fast-food restaurant.

We conducted a user study involving 21 pairs of participants to compare the proposed system with a baseline system in which all remote workers could communicate freely with customers (see Fig. 9.4). For each pair, we conducted four sessions of order-taking using CA. One participant played a competent and subpar remote worker under both conditions (proposed system or baseline system), whereas the other played different types of customer orders. After each session, participants playing the customer completed a questionnaire to rate the service provided by the CA in terms of satisfaction (Chung et al. 2018) and politeness (Berry et al. 1988; Nakamura 2007). After completing four sessions, the participants were interviewed. The system accurately classified remote workers (94% accuracy), and the intent classification during the support mode had an accuracy of 90.6%.

The findings suggest that the proposed system significantly improves the quality of interactions for remote subpart workers without disrupting the efficiency of competent remote workers. For subpar remote workers, customer service quality was significantly higher in terms of politeness and customer satisfaction than in the baseline system. However, we did not find a significant difference among competent remote workers. In addition, 17 of the 21 participants (approximately 81%) preferred to interact with the CA controlled using the proposed system.

Our study was conducted in a simulated robotic burger restaurant, which may not represent the situation in all service industries. The study's practical application may be restricted because the participants were only role-playing operators with reduced structured dialog scenarios. However, we believe that it shows the importance of adapting the system controlling CAs to help remote workers who need it to guarantee that CAs abide by "moral rules" of conversation. The proposed system did not face any challenges. Remote workers' freedom of speech and professionalism are in tension, creating potential ethical issues. The system also occasionally hampered fluid interactions when the remote workers' intentions were not in line with the preset categories.

9.3.1.3 Discussion

Our investigation of moral interactions with CAs suggests that recent machine learning techniques, such as intent recognition, could be used to support remote workers' conversational capabilities. Our results show that an inexperienced remote worker can interact politely and appropriately without an excessive burden when supported by the system. This aspect could extend to proactive, positive interaction with customers, such as hospitable engagement.

Because CAs are expected to be mobile, a further area of exploration is spatial interactions. We must study how CAs navigate physical spaces, respond to their environments, and adapt their behaviors to the constraints of their physical surroundings. This finding has significant implications for the effective utilization of CAs in the service industry. We are particularly interested in developing moral codes that govern these spatial interactions.

We expect society to change with the widespread adoption of CAs equipped with the moral computing techniques we envision. CAs could potentially serve as tools for remote work and as extensions and augmentations of human capabilities. In this view, CAs may be considered a significant step toward integrating robotic assistance into daily life.

However, the implications of this new technology need to be evaluated. As highlighted in our case studies, although the proposed techniques improve the efficiency of service provision, they can also affect the autonomy of remote workers. Balancing effectiveness and ethical considerations is crucial for developing sustainable and fair CA practices. Questions such as "How much autonomy should be given to remote workers?" and "What impact does regulating autonomy have on user experiences?" necessitate careful investigation.

9.3.2 Socio-cultural Aspects in Designing Avatar Behaviors

9.3.2.1 Motivation

CAs can be used in various applications, such as receptionists, shop clerks, and helpdesks. Regardless of the type of application, to make CAs accepted in society, the avatar's behavior must be natural, consistent, and meaningful for users. However, the user's impression of avatar behavior may differ for multiple reasons, and culture is a critical factor affecting the acceptability of avatar behaviors.

Nass et al. (2000) discussed whether the ethnicity of computer agents affects user attitudes and behaviors. Their experimental results demonstrated that when subjects interacted with an ethnically matched agent, they perceived the agent to be more similar to themselves, socially attractive, and trustworthy. The participants also conformed more to the decision of the ethnically matched agent and perceived the agent's arguments to be better.

In human communication science, Ting-Toomey and Dorjee (2018) described the differences in non-verbal behaviors across different cultures. For example, Italians use broader full-arm gestures than US Americans, and most of their hand gestures are expressive. Generally, southern Europeans tend to employ more animated hand gestures than northern Europeans. Concerning the perceived credibility aspect, Ting-Toomey and Dorjee (2018) described that facial composure and body posture influence judgments of credibility (i.e., whether a person has social influence power). In some Asian cultures (e.g., South Korea and Japan), influential individuals tend to maintain restrained facial expressions and rigid postures. However, in the US culture, relaxed facial expressions and postures are associated with credibility and positive impressions. In addition to bodily behavior, speaking style is a type of non-verbal information. Ting-Toomey and Dorjee (2018) also described that interrupting a conversation partner is perceived as impolite in many cultures while interrupting is accepted positively in other cultures as a way to express interest in the conversation.

Therefore, to develop culturally adaptive CAs, it is essential to understand how non-verbal communication styles vary across cultures and how to design CA that consider these cultural differences.

9.3.2.2 Culturally Adaptive Agents and Robots

Several attempts have been made to develop culturally adaptive virtual agents and communication robots. In a study on virtual agents, the CUBE-G project between Germany and Japan collected a comparable corpus in three prototypical social interaction scenarios: a first-time meeting, negotiation, and conversation with someone of higher social status (Rehm et al. 2009). Endrass et al. (2013) analyzed the differences between German and Japanese speakers. They found that usage of body postures categorized by Bull (1987) differed between the two cultures. The most frequent posture of German speakers was putting their hands into their pockets, but joining their hands was most frequently observed in the Japanese data. The three most frequent posture categories did not overlap between the two cultures. They found that the gesture expressivity differed between the two countries. In particular, gestures were performed faster, more powerfully, and more fluently by German speakers than by Japanese speakers. In addition, German participants used a wider space for their gestures than Japanese participants, while Japanese speakers used repetitive gestures more frequently. Based on these analyses, they developed and evaluated virtual agents that displayed prototypical or not-prototypical non-verbal behaviors in their culture. The experimental results showed that users preferred agent dialogues that reflected the behavioral patterns observed in their cultural backgrounds (Lugrin and Rehm 2021).

Research on Social Robotics has discovered that attitudes toward robots differ across cultures (Bartneck et al. 2005). It was also found that social signals displayed by robots were more accurately interpreted by native English speakers, and the interpretation of the robots' social signals differed depending on the culture (McKenna et al. 2018). The CASESSES Project between European countries and Japan aims to design culturally competent social robots for elderly care (Battistuzzi et al. 2018). They extracted key concepts from existing ethical guidelines for assistive technologies for people with dementia and applied them to scenarios describing how robots interact with the elderly belonging to different cultures.

9.3.2.3 Toward Designing Cultural and Ethical CAs

As described in the previous subsections, non-verbal behaviors are displayed and interpreted subconsciously rather than verbal information and are influenced by the user's cultural background. To consider this point, previous studies on virtual agents and social robots have attempted to propose non-verbal behavior models capable of producing culturally appropriate behaviors such as facial expressions, gesture expressivity, posture, and gaze.

Although the studies mentioned above focused on autonomous virtual agents and social robots, similar approaches can be used to design and develop culturally adaptive CAs. Furthermore, as discussed in Battistuzzi et al. (2018), it is also important to discuss cultural influences on ethical issues when designing avatars.

9.3.3 Cooperative Social Perception–Intention–Action (PIA) Model for Cybernetics Avatars

The term robot has been applied to automatons that execute simple and repetitive tasks to prevent humans from performing them. These early robots used an architecture composed of different layers (perception, modeling, planning, task execution, and motor control), where the robot sensed its surroundings before planning how to execute a particular task (Brooks 1986), and followed what was denominated in the perception–action (PA) model. The PA model does not account for humans' involvement in the process and does not allow establishing social interactions with them. In human–robot interaction (HRI) or Human–Robot Cooperation (HRC), the robot should understand human behavior and intentions to make these interactions safe, reliable, comfortable, and easily understandable to users, and the outcome of these interactions should be as productive as possible (Duchaine and Gosselin 2009).

In HRC tasks with humans, a robot must identify human intentions in a specific context and generate future predictions of human behavior to create one or several robot plans that anticipate user actions (Ferrer and Sanfeliu 2014). In this manner, the robot can adapt to changes in individual behavior and facilitate safe and comfortable interactions. Notably, human intention is the key issue in any HRI or HRC model. Examples of human–robot cooperative tasks include accompanying people (Garrell and Sanfeliu 2012), transporting a table between a robot and a human (Mörtl et al. 2012), dancing with a robot (Kosuge et al. 2003) and collaborative search between a robot, and a human (Dalmasso et al. 2023), among others.

9.3.3.1 The Social PIA Model

To describe the Social PIA paradigm (Domínguez-Vidal et al. 2023, 2024), we discuss a cooperative task example called the handover task. Two agents, a robot and a human—are four meters apart and tasked with a handover assignment in which the robot will deliver a box to the human, who will subsequently pick it up (see Fig. 9.7). Both agents must perform several tasks to comprehend the mission context (e.g., proximity), forecast their future location, device an optimal plan, commit to it, and execute it. The object must be within a certain proximity to each other to enable switching. After switching, the user must raise their hands, reach for the object, and move away from the hand. To achieve these subtasks, in the case of a robot, the robot has to perceive the environment and localize the human hand and obstacles. Moreover, it has to identify the human intention; for example, the human wants to grab the object or prefers to go away. If a human wants to grab an object, he will move toward the robot and raise his hand to take the object. The motion of the human skeleton can be predicted by a robot, which can anticipate the location of the human arm in the future. The robot then planned its forward motion and delivered the box. The robot anticipates how much it has to move when it has to raise its arm and when it has to deliver the box. If the human does not want to grab the box, it stops in the middle of the path. For example, when a robot is immobile, it should request clarification from a human regarding their intends goal. Depending on the human answer, the robot can be anticipative or simply ask the person what to do next.

Figure 5b depicts the SOCIAL-PIA model. This instance pertains to two agents (a robot and a human) who want to perform cooperative tasks. Each agent has a perception module for detecting objects, humans, and the environment. They also have an intention module that infers human intention (implicit intention) (Mutlu et al. 2009) or receives a verbal command or signal (explicit intention) (Domínguez-Vidal and Sanfeliu 2023). These two modules are the inputs for the Situation Awareness module (Endsley and Garland 2000), which compares and analyzes the current situation with what was expected (because the cooperative task and the steps to execute the task are known in advance) and computes the 2D or 3D prediction of the other agent's motion (Laplaza et al. 2022). These outputs are inputs to the decision-making module. In this module, the cooperative plan of each agent is elaborated, and the robot and the human create their plans, share their plans, and finally negotiate when the plan is unapproved by one of the agents. In certain scenarios, agents may negotiate their roles, with one agent assuming the role of leader, and the other that of follower. The intention module can also serve as an input to the decision-making process, influencing negotiation or role distribution. The output of this module is the subtask plan of each agent, for example, to move the robot and raise the hand in the cooperative case or not to move and ask the person what he wants to do if the human does not follow the expected cooperative subtasks. This cycle—prediction intention, situation awareness, and decision-making—is repeated until the task is completed or stopped by one of the agents.

A schematic of the Social PIA is shown in Fig. 5a, where two agents (AG) perform a cooperative task. Each agent performs the appropriate actions (AC) and uses the perception and intention modules (P–I) to understand the current situation and the intention of the other agent and/or scenario agents (for example, bystanders in the scenario). Moreover, they negotiate the subtasks to be performed using the PIA model, reorganize the roles of each partner, and create a joint plan.

9.3.3.2   The Social PIA Model Extended to Cybernetics Avatars

In the case of the Cybernetics Avatars (CAs), the Social PIA model can be extended, as shown in Fig. 9.6. Figure 6a shows a case of a CA interacting with an agent (AG, e.g., human). The CA is partially controlled by an operator (OP) who also perceives the actions and intentions of the CA and/or scenario agents (for example, bystanders) and performs the CA task. In some cases, it can negotiate CA tasks. The task execution between the CA and AG is the same as that explained for the PIA model. Although the operator (OP) can arbitrate (arbitration module), the dispute between the CA and AG does not arrive at a commitment consensus.

The previous diagram can be further extended when two CAs perform cooperative tasks. In this case, each CA is controlled by one operator (OP), and both OPs have the same interaction mechanism as their CAs. However, in this case, both operators could arbitrate the dispute over the CAs.

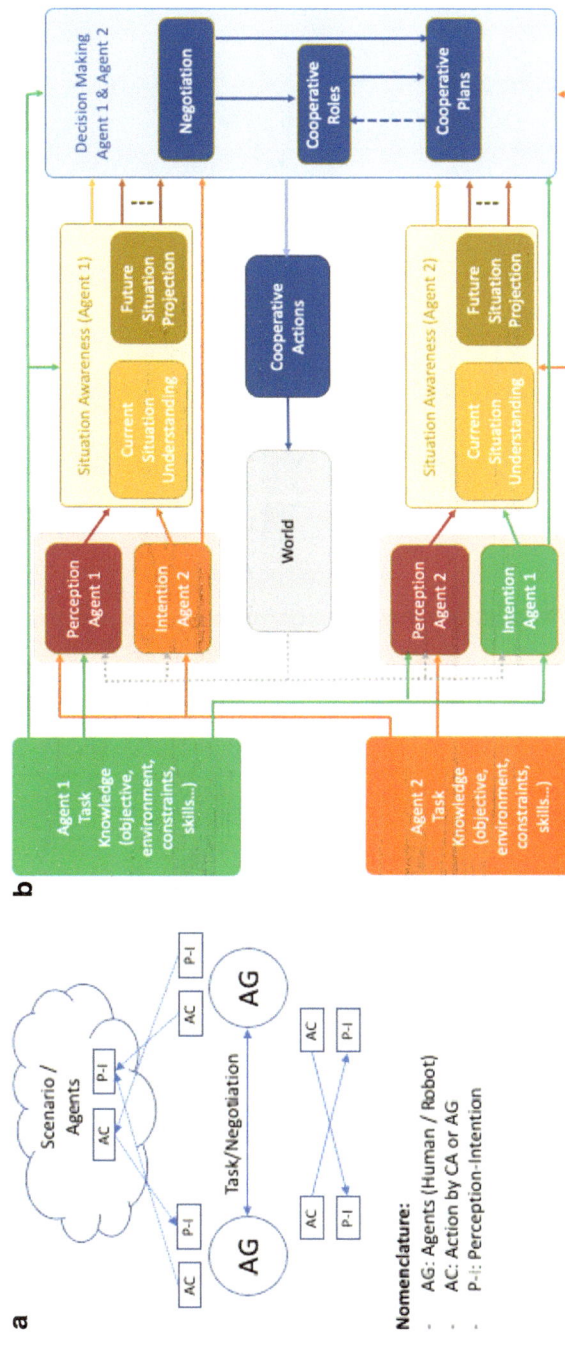

Fig. 9.5 **a** Schematic of the interaction between a robot and a human using the Social PIA model; **b** Social PIA model for two agents

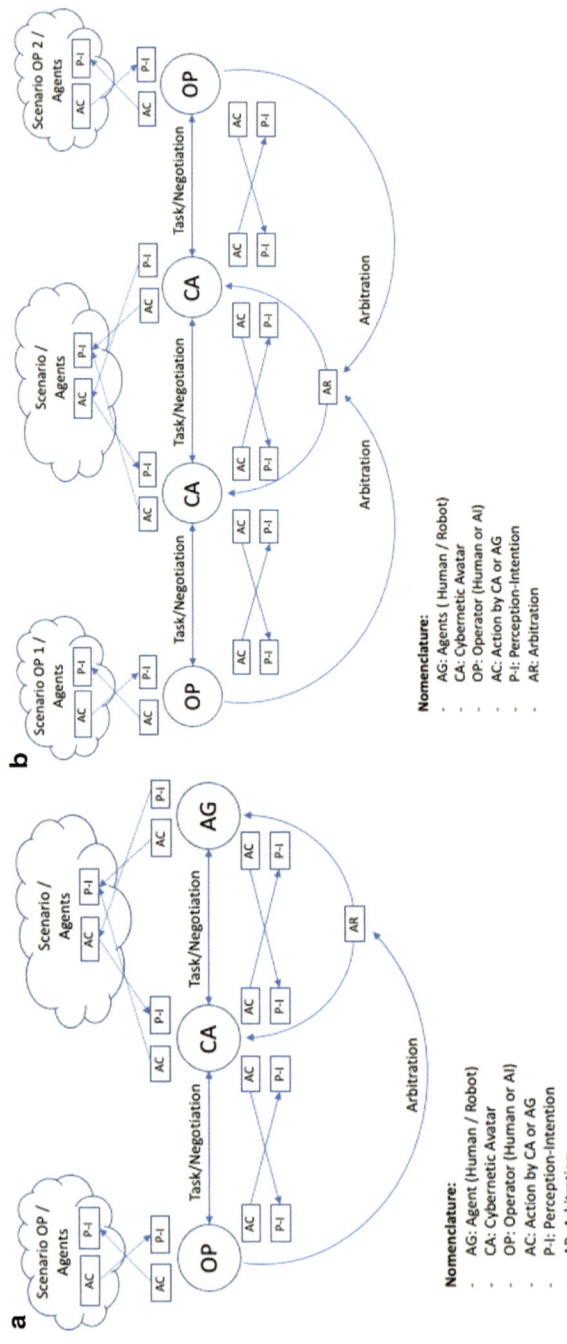

Fig. 9.6 **a** Schematic of the interaction between a CA and an agent (AG: human or robot) using the Social PIA model; **b** schematic of the interaction between two CAs using the Social PIA model

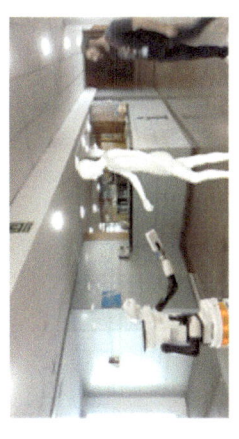

3- **Robot anticipates and moves** its arm to approach human delivery point

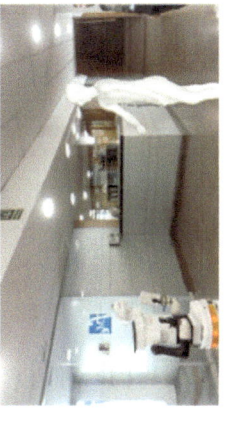

2- **Robot perceives** human **intention** and **predicts** human motion

1- Robot is waiting to deliver a box

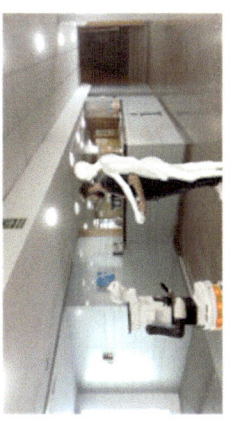

6- Human takes the box and **robot arm returns** to the rest position

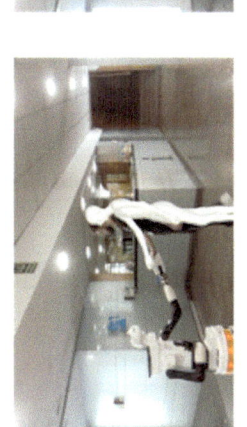

5- Human arrives to delivery point and **robot release the box**

4- **Robot anticipates and moves** its arm to approach human delivery point

Fig. 9.7 Different steps of the handover use case using the Social PIA model

9.3.3.3 The Use Case of Social PIA in a Handover Collaborative Task

We introduce the handover use case in Sect. 9.3.3.2. We use the Social PIA model to illustrate and describe this use case. Figure 9.7 shows a robot that must deliver a box to a human. In Photogram #1, the robot awaits delivery of the box. In Photogram #2, the robot senses that there is a clear way to proceed and uses its sensors (a stereo camera and lidar) to identify the person who needs to deliver the package (the robot uses the perception module). However, in Photogram #2, the robot perceives the intention of the person who will receive the box (the robot uses the intention module). The perception of the environment (there is a free path to deliver the box) and the perception of the human intention allow the robot to predict human motion in front of it, as shown in Photogram #2 (the robot analyzes the current situation and predicts human motion using the Situation Awareness module). The photogram human white model is a future projection of the human 2.5 s in advance. Then, the robot plans to anticipate human motion and moves its arm to reach the contact point with the human, Photogram #3 (the robot uses the decision-making module to plan in coordination with the human, and Cooperative Actions are executed). The human moves toward the robot; however, the human prediction has already arrived at the destination, that is, Photogram #4 (the Cooperative Action module is executed). Then, the human arrives at the contact point and grabs the box, and the robot releases the box at the same time, as in Photogram #5 (the Cooperative Action module is executed). Finally, the human takes the box out of the robot, and the robot moves its arm back to the rest position, as in Photogram #6 (the Cooperative Action module is executed). If the robot detects a different intention, its plan will be different.

9.4 Ethical Issues Concerning Cybernetic Avatars

Historically, humanity has been constrained by the need to exist in a physical space, possess a body located in a single place, and bear an identity tied to that physicality. Cybernetic Avatars (CAs) are technologies that aim to free humans from these constraints. Therefore, they have the potential to alter human existence fundamentally. This prospect offers immense hope but also raises a host of ethical concerns.

In our research, we used a variety of methods to gather views from a wide range of sectors and fields, focusing on the desired visions of the future and the ethical concerns associated with CA. At the symposia, we invited speakers, including Artificial Intelligence researchers, legal scholars, ethicists, philosophers, aestheticians, anthropologists, and science fiction writers, for lectures and discussions. In our workshops, we invited members of the public to experience CA and participated in discussions. We also conducted surveys to gauge public awareness, impressions, expectations, and concerns regarding CA.

In this section, we discuss the ethical concerns and desirable developments related to CA, based on the opinions we collected. The impact of CA technology on human life and society is far-reaching, resulting in a wide range of ethical issues. Therefore, it is not possible to comprehensively discuss all these issues. Instead, we focused on issues related to work, life, and gender.

9.4.1 Work and Life in Avatar Symbiotic Society

This subsection examines the impact and ethical considerations of CAs in individual workplaces and lives.

The extensive implementation and advancement of CA technology are anticipated to considerably alter the way people work. With the possibility of remote work gaining acceptance, job opportunities have broadened. Furthermore, CAs can enable individuals to perform previously impractical tasks without assistance. The removal of physical movement requirements allows the utilization of short free-time intervals for various tasks. Furthermore, operating multiple CAs concurrently can increase earnings for the same duration of work.

Consumers are likely to access diverse services more affordably and conveniently with advancements in CA technology. For example, if a household possesses a multifunctional digital assistant, various individuals can operate it at different times to fulfill various roles such as household assistants, caregivers, tutors, babysitters, and pet sitters. Additionally, the parallel use of digital assistants by experts, such as doctors, counselors, educators, consultants, and trainers in diverse locations, could broaden the scope of people benefiting from their specialized expertise.

With the development of CAs, daily life is likely to change significantly. By using shared CAs located worldwide, we can travel globally with ease. It may be easier to have richer interactions with loved ones who are far away through CAs, thus providing a more immersive experience than videoconferencing.

However, such transformations are not without concerns. One of the most obvious concerns is that while CAs may create new opportunities, they may also lead to unemployment and increased economic inequality. For example, if powerful CAs are only accessible to a limited number of people, a significant opportunity gap could emerge between those who can use superior CAs and those who cannot. This could exacerbate and entrench the existing economic inequalities. Such a scenario is not only ethically undesirable but also results in a loss by failing to utilize beneficial talent in society. It is desirable to ensure that as many people as possible have access to CAs. It will also be necessary to educate everyone on the skills required to operate CAs.

Furthermore, the ability of individuals to work more with the help of CAs could lead to situations in which some people are overworked, while others lose their jobs. Issues of work sharing and maintaining an appropriate work–life balance will become more critical than ever.

In the context of home services, security, surveillance, and education that use CAs, there is a potential risk of privacy violations. Furthermore, there may be cases in which individuals use CAs to impersonate others or employ AI to operate CAs and engage in work activities. It is essential that CAs are equipped with appropriate security measures and that operators are properly trained in professional ethics. It is also necessary to consider responses to potential issues that arise during CA operations. It is irresponsible for companies operating CA services to place all responsibilities for any damage caused solely on CA operators or users. For example, it is important for companies to provide remedies such as insurance for harm measures.

However, as domestic services and other professions that use CAs evolve into gig-work-like scenarios, ensuring compliance with professional ethics is likely to become more difficult. This could also lead to an increase in insecure employment, exacerbating inequality and poverty. It is essential to ensure that professionals that use CAs are stable, fulfilling, and socially respected.

Travel using CAs may provide fewer benefits to local people than conventional travel. Therefore, it is necessary to consider the welfare of the destination communities. In addition, remote technologies such as CAs may lead to a "disinhibition effect" (Suler 2004), potentially increasing the tendency to behave without restraint at the destination. Appropriate regulations on the number of CAs and their permitted areas of movement, speed, and data collection are necessary to ensure that such travel does not involve disturbing local people.

In the context of using CAs for educational purposes, it is crucial to evaluate their effectiveness carefully. Most of all, it is necessary to ensure that no child is disadvantaged through the use of CAs. If the cost of introducing CAs remains high, not all families in need of CA-based educational services will have equitable access. Consequently, those who cannot afford these services may be disadvantaged, leading to the creation or exacerbation of inequalities based on socioeconomic status.

As the society becomes increasingly dependent on CAs for various activities, individuals and communities may become vulnerable to system failures, technical malfunctions, and cyberattacks. In particular, if CAs malfunction or are hacked, not only could security be compromised, but vulnerable individuals (such as the elderly who rely on avatar assistance) could also experience disruptions in their daily lives and be unable to access necessary goods and services.

9.4.2   Gendering and Its Ethical Implications

This subsection examines gender embodiment in CAs. We begin with an overview of how gender roles are assigned to artificial entities and then explore the common concerns associated with such assignments. The following discussion is based on AI systems and robots. However, the entire point should also apply to CAs.

Until recently, AI systems and robots were not considered social agents with which ordinary people could communicate. AI systems did not speak a natural language. Robots were limited to industrial factories for mass production. However, interaction with AI assistants via natural language has become increasingly common. Social robots, some of which are remote-controlled CAs, are deployed in public places such as airports, shopping malls, restaurants, hotels, and hospitals to assist customers and provide helpful information.

The rise of human–robot interaction (HRI) has been a consequence of this shift. HRI aims to explore the interaction between humans and robots by utilizing a range of disciplines such as robotics, engineering, psychology, and design. A key finding of HRI research is anthropomorphism. Humans interact with AI systems and robots as social agents even though they are known to be inanimate artifacts. We tend to treat AI systems and robots with politeness and respect, rather than as mere tools. Through repeated interaction, individuals may develop friendships or, in some cases, even intimate relationships with them. Reports state that every day, hundreds of thousands of people say "good morning" to Amazon's Alexa and some have confessed their love for her (Cox 2018). Hiroshi Ishiguro designed Erica, a realistic humanoid robot that resembles a Japanese woman. Ishiguro commented that she was the most beautiful woman he saw (Nyholm 2023).

Anthropomorphism indicates that humans are inclined to project social cues onto artificial entities. If this is the case, it is expedient for engineers to have artificial entities that elicit this tendency, enabling users to engage more effectively. A frequently adopted approach is the incorporation of gender in AI systems and robots. Female coding is frequently employed in AI systems and social robots. AI assistants typically have feminine names such as Alexa, Siri, or Cortana. Their voices were soft, high-pitched voices associated with women. This result was intentional. Microsoft conducted an extensive survey of individuals' preferences for AI assistants, revealing that respondents from around the world favored a female assistant over a male assistant, ideally in their twenties or thirties at oldest (Kedmey 2015).

This apparent convergence of preferences for female voices may reflect the expectation that women are particularly adept at tasks requiring empathy and conversational skills. However, in other conversational settings, users may favor agents coded as males. For instance, Bayerische Motoren Werke (BMW) initially marketed automobiles using a navigation system programmed to communicate through a female voice. Some customers raised concerns about receiving directions from females. Consequently, BMW decided that the voice should suggest a male who was slightly dominant, somewhat friendly, and highly competent (Nass and Brave 2005).

This discrepancy may arise from the fact that certain roles in human society are gender-specific. Some tend to be female centric, whereas others tend to be male-centric. In female-centric domains, a female-coded robot may perform better, because it matches human expectations. In male-centric domains, users may prefer male-coded robots. This hypothesis predicts that aligning robots with human expectations will increase their acceptance in society.

However, the gendering of AI systems and robots has also raised concerns about the perpetuation of gender stereotypes and reinforcement of gender divides. According to critics, the conflation of female-voiced AI assistants with real women may propagate gender stereotypes and normalize one-sided command-based verbal interactions with women, as these AI assistants are expected to act submissively (West et al. 2019). Furthermore, associating social robots with subordinate tasks traditionally performed by women can reinforce social inequalities, as these tasks tend to be paid poorly (Tannenbaum et al. 2019). It is currently uncertain whether this alleged conflation of female-voiced AI assistants with real women occurs regularly. Similarly, the extent to which gendering of AI systems and robots contributes to social inequality remains unclear. Nevertheless, these concerns are widely shared among journalists and humanities scholars.

A related but different concern is the symbolic significance of gendering. Recent research on the generation of shared laughter has illustrated this phenomenon. Inoue and his colleagues attempted to teach the art of conversational laughter to a humanoid robot in the hope of improving the natural conversation between humans and robots (Inoue et al. 2022). They obtained speed-dating dialogs between male university students and the CA, which were remotely controlled by one of four amateur actresses. The dataset was annotated with different laughter types and was used for machine learning. This research was immediately acknowledged for its contribution to affective computing, winning the NETEXPLO Innovation 2022 Award. Despite its technological success, however, critics have focused on the ethical implications of this research. This research employed Erica, a realistic humanoid who has long been known for her highly feminine beauty. Critics have suggested that this research on shared laughter generation implies that Erica learned submissive female behavior during Japanese speed-dating events.

If gendered AI systems and robots have some symbolic significance, as shown in the previous concern, they could serve as positive role models for both boys and girls. The idea is that female-coded systems can be programmed to be assertive and confident, whereas those coded as males can possess nurturing and empathetic traits. This approach can assist in challenging traditional gender stereotypes and encourages children to develop a broader view of what it means to be a boy or girl. It is also noteworthy that gendering is only one approach for designing AI systems and robots that elicit a human tendency to project social cues onto artificial entities. In principle, designers can focus on creating AI systems and robots that are perceived as gender neutral. This could involve using gender-neutral names, avoiding stereotypical design features, and programming these systems with a variety of behaviors that do not conform to traditional gender roles (Schiebinger et al. 2011–2020).

This is not a place to resolve the controversy but merely to highlight the complexity of introducing AI systems and robots into society. It is crucial to examine the ethical implications and to design AI systems and robots that promote inclusivity, diversity, and accountability.

9.4.3 Summary

Technological progress does not automatically improve the well-being of all people. Technology has historically worsened the lives of certain groups, exacerbated inequalities, and threatened social security. Often, socially vulnerable people are harmed, rather than benefiting from technological progress.

Truly valuable innovations increase the overall utility and well-being of society, improve the quality of individual lives, contribute to human prosperity and peace, and help solve societal challenges. Therefore, to promote innovation, individuals' rights, health, property, and dignity should be respected.

Technology must advance equitably. Special care must be taken to avoid worsening the situation for minorities and disadvantaged people. Thus, innovation should be a means of redressing discrimination and inequality. Driving innovation must protect the needs and rights of vulnerable people whose voices are not adequately represented and provide them with opportunities to participate in decision-making processes.

People are often fearful and cautious of new technologies. Those who promote technology must fulfill their duty to provide adequate explanations, increase the transparency of the technology itself and its promotion process, and address people's concerns. The public's understanding of technology is a critical prerequisite for its adoption. Risks are inherent in new technologies, and even experts may fail to anticipate potential harm. Therefore, it is necessary to prepare flexible responses to unforeseen negative effects. Moreover, all stakeholders must understand such uncertainties before technology can be introduced into society.

9.5 Legal Issues Concerning Cybernetic Avatars

9.5.1 What is an Avatar?

The word "avatar" is derived from the Sanskrit word "Avatāra" (अवतार), meaning "a representation of a divine principle" (a symbolic representation) or "descent" (a figure that has descended to earth), which is the object of people's faith (Lenoir and Drucker 2011). The term usually refers to the incarnation of the god, Vishnu.

In the Ramayana and the Mahābhāratam, the great Hindu epics, and in the Bhagavadgītā, one of the Hindu scriptures, Rāma appears as a superhuman protagonist and Krishna is the avatara (incarnation) of Vishnu. James Cameron's movie *Avatar* was inspired by this word (Lenoir and Drucker 2011).

Mandalas in Japan, India, and China depict Buddhas and deities descending to earth in temporary forms, and these are called "gongen" [alter-ego Buddhas] in Japan. In American museums, alter-ego Buddhas in the mandalas are described as avatars (Ikegami and Tanaka 2020). Therefore, the term "avatar" is often used to mean "alter-ego of oneself," extending the original meaning of alter-ego (incarnation) of God, to that of a human being.

9.5.2   What is a Cybernetic Avatar?

The term "cybernetics" has several definitions (Umpleby 1982). It was originated from the ancient Greek word "Κυβερνήτης" (*kybernetikos*) [helmsman (good at steering)].

In the first half of the nineteenth century, French physicist André-Marie Ampère, in his classification of sciences, suggested that the nonexistent science of government control is called cybernetics. The term was soon forgotten; however, it was not used again until the American mathematician Wiener (1948) published his book Cybernetics in 1948. In that book, Wiener referred to an 1868 article on governors by British physicist James Clerk Maxwell and pointed out that the term governor is derived, via Latin, from the same Greek word that gives rise to cybernetics. The date of Wiener's publication is generally accepted to mark the birth of cybernetics as an independent science. Wiener defined cybernetics as "the science of control and communications in the animal and machine." This definition is closely related to the theory of automatic control and physiology, particularly the physiology of the nervous system (Britannica).

A derivative word of "cybernetics" is "cyberspace," coined by William Gibson in his Neuromancer (1984) by combining the words "cybernetics" and "space."

There is, however, no clear definition of "Cybernetic Avatar" (CA) to date.

The Japanese Council for Science, Technology, and Innovation (CSTI) and the Strategic Headquarters for the Promotion of Health and Medical Care within the "moonshot-type R&D projects" have been promoting the "Moonshot Goal 1:" to realize a society in which people are free from the constraints of body, brain, space, and time by 2050. They explain the concept of CA in relation to this goal, including ICT and robotics technologies, which extend people's physical, cognitive, and perceptual abilities, in addition to robots and avatars that show tri-dimensional (3D) images as a substitute for the person.

Therefore, to start considering the legal status (proof of existence) of CAs in the future, I classify them as shown in Table 9.2.

"Tangible object CAs" are not limited to robots, such as "geminoids" and "humanoids," but also include other tangible objects (solid, liquid, or gas). Stealing liquid CA constitutes theft under Article 235 of the Japanese Penal Code, which is similar to stealing water. In the case of projection mapping, where a CA is displayed onto an area filled with gas, the gas becomes a tangible CA, and this projected image becomes intangible in the same way as a hologram.

Although electricity is an intangible substance, it is positioned as a property (tangible object) under the current Japanese law, and Article 245 of the Japanese Penal Code provides that "electricity shall be deemed to be a property."

Table 9.2 Classification of CAs

Tangible objects CAs	Physical avatars, such as robots
Intangible CAs	Computer graphics avatars and software agents that are complete in the virtual world

The following is a summary of the current theories regarding this property. Until this provision was established, the electric theft judgment (May 21, 1903) deemed them to be tangible objects. There are three theories regarding the meaning of goods: (1) corporeality (goods are tangible objects), (2) manageability (intangible objects are good as long as they are manageable), and (3) physical manageability (objects with manageable materiality), with (1) corporeality being the most common theory (Otani 2019).

Article 85 of the Japanese Civil Code also provides that "in this Act, 'thing' means a tangible object."

The "intangible CA" could be a "digital twin," (the avatar of a real or deceased person), a virtual fictional person, an avatar of a character, and an electronic agent or bot.

9.5.3   Usage Aspects of CAs

Because a CA is intended to carry out social activities on behalf of the individual, we attempt to classify the phases of its use.

(1)   Situations in which CA can be used include (a) substitution of a real person, (b) reproduction of a deceased person, and (c) representation of a non-existent person.
(2)   The following methods of use are envisioned: (a) remote operation, (b) use within the scope of automatic program processing, and (c) autonomous operation.
(3)   With regard to the form of use, Moonshot Goal 1 involves developing a CA infrastructure that enables everyone to participate in diverse social activities by (a) combining a large number of avatars and robots remotely controlled by multiple people by 2050, (b) developing a CA infrastructure that can be used in a wide variety of social activities by the end of 2050, and (c) by 2030, developing technology that enables a single person to operate 10 or more avatars for a single task with the same speed and accuracy as a single avatar, and building the infrastructure necessary to operate such a system. The following is an example of a type of work that can be performed: Specifically, the concept of using a CA is "living a CA life", which means that (1) by 2050, technology will be developed to enable anyone who wants it to expand their physical, cognitive, and perceptual abilities to high levels, and a new lifestyle based on socially accepted ideas will be popularized; and (2) by 2030, anyone who wants it will be able to live through a CA.

9.5.4 Legal Status of CA

Legally, the notion of "person" includes not only natural persons but also legal entities. Because avatars are neither natural nor legal persons, they only have the legal status of objects (tangible or intangible) such as machines or software. Therefore, unless a legal personality is assigned to an avatar, its legal status will merely be that of the object. Therefore, various aspects must be clarified when performing legal acts using avatars.

Regarding tangible CAs, legal status theory (Saito 2017) is a future issue (Shimpo 2017), because legal status other than as a thing cannot be considered at this point.

The legal concept of CA for intangible objects such as electronic agents has been examined since the late 1990s. For example, Kerr and Shimpo (2000) presented three proposals for contracts using electronic agents: (1) a method that recognizes the juridical personality of the agent software; (2) a method that regards the action of the agent software as the act of a juridical person who uses the agent software; and (3) a method that refers to the legal status of slaves in ancient Roman times (Kerr and Shimpo 2000).

(1) Allows the agent software to have a juridical personality by equating it with other artificial juridical entities created by humans, such as corporations.

(2) Is a method in which an electronic device is regarded as the act of a juridical person who uses the action of agent software instead of being regarded as an independent juridical person. This method ignores the fact that agent software operates voluntarily in the process of concluding an agreement and pretends that it is nothing more than a mere means of communication. This is discussed as an application of "legal fiction." For example, the United Nations Commission on International Trade Law (UNCITRAL) Model Law Enactment Guide states that "data messages automatically generated by a computer without human intervention should be regarded as 'originally executed' by the legal entity, instead of being accomplished by the computer." The court indicated that the Uniform Commercial Code, Part 2 B, in its Compilation and Commentary, also states that electronic agent software is, in effect, merely an extension of the person who uses it and that his or her acts are constitutive of those of an individual. The Transactions Act also points out that it clearly recognizes electronic devices as an extension of human actions but also as something that can operate independently of human control and explicitly endorses contracts entered into by electronic agents and software. Furthermore, he states that there are other ways to extend the contract principle in a broad sense besides the application of legal mimicry.

Example (3) refers to the fact that Roman slaves were considered to have no juridical personality, but many legal rules existed that allowed them to participate in transactions and enter into contracts. An electronic device that voluntarily executes transactions similar to Roman slaves will be granted a certain resemblance to legal status. Once the certainty of this technology is guaranteed and a high degree of autonomy and intelligence comes to the fore, this will be sufficient reason to treat it as a legal intermediary rather than a mere tool.

Pagallo (2013) examined these three perspectives. He points out that:

(1) "It makes no sense to treat a robot as a tool as a legal entity that has the capacity to conclude contracts based on its own rights" (Pagallo 2013) and then present an idea based on the Roman law with reference to the legal status of slaves in ancient Rome.

(2) At the national meeting of the Uniform State Law Commission within the annual meeting of the National Bar Association in May 2003, a proposal was made to recognize the validity of contracts made by electronic agents, even if no human action or knowledge intervened (Pagallo 2013). This led to efforts to amend the Uniform Commercial Code through the US Electronic Signature Act and the Uniform Computer Information Transactions Act of 1999 (UCITA), 15 U.S.C. § 7001(h), which provides that a contract "may be executed, created or delivered by one or more electronic agents, provided that such electronic agents are not involved in its formation, creation, or delivery". The legal effect, validity, and enforceability of a contract cannot be denied solely because of the involvement of an electronic agent even if such an agent is legally vested in the person bound by the contract. The following is a summary of the provisions of the law: Furthermore, Article 14 of the Uniform Electronic Transactions Act (UETA) of the USA states that a contract may be formed by the interaction of the electronic agents of the parties, even if the individual was unaware of or did not confirm the acts of the electronic agents or the resulting terms and agreements (Pagallo 2013).

(3) He suggests that today's robots can be likened to ancient Roman slaves and that the agency rights that ancient Roman law granted to "objects" indicate a way to address the inconsistencies in the robot-as-tool approach. While the majority of slaves did not have the right to demand their own patriarchs, some enjoyed considerable autonomy and even entered into contracts and managed assets on behalf of their own patriarch's family business. More specifically, they were "rulers" ("institores"). He presents a discussion based on Justinian's Digest (XIV, 3, 11, 3; XV, 1, 47) of the Justinian Code. Moreover, because the *Compendium of Learned Treatises* recognized slavery as a system of private property and allowed slaves to work as property managers, bankers, or merchants, although the juridical personality, which was the basis of private rights, was removed, it is also possible that a certain property value for robots is a factor that should be considered with respect to such machines.

9.5.5   Corporate CAs

To discuss the legal status of CAs, I divided them into corporate and individual avatars for tangible and intangible CAs.

The corporate avatar does not mean an avatar with a juridical personality, but simply an avatar of a corporation. In both tangible and intangible objects, the local characters and the characters of various corporations and other entities are present and active. A wide variety of avatars, from the so-called "stuffed animals to virtual avatars", are used daily, and Japan is one of the most avatar-oriented societies.

However, these corporate avatars are not expected to be used as entities that perform legal acts, and it is unlikely that legal acts by corporate CAs will be discussed in the future development of CAs.

For example, the Chiba Prefecture mascot CHI-BA+KUN is an avatar that personifies the Chiba Prefecture. However, it is inconceivable that this avatar performs legal acts related to Chiba Prefecture on behalf of the government. More specifically, when the governor is a representative of an organization that is eligible for a subsidy, it is not expected that they would delegate the authority of representation as the chairperson of the organization to Chi-Ba-kun to dissolve the relationship of mutual representation, as this would fall under mutual representation under Article 108 of the Civil Code.

Therefore, although there are situations where a corporate avatar is used for factual acts, it is considered unnecessary to use a corporate avatar to perform legal acts related to the corporation.

9.5.6   Proposals for New ELSI Studies

9.5.6.1   Establishing a Research System to Ensure and Maintain the Social Acceptability of CA in Avatar Life

To realize a life with avatars, ethical, legal, and social implication (ELSI) issues in human society cannot be avoided. However, the study of ELSI issues to find ways to resolve them has been confined to research in the humanities and social sciences. It is necessary to show how these new issues can be resolved using technology and establish a comprehensive research system for this purpose. It is also necessary to develop a new research vision that considers the complex and diverse ELSE issues and responds to the arrival of an avatar-deployment society beyond expectations.

However, it is practically impossible to examine all issues that may or may not arise in the future within the framework of current knowledge, society, and institutions. Although research on ELSI issues associated with new technological developments has been conducted by enumerating and exemplifying the issues to be considered, and research has been conducted assuming new issues, it is partly because of research methods in the humanities and social sciences, but it is also because of research methods that are vertically divided into various fields and do not allow for cooperation between each field, and thus fragmented issues are being studied individually and The situation is that research is being conducted in a whack-a-mole fashion. Consequently, the lack of a research system and methods to fundamentally examine essential issues has contributed to the lack of significant progress in ELSI research.

9.5.6.2 ELSI Issues

With regard to ELSI issues, it may be a good idea to indicate that research will be conducted on issues that cannot be envisaged within the current framework or on tentative issues that require the presentation or construction of new concepts in the future, not within the existing ELSI framework, but from the four disciplines of ethics (E), law (L), society (S), and economics (E). We believe that this is one possible way to make further progress in ELSI research, which has not advanced well, by clarifying the areas of research responsibility.

We also proposes a research structure (sustainable ELSE research) for conducting new ELSI research that is comprehensive (encompassing), Looking Forward (looking forward), systematic (systematical), and integrated (integrated), rather than piecemeal research in individual fields (management systems) to conduct new ELSI research.

From the perspective of extending the functions of CA and improving productivity, ELSE research is a prerequisite for CA acceptability and promoting its use in society, which also contributes to indirect productivity improvements. The extension of individual activities as a community can also lead to increased productivity in the organization to which they belong, and in society in general.

By examining the possibilities of using CA as an extension of individual activities, noncontact technology, and teleoperation technology from an ELSE perspective, business continuity and ensuring (maintaining) productivity can be achieved through the use of CA as a continuation of activities, even in environments where people have to live with infectious diseases. The fulfillment of the statutory employment rate based on the Law for the Employment Promotion of Persons with Disabilities in the workplace by using CA is another example of ensuring productivity through the use of CA.

Because research in the humanities and social sciences does not produce inventions, but new knowledge can be gained through the accumulation of knowledge, we plan to construct an ELSE research management system to accumulate knowledge for this purpose.

9.5.7   Economic Security and CA R&D

9.5.7.1   Policy Development for CA R&D (Policy Advocacy, IP Protection, International Strategy, and Standardization)

To use CAs safely, securely, and reliably in daily life, it is necessary to build a certification infrastructure to guarantee their reliability and develop new continuous certification technologies to guarantee connectivity between the CA user and the main CA. It is also necessary to take institutional measures, such as standardization, building certification infrastructure, and implementing domestic and international policy development.

Recognizing the IP protection strategy from the perspective of economic security as a sensitive issue in CA development, international collaboration should be actively promoted as an opportunity to showcase the response to the challenging issues of the ELSE issue study, while also considering that a sensitive technology management mechanism based on the international security trade export control system and careful international deployment based on international collaboration must be promoted. At the same time, it is necessary to closely monitor the trends in studies in the international community in areas that cannot be addressed by the studies in this proposal, such as applications to autonomous weapons (LAWS).

9.5.7.2   Leading International Rulemaking

Proposals should be made for the legal framework necessary for the establishment of a CA certification infrastructure and for the international strategic policy development of CA R&D in the context of emerging international sensitive technology management systems, such as economic security and security trade export control.

It is also important to make the necessary recommendations for Japan's participation in strategic rulemaking to take the initiative in international rulemaking in CA infrastructure and CA life, based on trends in international legislation, and to maintain its advantage in terms of international competitiveness, even at the stage where CAs have spread internationally in 2050. The following are some of these recommendations:

9.5.8   CA R&D and the Legal System

In areas such as data management, information security, quality assurance, and environmental protection, initiatives based on a bi-design philosophy have been implemented, and standards for the establishment of management systems have been developed. Privacy Impact Assessment (PIA) was introduced as a specific personal data protection assessment tool under the Number Utilization Act.

However, mechanisms such as privacy by design and default as specified in the EU General Data Protection Regulation (GDPR) have not been implemented in Japan.

Although it is conceivable that knowledge of these management systems could be used in the construction of a certification infrastructure to prove the genuine existence of CAs, it is insufficient for their continuous certification as it is only an initiative in a discrete field. The most important issues are inclusiveness, privacy protection, responsibility for avatar actions, and the prevention of misuse, such as impersonation. Various research groups and organizations have individually considered and proposed ethical principles and regulations. It is important to create an international organization and network to discuss ELSI policies on avatars and their technical standards.

With regard to CA research and development and the legal system, proposals for a legal system are necessary for the establishment of a CA certification infrastructure, overcoming the disadvantages caused by the absence of a legal code to regulate and discipline the use of new technologies, issues associated with information acquisition and analysis, the development of guidelines, other rules and regulations, and transparency. Therefore, a mechanism ensuring transparency must be considered.

Ensuring the portability of CAs (portability (seamless transfer) of data required for CA use) and considering systems for portability and sharing mechanisms in the use of the CA itself are also needed.

9.6 Use of CA for Election Campaign

9.6.1 Can CA Be Used for Campaigning?

This section examines whether communicative avatars (CAs) can be used in election campaigns.

In conclusion, the CA cannot be used for election campaigning in Japan. The Public Offices Election Law regulates election campaigning. Although the Public Offices Election Law contains no provisions for CAs, their use is prohibited.

9.6.2 New Technologies and Election Campaigns

9.6.2.1 Use of the Internet in Election Campaigns

Today, with the proliferation of the Internet, it is being used extensively in election campaigns.

Politicians began using computer networks in the late 1980s. At that time, networks such as CompuServe and AOL were popular in the USA, and networks such as Nifty-Serve began to spread in Japan. The network consists of a modem connected to a computer and a dial-up connection to the host computer via a telephone line.

In the late 1980s, politicians in Japan began using e-mails to report to their supporters. At that time, e-mail was not yet in the format used today but was sent and received via a telephone line connected to a host computer.

The year 1995 was marked by two major events in terms of the spread of the Internet: first, commercial use of the Internet became widespread, and second, Microsoft released Windows 95, allowing many users to easily connect to the Internet at home.

After 1995, some Diet members and political parties began to establish homepages, and after 2000, politicians began to use homepages in earnest. 60% or more of the candidates in the 2003 House of Representatives election established their own homepages. Currently, a minority of candidates do not have a homepage.

Perhaps the first country where the use of the Internet also had a significant impact on election campaigns was South Korea, symbolized by the 2002 presidential election. In the 2002 election for South Korea's 16th president, supporters of candidate Roh Moo-hyun formed an advocacy group through the Internet. Initially, Roh Moo-hyun was not considered a strong candidate. However, Internet advocacy groups helped Roh eventually win elections.

In South Korea, Internet-based election campaigns have transformed political traditions. In the 2012 presidential election, the ban on third parties other than political parties and candidates using the Internet to campaign during the election period was lifted. In the 2012 presidential election, the ban on third parties other than political parties and candidates using the Internet during the election campaign period was lifted.

9.6.2.2 Use of the Internet in Japanese Election Campaigns

Before discussing Internet election campaigns in Japan, we must first explain their characteristics of Japanese election campaigns.

First, the Public Offices Election Law, a national law, regulates the campaigning of prefectural and municipal leaders and assembly members so that prefectures and municipalities cannot set their own campaigning methods. This is partly because of the fact that Japan is not a country with a federal system.

The second was public election campaigning, which began in 1925 (Quigley 1926). To ensure the fairness of election campaigns, the Public Offices Election Law stipulates that national or local governments bear the costs of using campaign vehicles, distributing postcards and leaflets, posting posters, newspaper advertisements, political and biographical broadcasts, and speeches. Instead, the law regulates in detail the period of campaigning, who may campaign, the method of campaigning, and the maximum expenses (Hayashida 1967). Distribution and posting of leaflets, flyers, documents, and graphics other than those stipulated by law are prohibited.

Third, a distinction is made between election campaigns conducted as political activities and those conducted for a specific election. The election campaign period is limited to the period from the day a candidate's candidacy is announced to the day before voting. Election campaigning conducted prior to notification is illegal, as advance campaigning is prohibited. In reality, however, the prohibition of advanced campaigning has been criticized for having no practical effect, since activities such as policy propaganda and party expansion by political parties and other political organizations can be conducted before the election campaign period as political activities.

Until now, websites were considered to be "documents and drawings." If a candidate updated his/her homepage or posted it on a blog during the election campaign period, it was considered a violation of the Public Offices Election Law because it would fall under the prohibition of distributing or posting documents and drawings other than those stipulated in the Public Offices Election Law. Therefore, candidates were not allowed to use the Internet for election campaigns during the campaign period. Ordinary voters other than candidates were also prohibited from using their homepages or blogs to call for support for candidates, which could violate the prohibition of popular votes for "dropout campaigns," as well as the Public Offices Election Law, because such campaigns could violate the prohibition of signature campaigns prohibited by the Law.

9.6.2.3 Amendments to the Public Offices Election Law

There have been strong calls for the Internet to be used for election campaigns, and an increasing number of political parties have begun to use the Internet as a political activity, even during election campaigns. Shinzo Abe immediately expressed his view that the ban on using the Internet for election campaigning should be lifted after the LDP won the Lower House election in December 2012, returning to the ruling party and becoming the prime minister for a second time. In 2013, the Diet amended the Public Offices Election Law to lift the ban on Internet use in election campaigns. As a result, candidates and political parties can update their websites during the campaign period and use X (Twitter), blogs, and other social networking services to conduct election campaigns.

Regarding e-mail, the ban on its transmission by candidates and political parties for campaigning purposes has been lifted, but its transmission by third parties for campaigning purposes remains prohibited.

9.6.3   CA and Campaigning

There is no provision for CA in the Public Offices Election Law. Why is the use of CA prohibited?

The Election Department of the Ministry of Internal Affairs and Communications has authority to interpret the provisions of the Public Offices Election Law.

According to the Elections Department of the Ministry of Internal Affairs and Communications, CAs fall under the category of billboards, signs, and puppets.

If the CA does not have a three-dimensional appearance, it may be used as a billboard or sign within the provisions of the Public Office Election Law. However, the quantity must not exceed three panels for a candidate, and the size must not exceed 350×100 cm.

If the CA has a three-dimensional appearance, it may be a puppet. Puppets made based on the likeness of candidates may not be used. Therefore, the CA cannot be used for campaigns or political activities.

9.6.4   Use of On-Screen Avatars

Avatars, which are not CAs in robot form but appear on the screen, can be used for election campaigns. Candidates and parties can update their websites during the campaign and use Twitter, blogs, and social networking services. Among them, avatars can be used.

However, their use as advertisements has also been regulated. This is an interesting example.

In recent years, Japanese cabs have installed tablets or liquid crystal display (LCD) displays on the headrests or backs of passenger seats. In these displays, moving and still image advertisements are shown and sound is played. Is the use of campaign-related avatar advertisements in cabs regulated by the Public Offices Election Law?

It is noteworthy that the Election Division of the Ministry of Internal Affairs and Communications has recently expressed its interpretation that such advertisements on displays in cabs are not considered paid Internet advertisements (Election Division, Ministry of Ministry of Internal Affairs and Communications 2021).

First, "if the distribution is to a certain number of cabs of a certain size, it is assumed that a cloud-based system, which is a type of network distribution system, will be used." As such, it is acknowledged that the Internet is being used. Then, regarding the relationship between the display of such monitor advertisements in cabs and the law, it states that they fall under the category of "documents and drawings" under the Public Offices Election Law. However, it states that they are subject to the Public Offices Election Law as either projections or billboards.

As mentioned earlier, this advertising method can display still or moving images, and if the advertisements are exposed to the eyes of many customers, the monitors on which the advertisements are displayed are considered to be subject to the regulations of the law as "projection, etc." if they display different screens one after another, and as "billboards and signs" if they display the same screen for a certain fixed period of time. If the monitor displays the same screen for a fixed period of time, it is considered to be a "billboard or signboard" and is subject to legal restrictions.

9.6.5 *Possibility of Using CA in Election Campaigns*

As mentioned previously, CAs cannot be used in election campaigns in Japan.

However, there are many advantages of using CA in election campaigns. Candidates with disabilities enjoy greater convenience. Candidates with speech impediments were unable to make campaign speeches because they were unable to speak for themselves. However, with the use of CA, candidates with speech impediments can make campaign speeches on the streets.

Candidates hospitalized because of serious disabilities or illness are not allowed to make campaign speeches on the streets. However, with the CA, they can make campaign speeches on the streets.

The use of CA also provides significant advantages for legislators with disabilities. In 2019, Eiko Kimura, a candidate with an extremely significant disability, was elected to the House of Councilors. Kimura was diagnosed with a spinal cord injury caused by an injury sustained when she was eight years old, leaving her with very little movement other than that of her right hand, which she used to operate a motorized wheelchair for transportation (Takenaka 2019). If she was able to use the CA, she would not have had to risk entering the chamber.

However, the use of CA in election campaigns is problematic. This can be distinguished from deep faking, which is a global problem. Fake information constitutes a major issue in election security.

In the USA, countermeasures against election faking are being taken from the perspective of Mis, Dis, and Malinformation (MDM). Foreign influencing tactics include leveraging misinformation, disinformation, and malinformation. The definitions of each are as follows (Cybersecurity and Infrastructure Security Agency 2020):

Misinformation is false but is not created or shared with the intention of causing harm.

Disinformation is deliberately created to mislead, harm, or manipulate a person, social group, organization, or country.

Malinformation is based on fact but used out of context to mislead, harm, or manipulate. An example of malinformation is the editing of a video to remove important contexts that harm or mislead.

It is also necessary to consider legal issues regarding the production and use of CA by a certain candidate or politician by a third party without consent. This issue must be considered from a wide range of perspectives including privacy, personal information, portrait rights, publicity, laws governing elections, and political activity.

References

Asilomar Conference (2017) Asilomar AI principles. Future of Life Institute

Bartneck C, Nomura T, Kanda T, Suzuki T, Kato K (2005) Cultural differences in attitudes towards robots. In: Proceedings of the symposium on robot companions: hard problems and open challenges in robot-human interaction

Battistuzzi L, Sgorbissa A, Papadopoulos C, Papadopoulos I, Koulouglioti C (2018) Embedding ethics in the design of culturally competent socially assistive robots. In: 2018 IEEE/RSJ international conference on intelligent robots and systems (IROS). IEEE, pp 1996–2001

Berry LL, Parasuraman A, Zeithaml VA (1988) SERVQUAL: a multiple-item scale for measuring consumer perceptions of service quality. J Retail 64:12–40

Britannica cybernetics. https://www.britannica.com/science/cybernetics. Accessed 16 Dec 2023

Brooks R (1986) A robust layered control system for a mobile robot. IEEE J Rob Autom 2:14–23. https://doi.org/10.1109/JRA.1986.1087032

Bull P (1987) Posture and gesture. Pergamon Press

Canca C (2020) Operationalizing AI ethics principles. Commun ACM 63:18–21. https://doi.org/10.1145/3430368

Chung N, Lee H, Kim J-Y, Koo C (2018) The role of augmented reality for experience-influenced environments: the case of cultural heritage tourism in Korea. J Travel Res 57:627–643. https://doi.org/10.1177/0047287517708255

Cox T (2018) Now you're talking: human conversation from the Neanderthals to artificial intelligence. Counterpoint

Dalmasso M, Domínguez-Vidal JE, Torres-Rodríguez IJ, Jiménez P, Garrell A, Sanfeliu A (2023) Shared task representation for human-robot collaborative navigation: the collaborative search case. Int J Soc Robot. https://doi.org/10.1007/s12369-023-01067-0

Daneshmand M, Even J, Kanda T (2023) Effortless polite telepresence using intention recognition. ACM Trans Hum-Robot Interact (THRI)

Domínguez-Vidal JE, Sanfeliu A (2023) Improving human-robot interaction effectiveness in human-robot collaborative object transportation using force prediction. In: 2023 IEEE/RSJ international conference on intelligent robots and systems (IROS)

Domínguez-Vidal JE, Rodríguez N, Sanfeliu A (2023) Perception-intention-action cycle as a human acceptable way for improving human-robot collaborative tasks. In: Companion of the 2023 ACM/IEEE international conference on human-robot interaction. ACM, New York, pp 567–571

Domínguez-Vidal JE, Rodríguez N, Sanfeliu A (2024) Perception-intention-action cycle in human-robot collaborative tasks: the collaborative lightweight object transportation use-case. Int J Soc Robot (accepted)

Duchaine V, Gosselin C (2009) Safe, stable and intuitive control for physical human-robot interaction. In: 2009 IEEE international conference on robotics and automation. IEEE, pp 3383–3388

Endrass B, André E, Rehm M, Nakano Y (2013) Investigating culture-related aspects of behavior for virtual characters. Auton Agent Multi-Agent Syst 27:277–304. https://doi.org/10.1007/s10458-012-9218-5

Endsley MR, Garland DJ (2000) Situation awareness analysis and measurement. CRC Press

EUR-Lex (2021) Proposal for a regulation of the European Parliament and of the Council on laying down harmonised rules on Artificial Intelligence (Artificial Intelligence Act) and amending certain Union Legislative Acts. https://eur-lex.europa.eu/legal-content/EN/TXT/?uri=celex%3A52021PC0206. Accessed 6 Dec 2023

European Commission (2019) European Commission's ethics guidelines for trustworthy AI. https://digital-strategy.ec.europa.eu/en/library/ethics-guidelines-trustworthy-ai. Accessed 6 Dec 2023

European Parliament (2017) European Parliament, report with recommendations to the Commission on Civil Law Rules on Robotics

Ferrer G, Sanfeliu A (2014) Proactive kinodynamic planning using the Extended Social Force Model and human motion prediction in urban environments. In: 2014 IEEE/RSJ international conference on intelligent robots and systems. IEEE, pp 1730–1735

Floridi L, Cowls J, Beltrametti M, Chatila R, Chazerand P, Dignum V, Luetge C, Madelin R, Pagallo U, Rossi F, Schafer B, Valcke P, Vayena E (2018) AI4People—an ethical framework for a good AI society: opportunities, risks, principles, and recommendations. Minds Mach (Dordr) 28:689–707. https://doi.org/10.1007/s11023-018-9482-5

Garrell A, Sanfeliu A (2012) Cooperative social robots to accompany groups of people. Int J Rob Res 31:1675–1701. https://doi.org/10.1177/0278364912459278

Gayet-Viaud C (2017) French cities' struggle against incivilities: from theory to practices in regulating urban public space. Eur J Crim Pol Res 23:77–97. https://doi.org/10.1007/s10610-016-9335-9

Gibson W (1984) Neuromancer. Ace

Hart SG (2006) Nasa-task load index (NASA-TLX); 20 years later. Proc Hum Factors Ergon Soc Annu Meet 50:904–908. https://doi.org/10.1177/154193120605000909

Hart SG, Staveland LE (1988) Development of NASA-TLX (task load index): results of empirical and theoretical research. In: Advances in psychology, pp 139–183

Huggins M, Alghowinem S, Jeong S, Colon-Hernandez P, Breazeal C, Park HW (2021) Practical guidelines for intent recognition. In: Proceedings of the 2021 ACM/IEEE international conference on human-robot interaction. ACM, New York, pp 341–350

IEEE SA (2019) IEEE global initiative on ethics of autonomous and intelligent systems. https://standards.ieee.org/industry-connections/ec/autonomous-systems/. Accessed 6 Dec 2023

Ikegami E, Tanaka Y (2020) Edo and avatar: our inner diversity. Asahi Shimbun Publishing

Inoue K, Lala D, Kawahara T (2022) Can a robot laugh with you? Shared laughter generation for empathetic spoken dialogue. Front Robot AI 9. https://doi.org/10.3389/frobt.2022.933261

Kedmey D (2015) Here's what really makes Microsoft's Cortana so amazing

Kerr IR, Shimpo F (trans) (2000) Intelligent software agents. In: Ibusuki M (ed) Cyberspace law, cyberlaw study group. Nippon Hyoronsha

Kosuge K, Hayashi T, Hirata Y, Tobiyama R (2003) Dance partner robot—Ms DanceR. In: Proceedings 2003 IEEE/RSJ international conference on intelligent robots and systems (IROS 2003) (Cat. No.03CH37453). IEEE, pp 3459–3464

Laplaza J, Garrell A, Moreno-Noguer F, Sanfeliu A (2022) Context and intention for 3D human motion prediction: experimentation and user study in handover tasks. In: 2022 31st IEEE international conference on robot and human interactive communication (RO-MAN). IEEE, pp 630–635

Lenoir F, Drucker M (2011) DIEU. Robert Laffont

Lugrin B, Rehm M (2021) Culture for socially interactive agents. The handbook on socially interactive agents. ACM, New York, pp 463–494

McKenna PE, Ghosh A, Aylett R, Broz F, Rajendran G (2018) Cultural social signal interplay with an expressive robot. In: Proceedings of the 18th international conference on intelligent virtual agents. ACM, New York, pp 211–218

Morley J, Floridi L, Kinsey L, Elhalal A (2020) From what to how: an initial review of publicly available AI ethics tools, methods and research to translate principles into practices. Sci Eng Ethics 26:2141–2168. https://doi.org/10.1007/s11948-019-00165-5

Mörtl A, Lawitzky M, Kucukyilmaz A, Sezgin M, Basdogan C, Hirche S (2012) The role of roles: physical cooperation between humans and robots. Int J Rob Res 31:1656–1674. https://doi.org/10.1177/0278364912455366

Mutlu B, Yamaoka F, Kanda T, Ishiguro H, Hagita N (2009) Nonverbal leakage in robots. In: Proceedings of the 4th ACM/IEEE international conference on human robot interaction. ACM, New York, pp 69–76

Nakamura Y (2007) An empirical study on service quality measurement scale. Yokohama Int J Soc Sci 11

Nass C, Brave S (2005) Wired for speech: how voice activates and advances the human computer relationship. MIT Press

Nass C, Isbister K, Lee E-J (2000) Truth is beauty: researching embodied conversational agents. In: Embodied conversational agents. The MIT Press, pp 374–402

Nyholm S (2023) This is technology ethics: an introduction. Wiley-Blackwell

OECD (2019) Recommendation of the Council on Artificial Intelligence. https://legalinstruments.oecd.org/en/instruments/OECD-LEGAL-0449. Accessed 6 Dec 2023

Otani M (2019) Lecture notes on criminal law, new 5th edn. Sei-bun-doh

Pagallo U (2013) The laws of robots. Springer, The Netherlands

Prem E (2023) From ethical AI frameworks to tools: a review of approaches. AI Ethics 3:699–716. https://doi.org/10.1007/s43681-023-00258-9

Rehm M, Nakano Y, André E, Nishida T, Bee N, Endrass B, Wissner M, Lipi AA, Huang H-H (2009) From observation to simulation: generating culture-specific behavior for interactive systems. AI Soc 24:267–280. https://doi.org/10.1007/s00146-009-0216-3

Saito K (2017) Granting juridical personality to artificial intelligence. IPSJ J 35:10–27

Schiebinger L, Klinge I, Paik HY, Sánchez de Madariaga I, Schraudner M, Stefanick M (eds) (2011–2020) Gendered innovations in science, health & medicine, engineering, and environment (genderedinnovations.stanford.edu)

Shimpo F (2017) Overview of issues by legal domain concerning robot law. J Law Inf Syst 65–78

Suler J (2004) The online disinhibition effect. CyberPsychol Behav 7:321–326. https://doi.org/10.1089/1094931041291295

Suzuki M, Takahashi R (2020) bert-japanese. In: github. https://github.com/cl-tohoku/bert-japanese. Accessed 13 Dec 2023

Takeuchi K, Yamazaki Y, Yoshifuji K (2020) Avatar work: telework for disabled people unable to go outside by using avatar robots. In: Companion of the 2020 ACM/IEEE international conference on human-robot interaction. ACM, New York, pp 53–60

Tannenbaum C, Ellis RP, Eyssel F, Zou J, Schiebinger L (2019) Sex and gender analysis improves science and engineering. Nature 575:137–146. https://doi.org/10.1038/s41586-019-1657-6

Ting-Toomey S, Dorjee T (2018) Communicating across cultures, 2nd edn. Guilford Press

Umpleby S (1982) Definitions of cybernetics. https://asc-cybernetics.org/definitions/. Accessed 16 Dec 2023

UNESCO (2021) Recommendation on the ethics of artificial intelligence. https://unesdoc.unesco.org/ark:/48223/pf0000381137. Accessed 6 Dec 2023

Velastin SA (2005) Intelligent CCTV surveillance: advances and limitations. In: Proceedings of the 5th international conference on methods and techniques in behavioral research—measuring behavior

West M, Kraut R, Chew HE (2019) I'd blush if I could: closing gender divides in digital skills through education

Wiener N (1948) Cybernetics: or control and communication in the animal and the machine. The MIT Press, Cambridge

Wilson JQ, Kelling GL (1982) BROKEN WINDOWS: the police and neighborhood safety. The Atlantic Monthly March, pp 29–38

Yamada K, Even J, Kanda T (2023) Enhancing avatar robot customer service through speech monitoring and filtering. In: The 2023 IEEE/RSJ international conference on intelligent robots and systems